BIOCHEMISTRY RESEARCH TRENDS

LIGANDS

SYNTHESIS, CHARACTERIZATION AND ROLE IN BIOTECHNOLOGY

BIOCHEMISTRY RESEARCH TRENDS

Additional books in this series can be found on Nova's website
under the Series tab.

Additional e-books in this series can be found on Nova's website
under the e-book tab.

BIOCHEMISTRY RESEARCH TRENDS

LIGANDS

SYNTHESIS, CHARACTERIZATION AND ROLE IN BIOTECHNOLOGY

PAULA GAWRYSZEWSKA
AND
PIOTR SMOLEŃSKI
EDITORS

nova publishers
New York

Library of Congress Cataloging-in-Publication Data

ISBN: 978-1-63117-143-7

Library of Congress Control Number: 2014930664

Published by Nova Science Publishers, Inc. † New York

CONTENTS

PREFACE

To meet the search for new therapeutic compounds this book summarizes the research on biologically active organic molecules (chapters 1, 2 and 3), metal complexes with biological activity (chapter 4), and shows the possibilities for coordination chemistry in the planning of metal complexes with interesting properties for application (chapters 5, 6, 7, 8 and 9). It should be remembered that in the design of a new potentially active metallodrug, beyond the nature of the metal, the choice of appropriate ligands which affect the thermodynamic and kinetic stability, as well as the solubility and lipophilicity of the complexes is of paramount importance. In Chapter 1 the interactions involving the members of the Nuclear Receptor superfamily and their ligands are discussed. The authors present an introduction about the structure-function relationship for nuclear receptors, the general features of the binding pocket, and a detailed discussion of three representative nuclear receptors.

Chapter 2 reviews the role of chemokine ligands and their receptors in normal and disease processes, and the emerging therapeutic approaches of using chemokine antagonists for appropriately targeted therapy.

Chapter 3 describes the results of a project focused on the synthesis, elucidation of biological activity, and biopharmaceutical application of the compounds C-97 C-98 and C-192 possessing one, two or four biophoric methylene bisphosphonic acid (MBPA) groups at the upper rim of the calix[4]arene molecular platform.

Chapter 4 describes a review of the biological properties of tris(pyrazol-1-yl)-borate and -metane homoscorpionate RX(R'pz)3 (pz = pyrazol-1-yl, X = C or B) ligands and their metal complexes. The main biological applications, enzyme modelling and antiproliferative and antimicrobial activity of such scorpionate-type complexes are classified by ligand and also by transition metal.

In chapter 5 the author exhibits the syntheses, structural determination, and the silver(I) discrete and polymeric coordination compounds bearing 1,3,5-triaza-7-phosphaadamantane (PTA) or its derivatives: oxide (PTA=O) and sulfide (PTA=S), which were recently reported in the literature. The ligands were also used as a new versatile P,N,O,S-building block towards engineering functional silver-organic frameworks (MOFs). Some biological properties of selected compounds - antibacterial and antifungal activities - are also presented. Furthermore, pronounced antiproliferative activity against human malignant melanoma is also described.

Chapter 6 deals with existing information on the synthesis of arylhydrazones of methylene active nitriles (AMANs) and their use as starting materials for the generation of new organic and coordination compounds. Reactions of AMANs with different nucleophiles lead to triazoloes, aminopyrazoles, arylazoaminopyrazoles, dicyanopyridazines and other compounds of pharmaceutical significance, depending on the reaction conditions and reagents used. The authors show that AMANs and their complexes and derivatives can be applied in different fields of modern chemistry, e.g., as pharmaceuticals, molecular magnets, switches and catalyst precursors.

In Chapter 7, in respect of searching for ligands allowing for coordination engineering towards biologically active metal-organic frameworks, the authors consider two types of organic ligands: carbacylamidophosphates (CAPh) – with the functional fragment -C(O)N(H)P(O)= and sulfonylamidophosphates (SAPh) - containing the -$S(O)_2N(H)P(O)$= fragment, their sodium salts and their lanthanide complexes. The determination of the ability of these ligands types to be applied as sensitizers of the lanthanide emission is also discussed.

Chapter 8 describes the coordination chemistry of N-acetylsalicylhydrazidate (H3ashz) as a versatile ligand for the synthesis of higher nuclearity complexes. The coordination flexibility and interesting donor properties of the ligand lead to the formation of various types of metal complexes, e.g., metallacrowns, 1D chain and mixed-valence complexes, which are illustrated herein, along with their applications.

In chapter 9 the authors represent an overview of the existing information on the chemistry and some applications of formazans, compounds with the R1NH-N=CR2-N=NR3 azohydrazone conjugated system. Representatives of the farmazan family are applied for the synthesis of different heterocycles, while the "formazan – tetrazolium salt" system is widely used as a redox-based staining component for different biochemical studies. Moreover, formazans are valuable spectrofotometric analytical agents and ionophors; they can be used for the simultaneous selective determination of several metal cations.

The authors hope that the compiled chapters will provide both beginners in the field and advanced researchers with comprehensive information on the subject.

The editors thank Peter Foulds, M.A. and Jerzy Sokolnicki, D.Sc. for help in editing the book.

In: Ligands
Editors: Paula Gawryszewska et al.
ISBN: 978-1-63117-143-7

Chapter 1

LIGAND INTERACTIONS IN NUCLEAR RECEPTORS

Ana Carolina Mafud, Jademilson C. dos Santos and Alessandro S. Nascimento*

Instituto de Fisica de São Carlos, Universidade de São Paulo, São Carlos, SP. Brazil

ABSTRACT

In this chapter the interactions involving the members of the Nuclear Receptors superfamily and their ligands are discussed. First, an introduction about the structure-function relationship for nuclear receptors is presented, and the general features of the binding pocket are shown.

Then, three representative nuclear receptors are discussed in more detail: i) the Peroxissome Proliferator-Activated Receptor gamma, a member of the thyroid subfamily; ii) NGFI-B, a member of the orphan nuclear receptor subfamily and, finally, iii) the mineralocorticoid receptor, a representative of the steroid hormone receptor subfamily. In the final section some perspectives for the design of new ligands are discussed.

Keywords: Nuclear receptors, PPAR, ligand, protein-ligand

* Universidade de São Paulo. São Carlos, SP. Brazil. Email: asnascimento@ifsc.usp.br.

INTRODUCTION

Nuclear receptors comprise a superfamily of ligand-regulated intracellular proteins able to recognize specific DNA sequences and regulate the transcription of target genes downstream to these sequences [1]. The proteins of this superfamily share a modular structure with four typical domains: i) an N-terminal domain with unknown structure, but probably disordered, since it was present in the crystal structure of an intact nuclear receptor but could not be defined in the electron density. The functional role of this domain is not well understood, but it is known to have a ligand-independent activation function (AF-1). It is also known to be involved in an interdomain interaction with the C-terminal domain in steroid receptors [2]. ii) A central DNA binding domain (DBD) folded as a typical zinc finger and highly conserved among the members of the family. This domain is responsible for DNA recognition and binding, and is the most conserved domain among the nuclear receptors. iii) A small *hinge* connecting the central DNA-binding domain and the C-terminal domain. This hinge has been shown to interact with the C-terminal domain, the DNA binding domains [3] and the DNA itself, playing a role in gene transcription regulation [4]. iv) A C-terminal domain, also called the ligand-binding domain (LBD), folded as a 3-layer α-helix sandwich with a buried and usually nonpolar binding pocket [5] (Figure 1). This domain typically contains 12 helices and serves as anchors for partner proteins and also for ligand recognition and binding. Helices 3 (H3) and 12 (H12) form a hydrophobic surface that recognizes coactivator proteins, while helices H3, H5, H10 and H12 form the ligand binding pocket. Helix H10 also serves as the dimerization surface. There is also a β-hairpin in the ligand binding domain, between helices H5 and H6, that does not have a clear functional role, but it has been shown to directly interact with the DBD of the partner receptor. Also, ligands shown to stabilize this β-hairpin of the LBD were shown to exhibit a different activation profile from ligands that stabilize the 'canonical' binding pocket close to helix H12, showing that this region has functional roles yet to be understood [6].

Nuclear receptors usually regulate gene transcription by forming dimers. Some receptors, such as the steroid receptors form homodimers, while thyroid hormone receptors, PPARs and vitamin D receptor form heterodimers. In the latter case, the nuclear receptor RXR (Retinoid X Receptor) is the usual dimerization partner. Due to these functional differences and the DNA sequences recognized by the receptors, they can be divided into subfamilies.

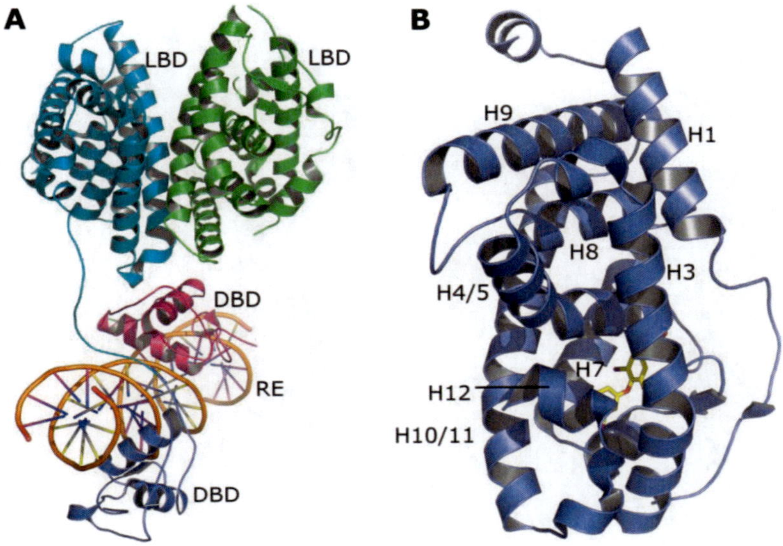

Figure 1. Nuclear receptors' canonical structure. (A) Crystal structure of PPARγ-RXRα dimer assembled on a responsive element (RE). (B) Typical structure of the ligand-binding domain.

The receptors in the steroid receptor subfamily are mainly cytoplasmatic proteins when unbound and, upon ligand binding, they are translocated to the cell nucleus, where they recognize responsive elements in the DNA with main sequence AGAACA in two half-sites [1,7]. In the unbound state, these proteins are typically found in a complex with heat shock proteins (HSPs). The HSPs are released with the structural rearrangements that follow ligand binding and the bound receptor is then capable of recruiting coactivator proteins, such as SRC-1 [7]. These coactivator proteins have histone acetyltransferase activity, and are also able to recruit general transcription factors, resulting in chromatin remodelling and gene transcription.

The receptors in the thyroid subfamily are found mainly in the cell nucleus bound to corepressor proteins such as the silencing mediator of retinoic acid and thyroid hormone receptors (SMRT) and N-CoR [8, 9], and sometimes bound to responsive elements. The interaction with high affinity ligands results in structural changes in the receptor, release of the corepressor proteins and recruitment of coactivator proteins. A complete structural characterization of the interaction between receptor and corepressors or coactivator proteins still awaits more experimental data.

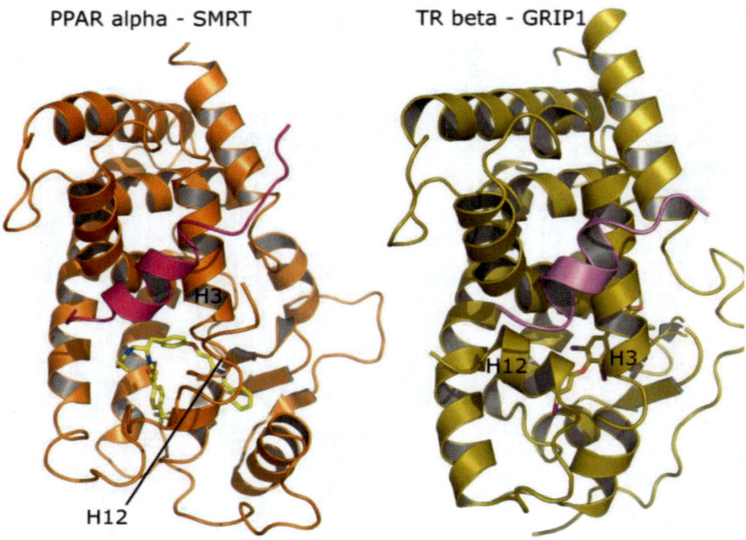

Figure 2. Crystal structure of PPARα in a complex with a corepressor peptide (left) and TRβ in a complex with a coactivator peptide (right). Corepressors and Coactivators use the same surface of the LBD, formed by helices H3, H12 and H5.

However, small peptides from those proteins have been crystallized in a complex with several nuclear receptors, revealing a conserved site for coactivator and corepressor binding. This site is located on the surface of helix 12. The interaction is directed by two charged residues located in helices H12 and H11, adding a polarity to the interaction between the peptide and the LBD [10] (Figure 2).

Additionally, Marimuthu et al. used an extensive mutational analysis to show that multiple sites can be explored for corepressor binding on thyroid hormone receptor (TR) LBD [11]. The deletion of helix 12 of TRβ LBD reduced corepressor binding but was not enough to abolish the interaction, showing that binding sites other than H12 could be used for corepressor protein interaction. Mutational analysis revealed a second site alongside helix H1 and a third site above helix H11, close to the dimerization interface [11].

Ligand Binding Pocket

Topologically, the LBD is highly conserved among the members of the nuclear receptor superfamily. However, despite the structural similarity, ligands found within the binding pocket can make precise contacts in each receptor-cognate ligand pair [12]. One important feature for the binding pockets in nuclear receptors is the volume of the pocket. This feature can substantially vary among the members of the superfamily. For example, the retinoic acid receptor (RAR) has a 420 Å^3 pocket, TRβ has a 600 Å^3 ligand pocket, the pregnane X receptor (PXR) has a 1,100 Å^3 pocket, and the Peroxissome Proliferator-Activated Receptors (PPARs) have the largest pockets, reaching 1,300 Å^3 [13].The pocket volume seems to be strictly associated with ligand selectivity, since TRs bind to very specific scaffolds with high affinity, while PXR and PPARγ bind to a wide range of molecules with a medium to low affinity [12, 14, 15]. It is important to highlight that some receptors can adapt the binding pocket volume upon ligand binding, as in the case of the vitamin D Receptor (VDR). This receptor can reduce its volume when bound to its natural ligand and expand by biding the Gemini agonist [16].

On the other hand, the subfamily of the so called 'orphan receptor' shows very interesting structural features in their pockets. This sub-family includes those nuclear receptors for which no ligand has yet been identified. Historically, after the cloning of the first nuclear receptors, a number of proteins were readily identified as nuclear receptors using their conserved DNA binding domain. However, some of these newly identified nuclear receptors had no ligand identified, and they were called orphan nuclear receptors [17]. Some of the orphan receptors were later 'deoprhanized' with the discovery of natural ligands able to bind and activate the receptor. PPARγ is an example of a deorphanized receptor. Examples of orphan receptor include NGFI-B (also called Nur77), Nurr1, Nor-1 and HNF-4. Interestingly, some of these receptors have their binding pocket occluded by side chains of residues locate in helices H3 and H8 (Figure 3).

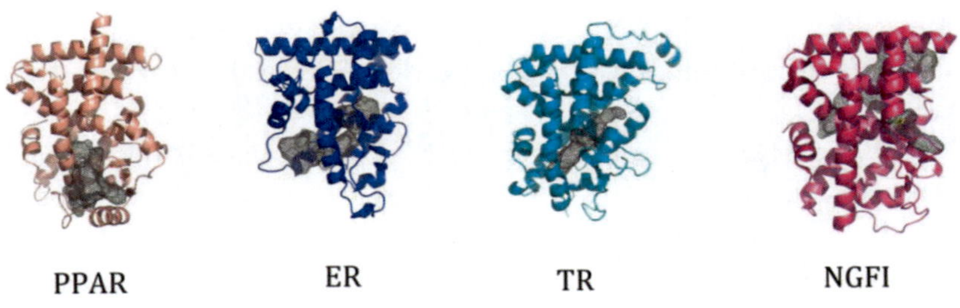

Figure 3. Buried cavities identified in some nuclear receptors. Note that for the orphan nuclear receptor NGFI-B, there is no cavity volume in the canonical binding pocket.

Consistent with a buried pocket, the ligand binding cavity in nuclear receptors is mainly comprised of hydrophobic residues with polar and charged residues in specific positions that dictate the polarity for ligand binding and determine ligand specificity. Contrary to what was previously thought, these polar residues can be very flexible in contributing to the ligand binding/unbinding dynamics, and also in creating a dynamic interaction pattern (e.g., [18, 19]). The interaction between these polar residues, the ligand and water molecules within the binding pocket were also shown to play a role in ligand selectivity and affinity [18, 20].

1. PPARs

The peroxisome proliferator-activated receptors (PPARs) are members of the thyroid-like subfamily of nuclear receptors and act as a transcription factor activated by endogenous and exogenous ligands regulating the expression of several target genes by binding to specific responsive elements, the PPREs [21]. These receptors have important roles in glucose uptake regulation, inflammation, and lipid and cholesterol metabolism. These roles make them targets for drugs used in the treatment of type 2 diabetes mellitus and dyslipidemia [22]. PPARs form heterodimers with RXR and, in contrast to classical steroid receptors that act as homodimers, the heterodimerization is essential for the recruitment of coactivator protein [23].

Peroxissome proliferation was observed in rodent liver as a response to drugs such as fibrate, phatalate and trichloroacetic acid herbicides [24], so these compounds were termed peroxisome proliferators [25]. In 1990, Isseman and Green hypothesized that the receptors for these peroxisome proliferators should be nuclear receptors, due to their ability to modulate gene transcription, which was already known. Screening a mouse liver cDNA library, they identified PPAR [26]. In 1992 Dreyer and co-workers cloned and identified the three PPAR isoforms in *Xenopus laevis*, identified as PPARα, PPARβ and PPARγ [24, 27].

The responsive elements recognized by PPARs (called PPREs) have a consensus sequence AGGTCAxAGGTCA [26] where X may be any nucleotide. The heterodimer assembly on the PPREs was structurally elucidated in 2008 [3] and will be discussed below.

The first crystal structure of a full length receptor was solved in 2008 for PPARγ in complex with RXR and assembled on its responsive element.

Coactivator peptides were also present in this crystal structure [3]. This structure revealed, for the first time, the arrangement of the DBDs with respect to the LBDs. First, the DBD domains of RXRα and PPARγ are bound on opposite sides of the DNA double helix, using the helix axis as a reference. Despite this organization of the DBDs, the LBDs form a classical dimerization interface using their helices H10. To permit this arrangement, the hinge domain of RXRα forms a long, flexible and disordered structure with several interactions with the responsive element (Figure 4). This structure also showed interdomain interactions involving RXRα DBD and PPARγ LBD. This interaction occurs through the β-hairpin region of the LDB, which may explain its role in receptor activation.

The PPARα gene was first cloned in 1990 from cDNA of mouse liver [26]. In humans, PPARα is expressed predominantly in the liver, adipose tissue and heart. This receptor controls the expression of genes involved in fatty acid oxidation, ketogenesis, lipid transport, inflammation and gluconeogenesis [28]. PPARα is the molecular target for the fibrates, a class of drugs used in the treatment of dyslipidemia (Figure 5). Activation of PPARα by fibrates can decrease triglyceride levels and increase high density lipoproteins (HDL), presumably through the metabolism of fatty acids and lipoproteins [29].

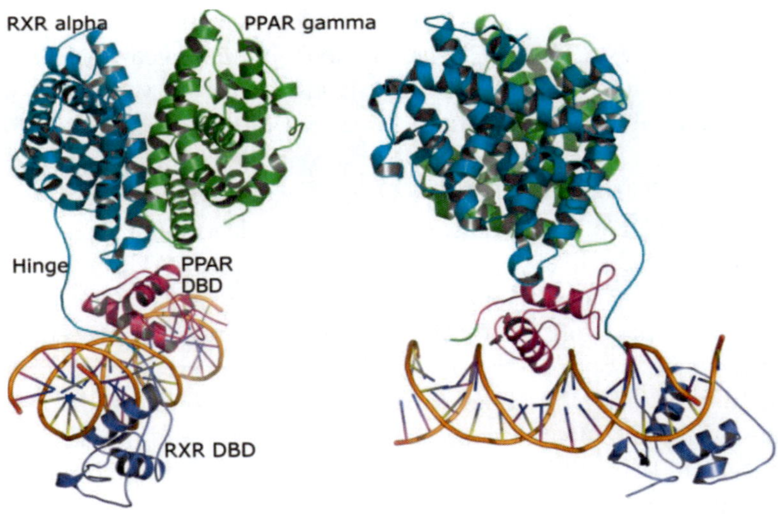

Figure 4. Crystal structure of full length PPARγ-RXRα complex (PDB 3DZY [3]). (Left) Arrangement of LBDs and DBDs with respect to each other, and (right) rotated view of the same structure.

Endogenous ligands are also able to activate this receptor. Some endogenous ligands include unsaturated fatty acids and eicosanoids, such as the inflammatory mediator leukotriene B4 and prostaglandins. It is believed that activation by endogenous ligands mainly occurs during fasting periods, when there is a large amount of free fatty acids in blood plasma. Dietary intake of specific fatty acids can also be an alternative for activation of PPARα [30].

PPARβ, also known as PPARδ, was originally cloned by Wahli and colleagues in *Xenopus laevis* [27]. The expression occurs in a wide range of tissues and cells, but at high levels in tissues related with metabolic activity, such as adipose tissue, liver and skin [23]. It is ubiquitously expressed, and the lack of a known ligand has caused PPARβ/δ to be less studied than the other isoforms. However, important advances have been made in genetic studies and in the synthesis of agonists selective for PPARβ/δ. These advances revealed the role of this receptor as a regulator of fatty acid catabolism and energy homeostasis [31]. Although no approved drug acting solely or mainly in PPARβ/δ exists to date, the metabolic processes regulated by PPARβ/δ, such as lipid metabolism, atherosclerosis, glucose metabolism and insulin sensitivity, make this receptor a potential candidate for drug development [32]. Some inflammatory processes are also attributed to PPARβ/δ, although the molecular mechanisms involved are not completely understood [33].

Three isoforms of the PPARγ gene have been identified in humans, with different distributions [34]. PPARγ1 is encoded by eight exons, comprising two specific γ1 exons in the 5'-end region not translated and six exons common to the three mRNA. PPARγ2 is encoded by seven exons with a specific exon encoding 28 amino acids at the N-terminal region. PPARγ3 encodes the same protein as PPARγ1 but is controlled by a separate promoter located in the 5'-end region [35]. PPARγ1 is expressed in a wide variety of tissues, including the heart, colon, small and large intestine, kidney, pancreas and spleen. PPARγ2 is expressed in adipose tissue; e PPARγ3 expression is restricted to macrophages and the intestines [34].

Some important roles of PPARγ in the human metabolism were readily recognized with the discovery of the thiazolidinediones (TZDs) as PPARγ high affinity ligands [36]. Lehmann and co-workers showed in 1995 that the insulin-sensitization effects of the TZD rosiglitazone were mediated by high affinity (40 nM) binding to PPARγ. Also, an ability to promote adipocyte differentiation was shown for TZDs, which explained the adipogenic effect of TZD therapy observed *in vivo*. PPARγ also plays a role in inflammation. Receptor activation can result in the inhibition of inflammatory cytokines including interleukin-6 (IL-6) and tumor necrosis factor α (TNF-α) by monocytes [37]. Since TNF-α is directly involved in insulin resistance, these cytokines may be key factors linking inflammation, improvement in insulin sensitization and obesity through a molecular mechanism involving this receptor. However, more experimental evidence is still necessary for a better understanding of the mechanisms involved in these processes.

1.1. PPAR Modulators

The PPARs are targets for several classes of drugs. As previously said, binding of fibrates to PPARα results in a significant reduction in serum triglycerides and a modest increase in HDL cholesterol levels (Figure 5). In addition, activation of PPARα increases insulin sensitization, reducing glucose tolerance in type 2 diabetic patients [38]. Fibrates are well tolerated by most patients, but some adverse effects are reported, mostly gastrointestinal. Other adverse effects include anxiety, headache, dizziness, sleep disorder and arthralgia. As fibrates are metabolized in the kidney, patients with improper kidney function are not recommended to use these drugs [30].

The TZDs are the only approved drugs acting on PPARγ (Figure 6). Three TZDs were initially approved for human therapy: troglitazone, rosiglitazone and pioglitazone. These are high affinity and selective ligands of PPARγ and promote insulin sensitization with beneficial effects in the treatment of type 2 diabetes mellitus. Troglitazone was withdrawn in 2000 due to hepatotoxicity [39].

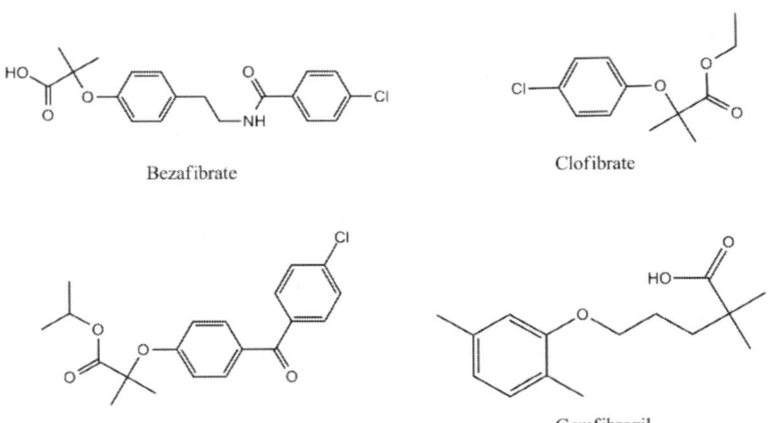

Figure 5. Chemical structures of representative fibrates.

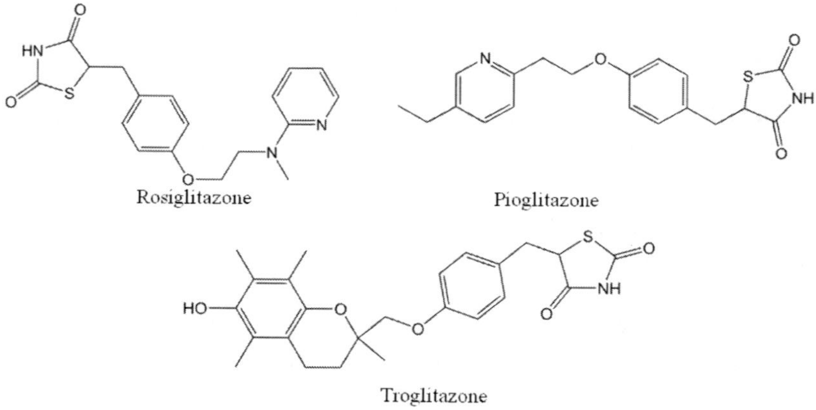

Figure 6. Chemical structures of thiazolidinediones (TZDs).

More recently, rosiglitazone had its use restricted or suspended because of the increased risk of death from cardiovascular events [40, 41], while pioglitazone is still used as an oral hypoglycemic agent. However, some important adverse effects have been reported, such as increased incidence of bone fractures, fluid retention and weight gain [42]. It is interesting to note that despite the similarity between rosiglitazone and pioglitazone, the latter does not cause the undesired effects to the same extent as rosiglitazone [43]. As noted by Bruning and co-workers, small differences in the interaction between ligand and receptor can result in important pharmacological profiles for these molecules [6].

Some of the mechanisms used by TZDs to enhance insulin sensitivity include an increased glucose disposal in adipocytes, stimulation of glucose disposal in skeletal muscle, reduced levels of TNF-α, reduced levels of leptin, reduced levels of free fatty acids (FFA), and increased levels of adiponectin [44]. It is well known that TZDs can induce genes directly involved in glucose metabolism, such as GLUT4 in adipocytes [45]. Also, TZDs increase the number of small adipocyte cells, which is believed to improve glucose uptake [46]. Acting on genes involved in glucose uptake or on the mediators of metabolic signaling (TNF-α, FFA, adiponectin), PPARγ activation by TZDs results in improved insulin sensitivity.

In terms of structural recognition rosiglitazone occupies roughly 40% of the ligand-binding site. In general the ligand is in a U-shaped conformation, wrapping around H3 with its central benzene ring directly behind H3 and the TZD head group and pyridine ring wrapping around H3.The TZD ring makes several specific interactions with amino acids in helices H3, H4, H10 and H12.

The carbonyl groups of the TZD form hydrogen bonds with two histidine residues, H323 and H449. Y473 in H12 forms a secondary hydrogen bond with H323. The electronegative nitrogen of the TZD head group is in a hydrogen-bonding distance of Y473 OH. A buried lysine residue, K367, forms another secondary hydrogen bond with the ligand, at residue H449. All of these primary and secondary hydrogen bonds result in a fixed conformation of the TZD head group and of the participating amino acids. The sulphur atom of the TZD ring is positioned in a hydrophobic region of the PPARγ ligand-binding pocket formed by F363, Q286, F282 and L469. The central benzene ring of the ligand occupies a very narrow pocket between C285 and M364. The bridging oxygen atom between the benzene ring and the pyridine ring provides vital geometry for the pyridine ring, which occupies the pocket between H3 and the β-sheet.

Since helix H12 is typically understood as the activation helix, i.e. the helix that directly interacts with ligands and coactivator proteins, the stabilization of this helix through ligand interactions became a goal in the development of new ligands with maximal potency and efficacy. The dynamics of the helix H12 will be discussed below, but it is interesting to note that TZDs are very efficient ligands in terms of H12 stabilization, via direct interactions of the TZD ring and Y473, as said above.

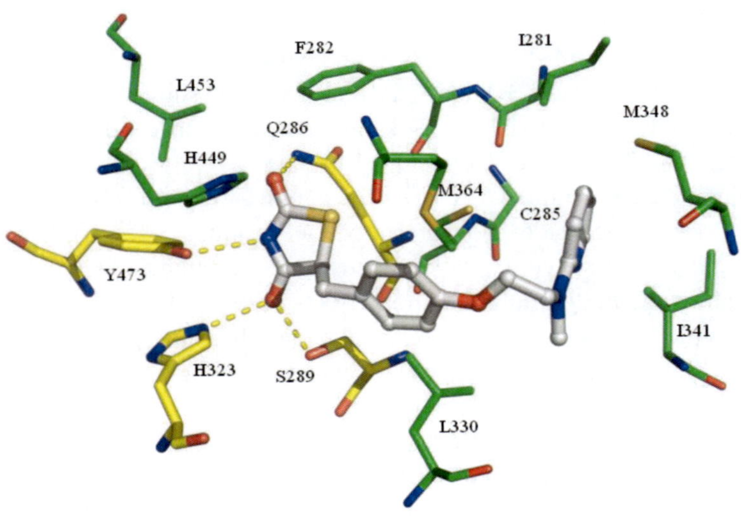

Figure 7. Interactions between rosiglitazone (show in grey sticks) and PPARγ. The polar residues of the PPARγ pocket interacting with rosiglitazone are shown in yellow sticks.

1.2. Full Agonists versus Partial Agonists for PPAR

According to the degree of receptor activation observed *in vitro*, the ligands can be classified as full agonists or partial agonists. Full agonists are those ligands that can cause maximal receptor activation. The TZDs are examples of PPARγ full agonists. On the other hand, a number of PPAR ligands have been identified that cannot cause maximal receptor activation, even when tested in high concentrations. Those ligands are termed partial agonists. The roles of full agonists *versus* partial agonists are better understood for PPARγ and will be briefly discussed in this section.

Three partial agonists have been described more recently, and their *in vitro/in vivo* profiles were compared with full agonists. MRL-24 binds to PPARγ with a 3 nM affinity, but results in only 20% of the receptor activation caused by rosiglitazone [47]. This difference in receptor activation was sufficient to restore glucose levels in a diabetic *db/db* mouse model, but reducing the weight gain as compared to rosiglitazone. Acton and co-workers showed that rosiglitazone (10mg/kg/day) resulted in a 10% weight gain in a mouse model, while MRL24 used at the same dose caused only a 3% weight gain [47]. GQ-16 is a TZD ligand that has its TZD ring as the B ring, not interacting with helix H12. As a result, GQ-16 causes only about 30% of the receptor activation caused by rosiglitazone and is also much less efficient in causing the differentiation of adipocytes than rosiglitazone, resulting in reduced weight gain [48]. Both GQ-16 and MRL24 can interact with a serine in position 273 in the LBD that has recently been shown to be the target position of CDK5. The phosphorylation of this serine residue is markedly associated with an adipogenic effect [49].

Balaglitazone, another PPARγ partial agonist, was also tested in diabetic animal models. The results showed that equi-efficacious doses of balaglitazone and rosiglitazone resulted in similar glucose control, but balaglitazone reduced the weight gain and the fluid retention effects observed for rosiglitazone [50]. These results showed that the beneficial and harmful effects of PPARγ activation can at least partially be dissociated by using submaximal receptor activation.

Structurally, important differences can also be observed for full agonists and some partial agonists. As a general rule, full agonists interact directly with H12, stabilizing the agonist conformation through a direct hydrogen bond to Y473, allowing H12 to dock against H3 and H11 and locking the helix in this position. In this conformation, H12 is itself a part of the coactivator-binding site, along with H3 and H5 [6] (Figure 2).

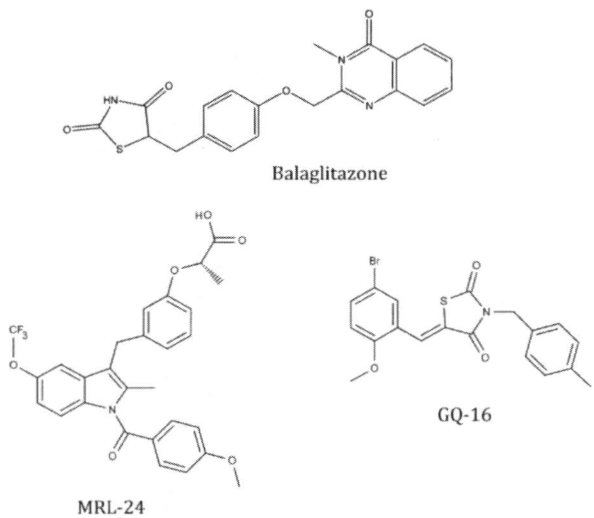

Figure 8. Chemical structures of GQ-16, MRL-24 and balaglitazone.

Several partial agonists have the capacity to activate PPARγ without a direct interaction with the H12 [6, 51, 52]. Due to the large volume of the LBP, these partial agonists can occupy a different region of the binding site with distinct biding modes. Alternatively, indirect interactions with H12 can be used by partial agonists. An example of this type of interaction is seen for the partial agonist INT131, a non-TZD ligand in phase III of clinical trials. INT131 is a highly selective ligand that can partially activate PPARγ and has little effect on lipid accumulation in 3T3-L1 adipocytes. Similar to other PPARγ ligands, INT131 binds to the same pocket but occupies a unique region in the pocket, lacking a direct interaction with H12, although two water molecules mediate an indirect interaction between INT131 and H12 [53].

Both GQ-16 and MRL-24 use a different region of PPARγ binding pocket for molecular interaction. These ligands bind to a site located between helix H3 and the β-hairpin, which is far from the helix H12, lacking any interaction with the latter. The position occupied by these ligands has been explained as the reason for their distinctive functional properties. In both cases, mass spectrometry and hydrogen-deuterium exchange experiments showed that the ligands stabilize the β-hairpin, in an interaction profile that significantly differs from the one observed with the TZD ligand. This binding site used by MRL24 and GQ-16 is also close to serine 273, blocking the phosphorylation of this residue by Cdk5.

1.3. PPAR Dual and Pan Agonists

PPARα is mainly involved in lipid metabolism, while PPARγ activation has effects for insulin sensitization. A global beneficial effect can in principle be achieved by simultaneously activating both receptors. PPARα activation would lead to the effects observed for the fibrates, such as the lowering of triglycerides and increase in HDL, while PPARγ activation could result in diabetic treatment, as observed for thiazolidinediones. A number of molecules were synthesized targeting this dual agonism. An important class of these dual agonists is the glitazars. The glitazars are molecules that combine structural features of the glitazones and the fibrates in single compounds able to activate both PPARα and PPARγ [54]. Two glitazars that entered in phase III were recently discontinued due to side effects. Muraglitazar and tesaglitazar increased the incidence of heart failure and decreased glomerular filtration. Although the expected beneficial effects were actually observed, the important side effects observed in phase III call into question the usability of PPAR α/γ dual agonists [55]. Structurally, a number of similarities can be readily identified among the PPAR α/γ dual agonists deposited in the protein data bank (PDB). In general, these ligands have an acidic group b that makes a hydrogen bond with Y314 (H4) in PPARα, or H323 in PPARγ (H4). The remaining residues involved in hydrogen bonds are conserved between these isoforms. For PPARα these residues are Ser280 (H4), His440 (H11) and Tyr464 in H12. For PPARγ the residues involved in polar interactions are Ser289 (H3) His449 (H11) and Tyr473 (H12), (Figure 9). This interaction pattern is observed for the dual agonists aleglitazar, AZ-242 and GW409544. One important difference between these ligands is the presence of water molecules surrounding GW409544, accounting for ligand stabilization [56].

Recent studies showed that PPARβ/δ can play a role in dyslipidemia. In contrast to PPARγ, the activated PPARβ/δ is not adipogenic and is associated with anti-obesity effects [57]. In addition, this receptor has recently been associated with a protective effect on glomerular filtration, which would be a goal for diabetic patients under pharmacological therapy [58,59]. Together, these effects suggest a beneficial effect of PPARβ/δ activated together with PPARγ and PPARα. Thus, the development of *pan* agonists, i.e. single molecules able to activate the three receptor isoforms, became a new paradigm in PPARs drug design. This strategy led to the discovery of new molecules with interesting effects in the treatment of type 2 diabetes mellitus and dyslipidemia [60]. However, due to the harmful side-effects, most of the PPAR pan agonists were discontinued during clinical trials.

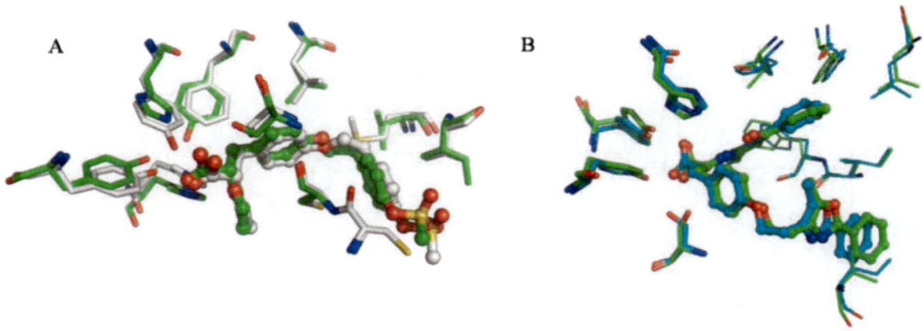

Figure 9. Structural details for the dual agonist (a) AZ 242 and (b) GW409544 to PPARα and γ. a) Superposition of the ligand binding sites in PPARα (white) and PPARγ (green), both occupied with AZ 242. a) Superposition of the ligand binding sites in PPARα (green) and PPARγ (blue), both occupied with GW409544.

The crystal structure of the pan agonist indeglitazar bound to the three receptor isoforms allows the comparison of the interactions between the ligand and the binding pockets [61].

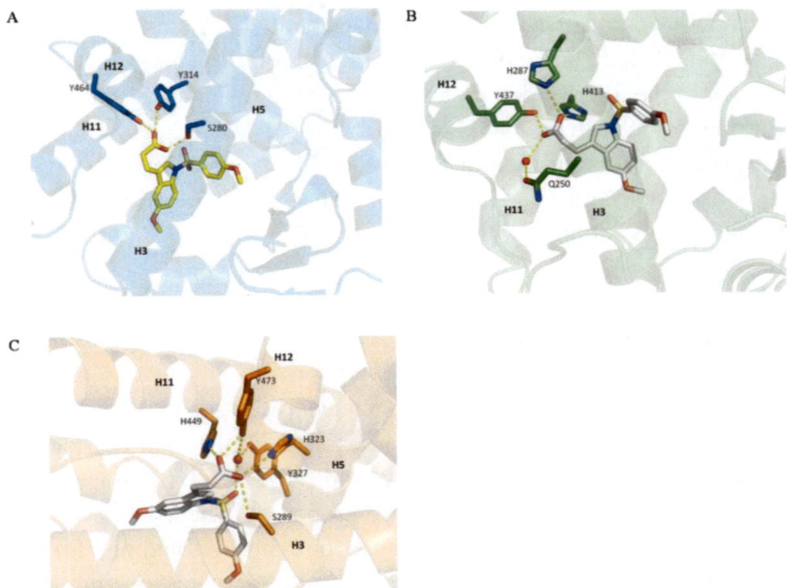

Figure 10. Structural details of indeglitazar bound with PPARα (a) PPARβ/δ(b) and PPARγ (c). PPARβ/δ and PPARγ require a water molecule for mediated hydrogen bonding.

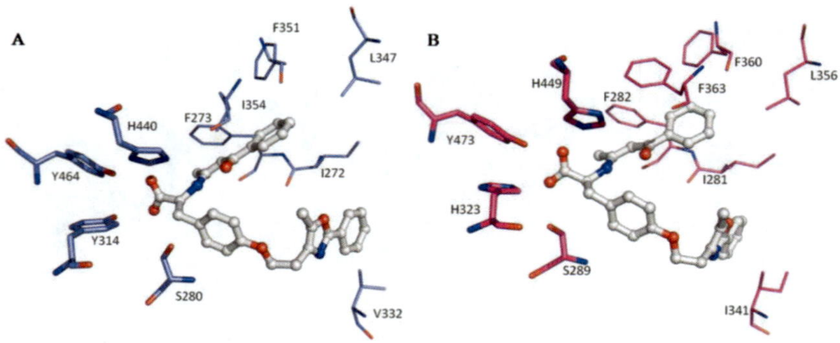

Figure 11. Interactions between GW409544 a PPARα/γ dual agonist (show in grey sticks) and (a) PPARα and (b) PPARγ. The polar residues of PPARα and γ pocket interacting with GW409544 are shown in sticks and nonpolar contacts in lines.

A direct interaction of indeglitazar with helix H12 was observed for three isoforms, although for PPARβ/δ and PPARγ a water molecule was found mediating the carboxylate interaction to one of the surrounding residues. It is believed that this molecule is involved in the partial agonism of indeglitazar. In the PPARγ complex the water interacts with Y327 and the carboxylate group of indeglitazar and within 2.8Å of one of the sulfonamide oxygens (Figure 10). In the PPARβ/δ complex a water molecule mediates an interaction between the ligand carboxylate and GQ250 [61]. A similar binding mode is also observed for the full agonist GW409544 (Figure 11). The interaction pattern is also observed for TIPP-03, another PPAR pan agonist. This ligand was crystallized in complex with PPARγ and PPARα [60].

1.4. PPARγ and H12 Dynamics

The crystal structure of the first two nuclear receptors LBD (RXR and TR) revealed an extended conformation for helix H12 in unbound RXR-LBD [62] and a closed conformation in the T3-bound TR-LBD [63]. Based on these observations a '*mouse trap*' model was proposed, where helix H12 would serve as a gate regulating ligand entry. Upon ligand binding, H12 would change from an extended conformation to a 'closed' state, locking the ligand in the pocket and forming a hydrophobic surface for the recognition of coactivator proteins. This model was further confirmed in molecular dynamics simulations [64] (Figure 12).

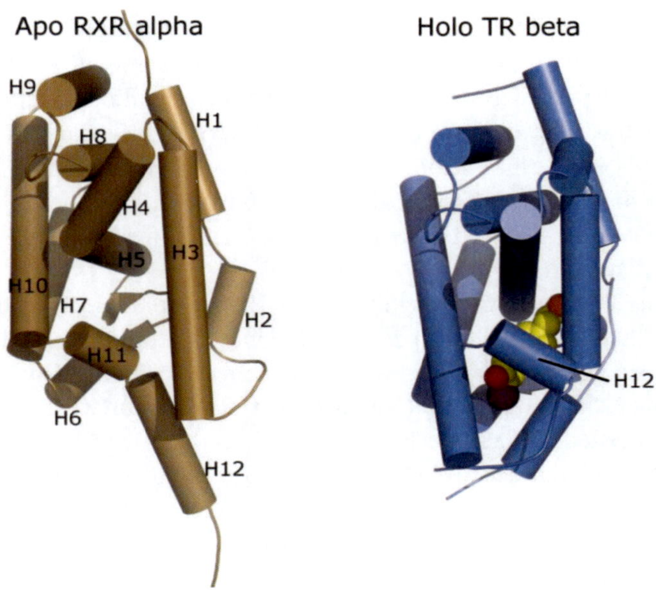

Figure 12. Mouse-trap model. (Left) Crystal structure of unbound RXRα and (right) bound TRβ. The helix H12 closes after ligand binding.

Molecular dynamics simulation of different ligand complexes with TR, ER, PPAR and MR suggested that helix H12 may have a more dynamic behaviour than previously suggested [65-67]. In line with the simulations, Kallenberger and co-workers measured the mobility of helix 12 using time-resolved fluorescence anisotropy of a fluorophore probe covalently attached to PPARγ-LDB [68]. They concluded that helix H12 is probably close to its usual position (i.e. docked against H3), but is very mobile in the absence of a ligand and, after ligand binding, it becomes immobilized on the surface of the receptor. These results suggest that the overall effect of ligand binding on H12 may be to increase the stability of the helix on the proper position to coactivator recruitment, reducing its mobility.

This altered mobility of H12 is also reflected in a functional role. Two PPARγ mutations known to cause insulin resistance, namely P467L and V290M, modified the dynamic behaviour of helix 12 and its response to the ligand, accounting for the loss of transcriptional activation observed for these mutants.

Hydrogen-deuterium exchange (HDX) experiments were used to probe the change in mobility for PPARγ full and partial agonists. Interestingly, rosiglitazone and MRL-20 showed a low rate of hydrogen exchange in helix

H12 when compared to the unbound receptor, indicating that this helix is highly stabilized by these full agonists. On the other hand, the partial agonists MRL-24, nTZDpa and BVT.13 did not show any significant stabilization of H12. In contrast, the region of the β-hairpin was significantly stabilized by these ligands, and the stabilization of this region correlates with the transactivation efficacy, showing that the stabilization of different regions of the binding pocket, far from helix H12, are structural mechanisms involved in receptor activation [6].

In terms of structural recognition, MRL-20 and MRL-24 show a 180° rotation in the ligand-binding pocket, allowing for different atomic contacts that are reflected in unique changes to the dynamics of the LBD, as determined by HDX analysis. In the partial agonist MRL-24 structure, the lactate group was found nestled against the β-hairpin, making hydrogen bonds as well as hydrophobic and van der Waal contacts (Figure 13). The acid portion of the lactate was positioned such that both oxygen atoms are in hydrogen-bonding distance with Ser342, as well as a water-mediated contact to Glu343. The lactate group was positioned in contact distance with Ile341, Leu340, and Met348. The observed interactions of MRL-24 with the PPARγ were consistent with the HDX data, with no significant stabilization of H12 but robust increased stabilization of the β-hairpin.

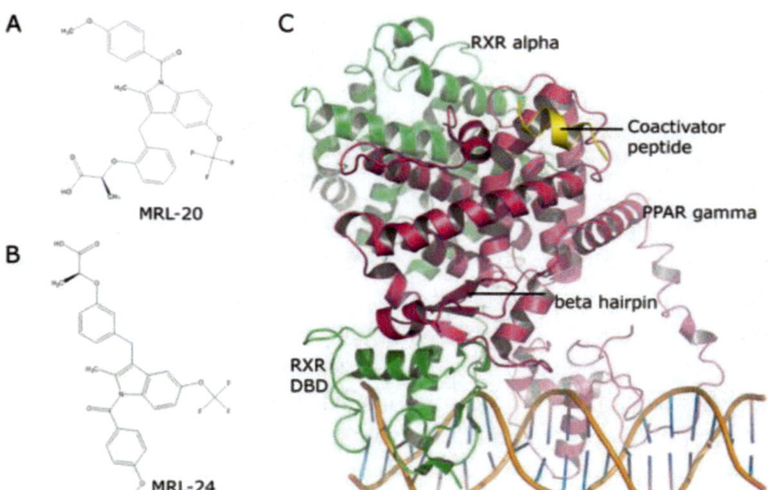

Figure 13. (A) Structure for MRL-20, (B) MRL-24 and (C) crystal structure of PPARγ (magenta) and RXRα (green) bound to the responsive element (PDB 3DZY [3]). The β-hairpin of PPARγ directly interacts with RXRα DBD.

The crystal structure of the full length PPARγ-RXR bound to the responsive element provided a structural explanation about the role of the β-hairpin on the receptor function. In this crystal structure the β-hairpin region of PPARγ LBD is in contact with RXRα DBD assembled on the responsive element (Figure 13) [69]. It is tempting to speculate that the ligand stabilization of the β-hairpin results in a better stabilization of DBD of the partner receptor on the DNA, favouring the transcriptional response.

1.5. PPAR Antagonism

A few molecules were identified with the ability to bind to PPARγ, inhibiting its activation and thus the gene transcription. GW9662 binds covalently to Cys285 located on helix H3 of PPARγ LBD. Despite the fact that this cysteine residue is conserved in all three PPAR subtypes, GW9662 displays higher affinity for PPARγ than for PPARα or PPARβ/δ. GW9662 antagonizes PPARγ activation in multiple cell types including adipocytes, macrophages and hepatic stellate cells [70]. GW9662 administration results in hypolipidemia, reducing the fatty acid level in the cell and also ameliorating insulin resistance.

The phosphonophosphate SR-202 was also identified as a PPARγ antagonist [71]. It inhibits both TZD-stimulated recruitment of the coactivator steroid receptor coactivator-1 and TZD-induced transcriptional activity of the receptor. SR-202 also efficiently antagonizes hormone- and TZD-induced adipocyte differentiation in vitro. The effects of this antagonist *in vivo* include reduction of both high fat diet-induced adipocyte hypertrophy and insulin resistance. These effects are accompanied by reduction of adipocyte size and reduction of TNF-α and leptin levels. Treatment with SR-202 also dramatically improves insulin sensitivity in diabetic *ob/ob* mice.

1.6. Conclusion

PPARs remain important targets for drug development, as both type 2 diabetes mellitus and obesity are becoming more important diseases in modern society. However, different approaches will have to be applied for the development of safer molecules, useful for therapeutics. First, the full activation of the receptors seems to be related to maximal beneficial response but with maximal chance of side effects. Submaximal activation is an interesting approach to pursue. A second strategy that needs more attention from basic research is the receptor antagonism. *In vivo* data from rodents show interesting results from the perspective of obesity treatment coupled to improvement in insulin sensitization.

2. ORPHAN RECEPTORS

As previously said, the orphan nuclear receptors are those nuclear receptors for which no ligand has been identified. One important group of orphan receptors is the NR4A subfamily of nuclear receptors. This subfamily comprises three members: NGFI-B (NR4A1), also referred to as NAK1, Nur77, ST-59, TR3 and eight other aliases [72]; NURR1 (NR4A2), and NOR1 (NR4A3). Like other NRs, members of the 4A subfamily regulate the expression of target genes binding to responsive elements in the DNA as monomers or as homodimers. Typical DNA sequences recognized by 4A receptors are AAAGGTCA (NBRE) and AAATG/CA/T CA (NurRE) [72, 73].

NGFI-B and Nurr1 can recognize direct repeat elements (AGGTCANNNAAAGGTCA) as heterodimers with RXR, whereas NOR-1 fails to form stable RXR heterodimers [74]. In heterodimers, agonist-bound RXR contributes to transcriptional activation.

Despite the lack of known ligands, NR4A receptors play key roles in the nervous, endocrine, and immune systems, and they are immediate-early response gene products that are induced after stimulation with serum, growth factors, and nerve growth factor [75-77]. In the next section, we describe the orphan receptor NGFI-B in more detail.

2.1. NGFI-B

NGFI-B was originally identified using differential hybridization in PC12 cells after induction with nerve growth factor (NGF) [78]. It was soon identified as a member of the nuclear receptors superfamily due to the high similarity among the DBD domain with the DBD of the steroid receptor. NGFI-B is expressed in various tissues, especially in the brain, and it plays a role in multiple cellular events such as cell proliferation, differentiation and apoptosis [79]. The importance of NGFI-B in the apoptotic process was demonstrated in knockout experiments in T-cell hybridomas [80]. Additionally, the receptor has been implicated in neurodegenerative pathologies, such as Parkinson disease, manic depression, and schizophrenia [81].

The recent crystallographic structures of NURR1 and NGFI-B ligand-binding domains revealed that these receptors have the canonical structure of nuclear receptors but they lack a ligand-binding pocket (LBP) [82, 83]. The structures showed a tight packing of the side chains of several bulky hydrophobic residues in helix H3, helix H5 and helix H10 within the pocket, in the same space usually occupied by ligands (Figure 3). Moreover, in contrast to what is generally observed in NR LBDs, in the orphan receptor NURR1 the coactivator cleft constituted by H3, H4, and H12 has a hydrophilic rather than a hydrophobic nature, with many charged residues on the surface. These results are consistent with the notion that orphan receptors cannot be modulated by ligands [83].

Although a clearly defined binding pocket is not found in NGFI-B, this receptor was demonstrated to be activated by synthetic ligands. Chintharlapalli and co-workers showed that 1,1-bis(3-indolyl)-1-(*p*-anisyl)methane can activate NGFI-B/GAL4 chimeras in pancreatic Panc28 cells[84]. After ligand administration the cells exhibited decreased survival and induction of cell death through a tumor necrosis factor-related apoptosis-inducing ligand and poly(ADP-ribose) polymerase cleavage [84]. Other ligands observed to bind to NGFI-B include cytosporone B [85] and, more recently, unsaturated fatty acids, arachidonic acid and docosahexaenoic acid [86].

The existence of ligands that can activate NGFI-B is intriguing. How can a protein receptor without a ligand pocket be activated by ligand binding? The crystal structure of NGFI-B bound to ethyl 2-[2,3,4 trymethoxy-6(1-octanoyl)phenyl]acetate (PDB 3V3Q, unpublished) showed for the first time a structural mechanism for ligand binding. In the crystal structure this ligand is docked on the surface created by helices H10, H12 and H5. Most of the polar interactions observed between the ligand and the receptor are mediated through atoms in the main chain, suggesting that the interaction might not be specific. A second ligand molecule was also observed on the surface formed by H5, H3 and the β-hairpin. For this second molecule polar interactions are observed with H41 and K125. Analysis of the temperature factors suggests that the first molecule (docked on the surface of helices H12, H10 and H5) is more stable than the second molecule. However, more experimental data of NGFI-B complexes of different ligands are still necessary for a better understanding on how this receptor can recognize natural and synthetic ligands.

Another intriguing point about NGFI-B and the orphan receptors is whether they can recruit coactivators after ligand binding. This question still awaits experimental answers. However, a piece of evidence was provided for the orphan nuclear receptor Nurr1. In the absence of a ligand, Nurrl is found bound to corepressor proteins such as SMRT and NCoR. Codina and co-workers used NMR fingerprints to show that peptides derived from these proteins interact with Nurrl through a region between helices H11 and H12, differently from what was shown for canonical nuclear receptors, such as TR [87].

This discovery opens new possibilities in the field. It is possible that orphan receptors have an entirely different way of ligand recognition and corepressor/coactivator recruitment or, alternatively, these receptors may be indicating that the entire nuclear receptor superfamily is functionally more plastic than previously thought. In any case, more structural and functional data will be necessary to unravel the mechanisms involved in orphan nuclear receptor activation by ligands.

The role of NGFI-B on apoptosis and cell death control makes this receptor an interesting target for the development of anti-cancer drugs. However, many molecular mechanisms still have to be better understood before the development of useful drugs. Basic cellular, functional and structural research can be important keys in achieving a better understanding of the NGFI-B structure-function relationship.

3. Steroid Hormone Receptors

The subfamily of the steroid hormone receptors (SHR) includes the probably most studied members of the nuclear receptors super family. The glucocorticoid receptor (GR), mineralocorticoid receptor (MR), progesterone receptor (PR), androgen receptor (AR), estrogen receptors α (ERα) and β (ERβ) and the estrogen-related receptors (ERRs) are members of this subfamily [88].

Structurally, these receptors have unique features when compared to the thyroid subfamily, for example. First, the steroid receptors have long N-terminal domains (also called A domain), with up to 600 residues. This long N-terminal domain can directly interact with the C-terminal LBD, enhancing the transcriptional response. Secondly, they usually bind to the responsive elements as homodimers. Third, in the unbound state, they are usually found in the cell cytoplasm in a complex with heat shock proteins (HSPs).

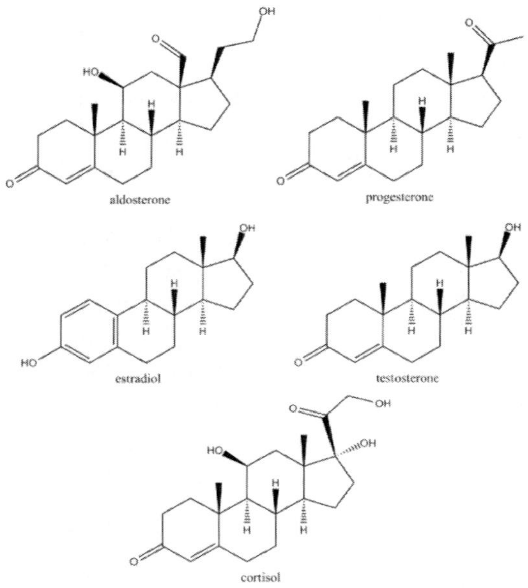

Figure 14. Chemical structures for representative steroid hormones. Aldosterone is the MR natural ligand. Cortisol binds with high affinity to GR and MR. Progesterone, testosterone and estradiol are the natural ligands for PR, AR and ER, respectively.

The steroid ligands have a common route of synthesis, derived from a cholesterol molecule [89]. The structure of representative steroid hormones is shown in (Figure 14).

Functionally, these receptors are fairly diverse. A large repertoire of molecular functions under hormone control is regulated by SHR, such as inflammatory responses, electrolyte balance, and reproductive cycle control.

3.1. Mineralocorticoid Receptor

The mineralocorticoid receptor (MR) is probably the last SHR to diverge in the phylogenetic tree. It shares a high degree of identity with the glucocorticoid receptor, and also binds to glucocorticoids with high affinity. MR has two natural ligands, synthesized in the adrenal gland. Aldosterone and cortisol can bind to human MR with similar and high affinity (0.5 and 0.8 nM, respectively), although the plasma concentration of these hormones differs significantly. The plasma concentration for aldosterone is 0.35 nmol/L and 400 nmol/L for cortisol, i.e. cortisol is found in concentrations 1,000 times higher than aldosterone [90]. The difference in concentrations suggests that the receptor is always occupied by cortisol, the ligand in higher concentration. However, in aldosterone target tissues, the enzyme 11β-hydroxyesteroid dehydrogenase 2 (11β-HSD2) is expressed and converts cortisol into cortisone, a metabolite inactive on MR [91]. This 'pre-receptor' mechanism reduces the cellular concentrations of cortisol but still leaves a hormone excess, suggesting that some other mechanisms may play a role in the receptor level [92].

Experimental evidence of a differential mechanism for recognition between aldosterone and cortisol came from dissociation kinetics. Hellal-Levy and co-workers showed that the dissociation half-time for aldosterone is 8 hours, while for cortisol this time is reduced to about 2 hours [93]. However, the structural basis of this mechanism is not yet completely understood.

MR regulates the transcription of several genes involved in ion balance. The main mechanism is related to sodium absorption and potassium elimination in the renal duct through Na^+/K^+ pumps. Some of the genes whose transcription is regulated by MR are the amiloride sensible sodium channel (*ENaC*), the gene for the Na^+/K^+ ATPase in the nefron and the gene responsible by *ENaC* traffic (*Sgk1*) [94]. The overall effects of MR on its target genes became clear in knockout experiments with rodents. The animals that had the MR gene deleted died 10 days after birth. They had a similar phenotype with primary pseudo hypoaldosteronism, loss of sodium retention and consequent water loss [95].

Aldosterone also plays a role in the development of cardiac hypertrophy and failure. Qin and co-workers showed that rodents with hyper expression of 11β-HSD2 in their cardiomyocites spontaneously develop cardiac hypertrophy, fibrosis, cardiac failure, and die early [96]. In these animals, the effect of receptor activation by cortisol is blocked, suggesting that aldosterone is a mediator of these effects. Additionally, the administration of eplerenone, a MR blocker, resulted in ameliorating effects [96].

Clinical studies also evaluated the effect of MR blockage in humans. 1,663 patients with heart failure received spironolactone, a MR blocker, for two years. The risk of death was reduced in 30%, with a 31% reduction in mortality by cardiac causes and a 29% reduction in sudden cardiac death. A second study with eplerenone administration for 16 months also showed a 15% reduction in death risk and a 17% reduction in the risk of death by cardiac causes [97].

Baxter and co-workers revised a series of studies and showed that in a non-selected population of hypertension patients there is an 8-13% prevalence of primary aldosteronism [98]. For these patients treatment with MR blockers would be of benefit. Two mineralocorticoid blockers (antagonists) are approved for human use. Spironolactone and eplerenone can bind with high affinity to MR and prevent the effects of aldosterone activation (Figure 15). As a result, sodium reabsorption and potassium elimination is reduced (hence the name '*potassium-sparing diuretics*'). Sodium elimination culminates in water loss and reduction of plasma volume blood pressure. The effect of MR on hypertension can also be observed in people harbouring the mutation S810L. This mutant receptor can be activated by cortisone, found in high concentrations, leading to an always-activated receptor. The phenotype observed is from severe hypertension [90, 99].

According to the World Health Organization (WHO) cardiovascular diseases account for 17 million deaths, and complications of hypertension account for 9.4 million deaths. Hypertension is responsible for at least 45% of deaths due to heart disease and 51% of deaths due to stroke (http://www.who.int). Since MR blockage can be effective in a number of cases, this nuclear receptor is an interesting therapeutic target.

The crystal structure of MR LBD was solved by different groups in 2005 [100,101]. The structure showed the typical arrangement of 12 helices with the ligand bound in the hydrophobic cavity formed by helices H5, H3 and H12. A comparison between the different ligands bound to this receptor showed that only subtle structural changes are observed when agonists or antagonists are bound to MR.

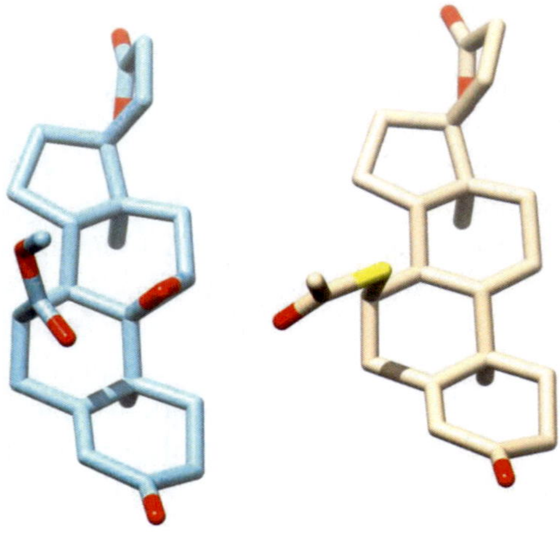

Figure 15. Eplerenone (left) and spironolactone (right) structures.

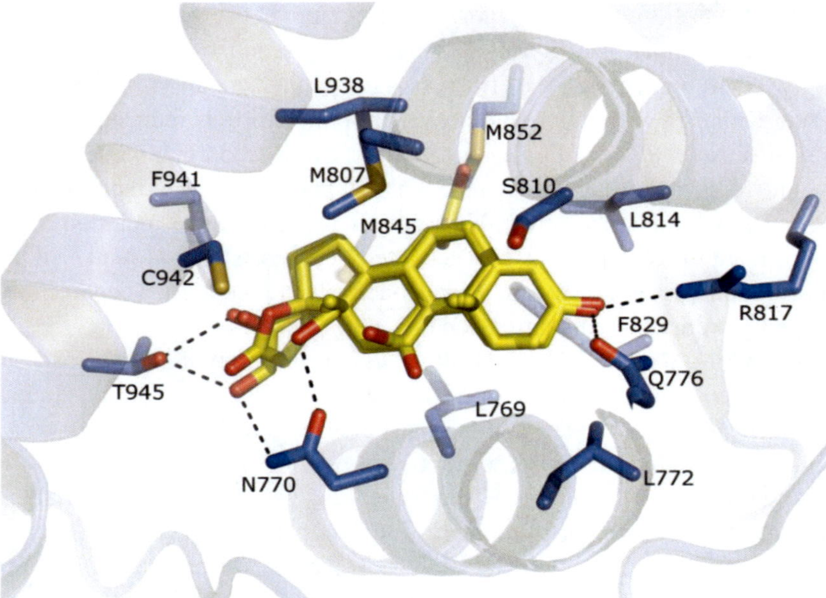

Figure 16. Interaction between MR and aldosterone, cortisone and spironolactone. The structure of these three ligands is shown in yellow sticks.

Aldosterone forms hydrogen bonds with R817 and Q776 through its carbonyl group in position 3. The oxygen atom at position 10 interacts with N770 and hydroxyl 21 interacts with T945. Similar interactions are observed for cortisone, except that this ligand has a carbonyl group in position 11, which would result in non-favourable interactions with N770 and cause repulsion with the carbonyl groups of helix 3 (Figure 16). This is the mechanism used to explain the low affinity between cortisone and wild type MR. Compared to cortisone, cortisol has a hydroxyl group in position 11 that can form favourable hydrogen bonds with N770 (through the oxygen of the amide) and with the main chain carbonyl groups.

A water molecule was modeled in the region close to the receptor β-hairpin for most of the ligands crystallized in a complex with MR. This water molecule participates in a hydrogen bond network that involves the carbonyl at position 3 of steroidal ligands, R817, Q776 and S810. A molecular dynamics simulation done in our group suggests that this network of polar contacts affects ligands' stability within the active site and mays also regulate ligand binding/unbinding kinetics (Nascimento, A.S., unpublished). The network of polar interactions, involving water molecules, has also been used to explain molecular selectivity at the receptor level [100]. Taken together, the set of structural data available to date provides only a few insights about the molecular mechanisms of ligand recognition by MR. More experimental data may deepen the current understanding of the structural biology of this receptor.

The CHEMBL database [102] lists over 1,000 compounds with known activity on human MR[1]. However, most of these compounds were designed as progesterone antagonists or glucocorticoid receptor antagonists, and were later found to be unspecific compounds (e.g., [103-105]). These data highlight two important aspects concerning the development of new MR blockers. First, the number of molecules proposed in recent years as specific MR blockers is small. Second, it is difficult to achieve specific ligands for steroid receptors. The ligand- binding domain within the steroid receptors is conserved, and so are their binding pocket residues. So, off-target interaction is a common feature between the members of this family. Also, these receptors evolved to bind similar molecules, derived from the same cholesterol scaffold.

For MR, this off-target interaction can result in important side effects. The approved blocker spironolactone is known to interact with the androgen receptor and progesterone receptor as well. This unintended interaction can lead to undesired effects such as gynecomastia, impotence and menstrual irregularities. Eplerenone, due to its 9,11-epoxide group, has a very low affinity for these receptors [106]. The side effects observed in spironolactone therapy are important causes for the low usage of this pharmacological treatment for hypertension.

3.2. Conclusion

Mineralocorticoid is probably the most recent steroid receptor to diverge in the evolutionary tree. As an inheritance, the receptor still recognizes glucocorticoids with high affinity. This receptor is a promising target for the pharmacological therapy of hypertension, and is a potential target for cardiac failure prevention. However, the usual off-target interaction of mineralocorticoid ligands with other steroid hormone receptors imposes technical difficulties in the development of new compounds that could be useful in clinical practice. More structural studies may clarify the mechanism at the receptor level for the recognition of its ligands.

[1] Access in May, 2013.

4. Perspectives for the Design of Novel Ligands

The interactions between the nuclear receptors briefly described in this chapter and its ligands leave some general lessons which are useful for the design of new ligands for nuclear receptors. First, the role of water molecules in the binding site should not be ignored. Water molecules were shown to play important roles in the partial agonism of INT131 on PPARγ [20], in the selectivity of the thyromimetic TRIAC to the TRβ isoform [107], and to participate in the network of hydrogen bonds that alter the role of cortisone in the presence of a mutation in MR [100], to number a few examples. Together, these results show that water molecules in the binding pocket region can play roles as structural elements of the receptor.

Second, all the possible interaction modes should be explored. The big binding pocket of PPARγ allows very different binding modes for the same receptors, and sometimes, for very similar ligands. This is the case for MRL-20 and MRL-24 [6]. This point is especially important for the so-called '*structure-based*' strategies for ligand design. Energetically favourable but uncommon interactions should not be ignored. In addition, as shown for the orphan nuclear receptor NGFI-B, the nuclear receptors may be flexible.

Third, sometimes more is less. Full activation of nuclear receptors can be an interesting strategy for maximizing the desired response. However, the side effects can also be maximal with receptor full activation. PPARγ shows that a submaximal response (10-20% of the maximal response) can result in the desired effect with minimal side effects in the individual.

Fourth, comparative analysis of ligand binding should be used to ensure specificity. Most of the reported mineralocorticoid antagonists have glucocorticoid/androgen/progesterone activity. The interaction with these SHRs must be avoided in most cases, due to the chance of important side effects. Even during the design stage, candidate molecules can be screened for their potential to bind to each of these receptors, eliminating those molecules with a higher chance of cross-interaction from the drug development pipeline.

References

[1] Ribeiro RC, Kushner PJ, Baxter JD (1995) The nuclear hormone receptor gene superfamily. *Annu. Rev. Med.* 46: 443-453.

[2] Rogerson F, Fuller P (2003) Interdomain interactions in the mineralocorticoid receptor. *Molecular and Cellular Endocrinology* 200: 45-55.

[3] Chandra V, Huang PX, Hamuro Y, Raghuram S, Wang YJ, et al. (2008) Structure of the intact PPAR-gamma-RXR-alpha nuclear receptor complex on DNA. *Nature* 456: 350-U333.

[4] Nascimento AS, Gomes Dias SM, Nunes FM, Aparicio R, Ambrosio ALB, et al. (2006) Structural rearrangements in the thyroid hormone receptor hinge domain and their putative role in the receptor function. *Journal of Molecular Biology* 360: 586-598.

[5] Ribeiro RC, Apriletti JW, Wagner RL, Feng W, Kushner PJ, et al. (1998) X-ray crystallographic and functional studies of thyroid hormone receptor. *J. Steroid. Biochem. Mol. Biol.* 65: 133-141.

[6] Bruning JB, Chalmers MJ, Prasad S, Busby SA, Kamenecka TM, et al. (2007) Partial agonists activate PPARgamma using a helix 12 independent mechanism. *Structure* 15: 1258-1271.

[7] Aranda A, Pascual A (2001) Nuclear hormone receptors and gene expression. *Physiol. Rev.* 81: 1269-1304.

[8] Varlakhanova N, Snyder C, Jose S, Hahm JB, Privalsky ML (2010) Estrogen receptors recruit SMRT and N-CoR corepressors through newly recognized contacts between the corepressor N terminus and the receptor DNA binding domain. *Mol. Cell Biol.* 30: 1434-1445.

[9] Jepsen K, Rosenfeld MG (2002) Biological roles and mechanistic actions of co-repressor complexes. *J. Cell Sci.* 115: 689-698.

[10] Feng W, Ribeiro R, Wagner R, Nguyen H, Apriletti J, et al. (1998) Hormone-dependent coactivator binding to a hydrophobic cleft on nuclear receptors. *Science* 280: 1747-1749.

[11] Marimuthu A, Feng W, Tagami T, Nguyew H, Jameson J, et al. (2002) TR surfaces and conformations required to bind nuclear receptor corepressor. *Molecular Endocrinology* 16: 271-286.

[12] Renaud JP, Moras D (2000) Structural studies on nuclear receptors. *Cell Mol. Life Sci.* 57: 1748-1769.

[13] Nagy L, Schwabe J (2004) Mechanism of the nuclear receptor molecular switch. *Trends in Biochemical Sciences* 29: 317-324.

[14] Puhl AC, Bernardes A, Silveira RL, Yuan J, Campos JL, et al. (2012) Mode of peroxisome proliferator-activated receptor γ activation by luteolin. *Mol. Pharmacol.* 81: 788-799.

[15] Waku T, Shiraki T, Oyama T, Maebara K, Nakamori R, et al. (2010) The nuclear receptor PPARgamma individually responds to serotonin- and fatty acid-metabolites. *EMBO J.* 29: 3395-3407.

[16] Molnár F, Peräkylä M, Carlberg C (2006) Vitamin D receptor agonists specifically modulate the volume of the ligand-binding pocket. *J. Biol. Chem.* 281: 10516-10526.

[17] Giguere V (1999) Orphan nuclear receptors: from gene to function. *Endocr. Rev.* 20: 689-725.

[18] Martinez L, Nascimento AS, Nunes FM, Phillips K, Aparicio R, et al. (2009) Gaining ligand selectivity in thyroid hormone receptors via entropy. *Proceedings of the National Academy of Sciences of the United States of America* 106.

[19] Bleicher L, Aparicio R, Nunes FM, Martinez L, Gomes Dias SM, et al. (2008) Structural basis of GC-1 selectivity for thyroid hormone receptor isoforms. *Bmc Structural Biology* 8.

[20] Motani A, Wang ZL, Weiszmann J, McGee LR, Lee G, et al. (2009) INT131: A Selective Modulator of PPAR gamma. *Journal of Molecular Biology* 386: 1301-1311.

[21] Hihi AK, Michalik L, Wahli W (2002) PPARs: transcriptional effectors of fatty acids and their derivatives. *Cell Mol. Life Sci.* 59: 790-798.

[22] Harmon GS, Lam MT, Glass CK (2011) PPARs and lipid ligands in inflammation and metabolism. *Chem. Rev.* 111: 6321-6340.

[23] Berger J, Moller DE (2002) The mechanisms of action of PPARs. *Annu. Rev. Med.* 53: 409-435.

[24] Dreyer C, Keller H, Mahfoudi A, Laudet V, Krey G, et al. (1993) Positive regulation of the peroxisomal beta-oxidation pathway by fatty acids through activation of peroxisome proliferator-activated receptors (PPAR). *Biol. Cell* 77: 67-76.

[25] Lalloyer F, Staels B (2010) Fibrates, glitazones, and peroxisome proliferator-activated receptors. *Arterioscler. Thromb. Vasc. Biol.* 30: 894-899.

[26] Issemann I, Green S (1990) Activation of a member of the steroid hormone receptor superfamily by peroxisome proliferators. *Nature* 347: 645-650.

[27] Dreyer C, Krey G, Keller H, Givel F, Helftenbein G, et al. (1992) Control of the peroxisomal beta-oxidation pathway by a novel family of nuclear hormone receptors. *Cell* 68: 879-887.

[28] Lee CH, Olson P, Evans RM (2003) Minireview: lipid metabolism, metabolic diseases, and peroxisome proliferator-activated receptors. *Endocrinology* 144: 2201-2207.

[29] Forman BM, Chen J, Evans RM (1997) Hypolipidemic drugs, polyunsaturated fatty acids, and eicosanoids are ligands for peroxisome proliferator-activated receptors alpha and delta. *Proc. Natl. Acad. Sci. U S A* 94: 4312-4317.

[30] Barter PJ, Rye KA (2008) Is there a role for fibrates in the management of dyslipidemia in the metabolic syndrome? *Arterioscler. Thromb. Vasc. Biol.* 28: 39-46.

[31] Evans RM, Barish GD, Wang YX (2004) PPARs and the complex journey to obesity. *Nat. Med.* 10: 355-361.

[32] Cariou B, Charbonnel B, Staels B (2012) Thiazolidinediones and PPARgamma agonists: time for a reassessment. *Trends Endocrinol. Metab.* 23: 205-215.

[33] Bensinger SJ, Tontonoz P (2008) Integration of metabolism and inflammation by lipid-activated nuclear receptors. *Nature* 454: 470-477.

[34] Fajas L, Auboeuf D, Raspe E, Schoonjans K, Lefebvre AM, et al. (1997) The organization, promoter analysis, and expression of the human PPARgamma gene. *J. Biol. Chem.* 272: 18779-18789.

[35] Fajas L, Fruchart JC, Auwerx J (1998) PPARgamma3 mRNA: a distinct PPARgamma mRNA subtype transcribed from an independent promoter. *FEBS Lett.* 438: 55-60.

[36] Lehmann JM, Moore LB, Smith-Oliver TA, Wilkison WO, Willson TM, et al. (1995) An antidiabetic thiazolidinedione is a high affinity ligand for peroxisome proliferator-activated receptor gamma (PPAR gamma). *J. Biol. Chem.* 270: 12953-12956.

[37] Jiang C, Ting AT, Seed B (1998) PPAR-gamma agonists inhibit production of monocyte inflammatory cytokines. *Nature* 391: 82-86.

[38] Guerre-Millo M, Gervois P, Raspe E, Madsen L, Poulain P, et al. (2000) Peroxisome proliferator-activated receptor alpha activators improve insulin sensitivity and reduce adiposity. *J. Biol. Chem.* 275: 16638-16642.

[39] Bailey CJ (2000) The rise and fall of troglitazone. *Diabetic Medicine* 17: 414-415.

[40] Friedland SN, Leong A, Filion KB, Genest J, Lega IC, et al. (2012) The cardiovascular effects of peroxisome proliferator-activated receptor agonists. *Am. J. Med.* 125: 126-133.

[41] Nissen SE, Wolski K (2007) Effect of rosiglitazone on the risk of myocardial infarction and death from cardiovascular causes. *New England Journal of Medicine* 356: 2457-2471.

[42] Henke BR (2004) Peroxisome proliferator-activated receptor alpha/gamma dual agonists for the treatment of type 2 diabetes. *J. Med. Chem.* 47: 4118-4127.

[43] Erdmann E, Spanheimer R, Charbonnel B (2010) Pioglitazone and the risk of cardiovascular events in patients with Type 2 diabetes receiving concomitant treatment with nitrates, renin-angiotensin system blockers, or insulin: results from the PROactive study (PROactive 20). *J. Diabetes* 2: 212-220.

[44] Reginato MJ, Lazar MA (1999) Mechanisms by which Thiazolidinediones Enhance Insulin Action. *Trends Endocrinol. Metab.* 10: 9-13.

[45] Wu Z, Xie Y, Morrison RF, Bucher NL, Farmer SR (1998) PPARgamma induces the insulin-dependent glucose transporter GLUT4 in the absence of C/EBPalpha during the conversion of 3T3 fibroblasts into adipocytes. *J. Clin. Invest.* 101: 22-32.

[46] Okuno A, Tamemoto H, Tobe K, Ueki K, Mori Y, et al. (1998) Troglitazone increases the number of small adipocytes without the change of white adipose tissue mass in obese Zucker rats. *Journal of Clinical Investigation* 101: 1354-1361.

[47] Acton JJ, Black RM, Jones AB, Moller DE, Colwell L, et al. (2005) Benzoyl 2-methyl indoles as selective PPAR gamma modulators. *Bioorganic & Medicinal Chemistry Letters* 15: 357-362.

[48] Amato AA, Rajagopalan S, Lin JZ, Carvalho BM, Figueira ACM, et al. (2012) GQ-16, a Novel Peroxisome Proliferator-activated Receptor γ (PPARγ) Ligand, Promotes Insulin Sensitization without Weight Gain. *Journal of Biological Chemistry* 287: 28169-28179.

[49] Choi JH, Banks AS, Estall JL, Kajimura S, Bostrom P, et al. (2010) Anti-diabetic drugs inhibit obesity-linked phosphorylation of PPARgamma by Cdk5. *Nature* 466: 451-456.

[50] Larsen PJ, Lykkegaard K, Larsen LK, Fleckner J, Sauerberg P, et al. (2008) Dissociation of antihyperglycaemic and adverse effects of partial perioxisome proliferator-activated receptor (PPAR-gamma) agonist balaglitazone. *European Journal of Pharmacology* 596.

[51] Amato AA, Rajagopalan S, Lin JZ, Carvalho BM, Figueira AC, et al. (2012) GQ-16, a novel peroxisome proliferator-activated receptor γ

(PPARγ) ligand, promotes insulin sensitization without weight gain. *J. Biol. Chem.* 287: 28169-28179.

[52] Weidner C, de Groot JC, Prasad A, Freiwald A, Quedenau C, et al. (2012) Amorfrutins are potent antidiabetic dietary natural products. *Proc. Natl. Acad. Sci. U S A* 109: 7257-7262.

[53] Zhang H, Xu X, Chen L, Chen J, Hu L, et al. (2011) Molecular determinants of magnolol targeting both RXRalpha and PPARgamma. *PLoS One* 6: e28253.

[54] Balakumar P, Rose M, Ganti SS, Krishan P, Singh M (2007) PPAR dual agonists: are they opening Pandora's Box? *Pharmacol. Res.* 56: 91-98.

[55] Conlon D (2006) Goodbye glitazars? *The British Journal of Diabetes & Vascular Disease* 6: 135-137.

[56] Xu HE, Lambert MH, Montana VG, Plunket KD, Moore LB, et al. (2001) Structural determinants of ligand binding selectivity between the peroxisome proliferator-activated receptors. *Proc. Natl. Acad. Sci. U S A* 98: 13919-13924.

[57] Evans JL, Lin JJ, Goldfine ID (2005) Novel approach to treat insulin resistance, type 2 diabetes, and the metabolic syndrome: simultaneous activation of PPARalpha, PPARgamma, and PPARdelta. *Curr. Diabetes Rev.* 1: 299-307.

[58] Yan Z, Ni Y, Wang P, Chen J, He H, et al. (2013) Peroxisome proliferator-activated receptor delta protects against obesity-related glomerulopathy through the P38 MAPK pathway. *Obesity* (Silver Spring) 21: 538-545.

[59] Lee EY, Kim GT, Hyun M, Kim S, Seok S, et al. (2012) Peroxisome proliferator-activated receptor-δ activation ameliorates albuminuria by preventing nephrin loss and restoring podocyte integrity in type 2 diabetes. *Nephrol. Dial. Transplant.* 27: 4069-4079.

[60] Oyama T, Toyota K, Waku T, Hirakawa Y, Nagasawa N, et al. (2009) Adaptability and selectivity of human peroxisome proliferator-activated receptor (PPAR) pan agonists revealed from crystal structures. *Acta Crystallogr. D Biol. Crystallogr.* 65: 786-795.

[61] Artis DR, Lin JJ, Zhang C, Wang W, Mehra U, et al. (2009) Scaffold-based discovery of indeglitazar, a PPAR pan-active anti-diabetic agent. *Proc. Natl. Acad. Sci. U S A* 106: 262-267.

[62] Bourguet W, Ruff M, Chambon P, Gronemeyer H, Moras D (1995) Crystal structure of the ligand-binding domain of the human nuclear receptor RXR-alpha. *Nature* 375: 377-382.

[63] Wagner RL, Apriletti JW, McGrath ME, West BL, Baxter JD, et al. (1995) A structural role for hormone in the thyroid hormone receptor. *Nature* 378: 690-697.

[64] Blondel A, Renaud JP, Fischer S, Moras D, Karplus M (1999) Retinoic acid receptor: a simulation analysis of retinoic acid binding and the resulting conformational changes. *J. Mol. Biol.* 291: 101-115.

[65] Sonoda MT, Martinez L, Webb P, Skaf MS, Polikarpov I (2008) Ligand dissociation from estrogen receptor is mediated by receptor dimerization: Evidence from molecular dynamics simulations. *Molecular Endocrinology* 22.

[66] Martínez L, Webb P, Polikarpov I, Skaf MS (2006) Molecular dynamics simulations of ligand dissociation from thyroid hormone receptors: evidence of the likeliest escape pathway and its implications for the design of novel ligands. *J. Med. Chem.* 49: 23-26.

[67] Martínez L, Sonoda MT, Webb P, Baxter JD, Skaf MS, et al. (2005) Molecular dynamics simulations reveal multiple pathways of ligand dissociation from thyroid hormone receptors. *Biophys. J.* 89: 2011-2023.

[68] Kallenberger BC, Love JD, Chatterjee VK, Schwabe JW (2003) A dynamic mechanism of nuclear receptor activation and its perturbation in a human disease. *Nat. Struct. Biol.* 10: 136-140.

[69] Chandra V, Huang P, Hamuro Y, Raghuram S, Wang Y, et al. (2008) Structure of the intact PPAR-gamma-RXR-alpha nuclear receptor complex on DNA. *Nature:* 350-356.

[70] Miyahara T, Schrum L, Rippe R, Xiong S, Yee HF, Jr., et al. (2000) Peroxisome proliferator-activated receptors and hepatic stellate cell activation. *J. Biol. Chem.* 275: 35715-35722.

[71] Rieusset J, Touri F, Michalik L, Escher P, Desvergne B, et al. (2002) A new selective peroxisome proliferator-activated receptor gamma antagonist with antiobesity and antidiabetic activity. *Mol. Endocrinol.* 16: 2628-2644.

[72] Paulsen RF, Granas K, Johnsen H, Rolseth V, Sterri S (1995) Three related brain nuclear receptors, NGFI-B, Nurr1, and NOR-1, as transcriptional activators. *J. Mol. Neurosci.* 6: 249-255.

[73] Maruyama K, Tsukada T, Ohkura N, Bandoh S, Hosono T, et al. (1998) The NGFI-B subfamily of the nuclear receptor superfamily (review). *Int. J. Oncol.* 12: 1237-1243.

[74] Castro DS, Arvidsson M, Bondesson Bolin M, Perlmann T (1999) Activity of the Nurr1 carboxyl-terminal domain depends on cell type and integrity of the activation function 2. *J. Biol. Chem.* 274: 37483-37490.

[75] Demura M, Wang F, Yoneda T, Karashima S, Mori S, et al. (2011) Multiple noncoding exons 1 of nuclear receptors NR4A family (nerve growth factor-induced clone B, Nur-related factor 1 and neuron-derived orphan receptor 1) and NR5A1 (steroidogenic factor 1) in human cardiovascular and adrenal tissues. *J. Hypertens.* 29: 1185-1195.

[76] Lammi J, Aarnisalo P (2008) FGF-8 stimulates the expression of NR4A orphan nuclear receptors in osteoblasts. *Mol. Cell Endocrinol.* 295: 87-93.

[77] Cho SD, Lee SO, Chintharlapalli S, Abdelrahim M, Khan S, et al. (2010) Activation of nerve growth factor-induced B alpha by methylene-substituted diindolylmethanes in bladder cancer cells induces apoptosis and inhibits tumor growth. *Mol. Pharmacol.* 77: 396-404.

[78] Milbrandt J (1988) Nerve growth factor induces a gene homologous to the glucocorticoid receptor gene. *Neuron* 1: 183-188.

[79] Liu ZG, Smith SW, McLaughlin KA, Schwartz LM, Osborne BA (1994) Apoptotic signals delivered through the T-cell receptor of a T-cell hybrid require the immediate-early gene nur77. *Nature* 367: 281-284.

[80] Woronicz JD, Calnan B, Ngo V, Winoto A (1994) Requirement for the orphan steroid receptor Nur77 in apoptosis of T-cell hybridomas. *Nature* 367: 277-281.

[81] Wahlberg LU, Lind G, Almqvist PM, Kusk P, Tornoe J, et al. (2012) Targeted delivery of nerve growth factor via encapsulated cell biodelivery in Alzheimer disease: a technology platform for restorative neurosurgery. *J. Neurosurg.* 117: 340-347.

[82] Flaig R, Greschik H, Peluso-Iltis C, Moras D (2005) Structural basis for the cell-specific activities of the NGFI-B and the Nurr1 ligand-binding domain. *J. Biol. Chem.* 280: 19250-19258.

[83] Wang Z, Benoit G, Liu J, Prasad S, Aarnisalo P, et al. (2003) Structure and function of Nurr1 identifies a class of ligand-independent nuclear receptors. *Nature* 423: 555-560.

[84] Chintharlapalli S, Burghardt R, Papineni S, Ramaiah S, Yoon K, et al. (2005) Activation of Nur77 by selected 1,1-Bis(3'-indolyl)-1-(p-substituted phenyl)methanes induces apoptosis through nuclear pathways. *J. Biol. Chem.* 280: 24903-24914.

[85] Zhan Y, Du X, Chen H, Liu J, Zhao B, et al. (2008) Cytosporone B is an agonist for nuclear orphan receptor Nur77. *Nat. Chem. Biol.* 4: 548-556.

[86] Vinayavekhin N, Saghatelian A (2011) Discovery of a protein-metabolite interaction between unsaturated fatty acids and the nuclear

receptor Nur77 using a metabolomics approach. *J. Am. Chem. Soc.* 133: 17168-17171.

[87] Codina A, Benoit G, Gooch JT, Neuhaus D, Perlmann T, et al. (2004) Identification of a novel co-regulator interaction surface on the ligand binding domain of Nurr1 using NMR footprinting. *J. Biol. Chem.* 279: 53338-53345.

[88] Evans RM (1988) The steroid and thyroid hormone receptor superfamily. *Science* 240: 889-895.

[89] Fardella CE, Miller WL (1996) Molecular biology of mineralocorticoid metabolism. *Annu. Rev. Nutr.* 16: 443-470.

[90] Rafestin-Oblin ME, Souque A, Bocchi B, Pinon G, Fagart J, et al. (2003) The severe form of hypertension caused by the activating S810L mutation in the mineralocorticoid receptor is cortisone related. *Endocrinology* 144: 528-533.

[91] Funder JW (2000) Aldosterone and mineralocorticoid receptors: orphan questions. *Kidney Int.* 57: 1358-1363.

[92] Pascual-Le Tallec L, Lombes M (2005) The mineralocorticoid receptor: A journey exploring its diversity and specificity of action. *Molecular Endocrinology* 19: 2211-2221.

[93] Hellal-Levy C, Fagart J, Souque A, Rafestin-Oblin ME (2000) Mechanistic aspects of mineralocorticoid receptor activation. *Kidney Int.* 57: 1250-1255.

[94] Bhargava A, Pearce D (2004) Mechanisms of mineralocorticoid action: determinants of receptor specificity and actions of regulated gene products. *Trends Endocrinol. Metab.* 15: 147-153.

[95] Berger S, Bleich M, Schmid W, Cole TJ, Peters J, et al. (1998) Mineralocorticoid receptor knockout mice: pathophysiology of Na+ metabolism. *Proc. Natl. Acad. Sci. U S A* 95: 9424-9429.

[96] Qin W, Rudolph AE, Bond BR, Rocha R, Blomme EA, et al. (2003) Transgenic model of aldosterone-driven cardiac hypertrophy and heart failure. *Circ. Res.* 93: 69-76.

[97] Kalidindi SR, Tang WH, Francis GS (2007) Drug insight: aldosterone-receptor antagonists in heart failure--the journey continues. *Nat. Clin. Pract. Cardiovasc. Med.* 4: 368-378.

[98] Baxter JD, Funder JW, Apriletti JW, Webb P (2004) Towards selectively modulating mineralocorticoid receptor function: lessons from other systems. *Mol. Cell Endocrinol.* 217: 151-165.

[99] Geller DS, Farhi A, Pinkerton N, Fradley M, Moritz M, et al. (2000) Activating mineralocorticoid receptor mutation in hypertension exacerbated by pregnancy. *Science* 289: 119-123.

[100] Bledsoe RK, Madauss KP, Holt JA, Apolito CJ, Lambert MH, et al. (2005) A ligand-mediated hydrogen bond network required for the activation of the mineralocorticoid receptor. *J. Biol. Chem.* 280: 31283-31293.

[101] Fagart J, Huyet J, Pinon GM, Rochel M, Mayer C, et al. (2005) Crystal structure of a mutant mineralocorticoid receptor responsible for hypertension. *Nat. Struct. Mol. Biol.* 12: 554-555.

[102] Gaulton A, Bellis LJ, Bento AP, Chambers J, Davies M, et al. (2012) ChEMBL: a large-scale bioactivity database for drug discovery. *Nucleic Acids Res.* 40: D1100-1107.

[103] Zhi L, Tegley CM, Pio B, Edwards JP, Motamedi M, et al. (2003) 5-benzylidene-1,2-dihydrochromeno[3,4-f]quinolines as selective progesterone receptor modulators. *J. Med. Chem.* 46: 4104-4112.

[104] Zhi L, Tegley CM, Pio B, West SJ, Marschke KB, et al. (2000) Nonsteroidal progesterone receptor antagonists based on 6-thiophenehydroquinolines. *Bioorg. Med. Chem. Lett.* 10: 415-418.

[105] Link JT, Sorensen BK, Patel J, Emery M, Grynfarb M, et al. (2004) Discovery of novel nonsteroidal glucocorticoid receptor modulators. *Bioorg. Med. Chem. Lett.* 14: 2209-2212.

[106] Goodman LS, Gilman A, Brunton LL, Lazo JS, Parker KL (2006) *Goodman & Gilman's the pharmacological basis of therapeutics*. New York: McGraw-Hill. xxiii, 2021 p. p.

[107] Martinez L, Nascimento AS, Nunes FM, Phillips K, Aparicio R, et al. (2009) Gaining ligand selectivity in thyroid hormone receptors via entropy. *Proceedings of the National Academy of Sciences of the United States of America* 106: 20717-20722.

In: Ligands
Editors: Paula Gawryszewska et al.
ISBN: 978-1-63117-143-7

Chapter 2

THE ROLE OF CHEMOKINE LIGANDS IN HUMAN HEALTH AND DISEASE

Ning Li,* Canhua Jiang, Feng Guo, Tong Su, Xinqun Chen, Lian Zheng and Xinchun Jian
Department of Oral and Maxillofacial Surgery, Xiangya Hospital, Central South University, Changsha, P.R. China

ABSTRACT

Chemokine ligands constitute a family of structurally related proteins. So far, the human chemokine superfamily includes at least 48 ligands, which bind to 18 functionally signaling G-protein-coupled receptors and 2 decoy or scavenger receptors. Chemokine ligands exert different functions under physiological conditions, such as the implantation of the early conceptus, the migration of cells during embryonic development, and the overall growth of the embryo. Chemokine ligands also play a significant role in homeostasis, development, tissue repair and angiogenesis. In addition, they have an important role in the development and maintenance of innate and adaptive immunity. When the physiological role of chemokine ligands is subverted or chronically amplified, disease often follows. Chemokine ligands and their receptors play a key role in the pathogenesis of chronic

* Corresponding author: Dr. Ning Li. Department of Oral and Maxillofacial Surgery, Xiangya Hospital, Central South University, Changsha, 410008. China. Email: liningbeta @hotmail.com.

inflammation, autoimmune diseases, tumorigenesis and cancer metastasis. Physicochemical properties of chemokine ligands and chemokine receptors confer the ability to homo- and hetero-oligomerize. A great deal of effort is currently being made to establish new therapeutic compounds able to target the chemokine ligand/chemokine receptor system. This article reviews the role of chemokine ligands and their receptors in normal and disease processes, and the emerging therapeutic approaches of using chemokine antagonists for appropriately targeted therapy.

Keywords: Chemokine ligand, human, disease, health, therapeutic approach

INTRODUCTION

To date, at least 46 chemokine ligands have been found in humans. Chemokine ligands have been grouped into four families, CXC, CC, CX3C, and XC, based on the arrangement of cysteine residues involved in the formation of disulfide bonds. In the CC family, the first two cysteines near the amino terminus are adjacent; whereas in the CXC and CX3C chemokine family, one or three amino acid residues are inserted between the first two of four cysteine residues, respectively; in the XC subfamily, the first and third cysteine residues are absent [1]. With the exceptions of CX3CL1/fractalkine and CXCL16/SR-PSOX, the chemokine ligands constitute a family of small soluble signaling molecules of approximately 70 amino acid residues with a molecular weight of 7–12 kDa. The human genome encompasses more than 50 different chemokine genes and pseudogenes. A large number of human chemokine genes are known to be clustered on specific chromosomal regions. There are two major gene clusters comprising exclusively either CXC or CC genes on chromosome 4q13.3-q21.1 and 17q12, respectively. These major clusters can be subdivided into two regions. For the CXC gene cluster, the regions are named GRO and IP10, while the regions of the CC gene cluster are called MCP and MIP [2]. The GRO region contains the CXCL1–CXCL8 genes, and the IP10 region contains the CXCL9–CXCL13 genes, respectively. In the CC major cluster, the MCP and MIP regions comprise 6 and 12 genes, respectively (CCL2, CCL7, CCL11, CCL8, CCL13, CCL1 versus CCL5, CCL16, CCL14, CCL15, CCL23, CCL18, CCL3, CCL4, CCL3L3, CCL4L1, CCL3L1, CCL4L2). In addition to the two major clusters, a CC "mini"-cluster is found on chromosome 7 (comprising the CCL26 and CCL24 genes), on chromosome 9 (CCL27, CCL19, CCL21), and on chromosome 16 (CCL22,

CX3CL1, and CCL17), respectively. Both XCL1 and XCL2 are also found in a "mini"-cluster on chromosome 1. Besides their structural classification, another organization of chemokine ligands has been proposed based on their expression and their functional activity. This classification groups chemokine ligands into three "families": pro-inflammatory, homeostatic, and mixed function [3]. Pro-inflammatory chemokine ligands are up-regulated under inflammatory conditions and are involved in leukocyte recruitment. Homeostatic chemokine ligands are expressed constitutively at non-inflamed sites, and are involved in the homeostatic migration and homing of cells in physiological conditions. Some chemokine ligands have both properties, and are thus called mixed-function chemokine ligands.

Chemokine ligands exert their effects by binding to chemokine receptors. Chemokine receptors are typically 340–380 amino acids in length with an extracellular acidic N-terminus which is important in tethering the basic chemokine ligand to the receptor. This facilitates a second interaction of lower affinity, in which the chemokine is delivered to the remainder of the receptor, leading to the activation of intracellular signaling. These chemokine receptors are also grouped into four families, CXCR, CCR, XCR, and CX3R, with CCR and CXCR being used to define receptors for CC and CXC chemokines, respectively [4]. The entire group of chemokine receptors belongs to the seven transmembrane G-protein coupled receptors (GPCRs) that usually combine the receptor to the Gai subunit of heterotrimeric G proteins. So far, around 25 human chemokine receptor genes have been identified (Table 1).

Interestingly, 12 of these receptors are found on human chromosome 13 and are stretched around 13.5 megabases. In addition, several decoy receptors have been reported to bind chemokine ligands without eliciting signal transduction. Promiscuity is common, with chemokine receptors typically interacting with several different ligands, although their ligand repertoire is typically class restricted. This means that CC chemokine receptors will interact with only CC chemokines and, likewise, CXC chemokine receptors will bind only CXC chemokines. Two exceptions to this rule of thumb are the chemokines CXCL9, 10 and 11, which have agonist activity at CXCR3, and also antagonist activity at CCR3 [5]. Another is the Duffy antigen receptor complex (DARC), which binds both CC and CXC chemokines [6]. As more and more chemokines were identified, intensive efforts were undertaken to match these with the many 'orphan' receptors previously identified as being expressed by leukocytes [7]. Chemokine receptors can be broadly divided into

Table 1. The chemokine superfamily of human

Human	Other names	Chromosome	Function	Cluster	Receptor
CXC family					
CXCL1	Groα	4q13.3	I	GRO	CXCR2, CXCR1
CXCL2	Groβ	4q13.3	I	GRO	CXCR2
CXCL3	Groγ	4q13.3	I	GRO	CXCR2
CXCL4	PF4	4q13.3	U	GRO	CXCR3B
CXCL4V 1		4q13.3	U	GRO	
CXCL5	ENA-78	4q13.3	I	GRO	CXCR2
CXCL6	GCP-2	4q13.3	I	GRO	CXCR1, CXCR2
CXCL7	NAP-2	4q13.3	I	GRO	Unknown
CXCL8	IL-8	4q13.3	I	GRO	CXCR1, CXCR2
CXCL9	MIG	4q21.1	I	IP10	CXCR3, CXCR3B
CXCL10	IP-10	4q21.1	I	IP10	CXCR3, CXCR3B
CXCL11	I-TAC	4q21.1	I	IP10	CXCR3, CXCR3B, CXCR7
CXCL12	SDF-1α/β	10q11.21	H		CXCR4, CXCR7
CXCL13	BLC, BCA-1	4q21.1	H	IP10	CXCR5
CXCL14	BRAK, Bolekine	5q31.1	I		Unknown
Unknown					Unknown
CXCL16		17p13.2	I		CXCR6
CXCL17	DMC	19q13.2	U		Unknown
CC family					
CCL1	I-309	17q11.2	I	MCP	CCR8
CCL2	MCP-1	17q11.2	I	MCP	CCR2
CCL3	MIP-1α,LD78β	17q11.2	I	MIP	CCR1, CCR5
CCL1	LD78β	17q12	I	MIP	
CCL3	LD78β	17q12	I	MIP	
CCL4	MIP-1β	17q12	I	MIP	CCR5
CCL1	AT744.2	17q12	I	MIP	

Human	Other names	Chromosome	Function	Cluster	Receptor
CCL2		17q12	I	MIP	
CCL5	RANTES	17q12	I		CCR1, CCR3, CCR5
CCL7	MCP-3	17q11.2	I	MCP	CCR1, CCR2, CCR3
CCL8	MCP-2	17q11.2	I	MCP	CCR1, CCR2, CCR3, CCR5
CCL11	Eotaxin	17q11.2	I	MCP	CCR3
CCL13	MCP-4	17q11.2	I	MCP	CCR1, CCR2, CCR3
CCL14	HCC-1	17q12	H	MIP	CCR1
CCL15	HCC-2	17q12	H	MIP	CCR1, CCR3
CCL16	HCC-4 LEC	17q12	H	MIP	CCR1, CCR2, CCR5, HRH4
CCL17	TARC	16q13	D	MIP	CCR4
CCL18	PARC	17q12	H		Unknown
CCL19	MIP3β, ELC	9p13.3	H		CCR7
CCL20	MIP3α, LARC	2q36.3	D		CCR6
CCL21	SLC	9p13.3	D		CCR7
CCL22	MDC	16q13	D		CCR4
CCL23	MPIF-1	17q12	I	MIP	CCR1, FPRL-1
CCL24	Eotaxin 2	7q11.23	I		CCR3
CCL25	TECK	19p13.2	H		CCR9
CCL26	Eotaxin 3	7q11.23	I		CCR3
CCL27	CTACK, ILC	9p13.3	H		CCR10
CCL28	MEC	5p12	U		CCR10,CCR3
Other classes					
XCL1	Lymphotacti n SCM-1α	1q24.2	D		XCR1
XCL2	SCM-1β	1q24.2	D		
CX3CL1	Fractalkine	16q13	I		CX3CR1

Functions are as follows:

I, inflammatory; H, homeostatic; D, dual (homeostatic and inflammatory); U, unknown.

GRO, GRO region of the CXC major gene cluster; IP10, IP10 region of the CXC major gene cluster; MCP, MCP region of the CC major gene cluster; MIP, MIP region of the CC major gene cluster.

those that are expressed exclusively by a particular subset of leukocytes or those which are more widely expressed. For example, CCR6 expression is largely restricted to immature dendritic cells and memory T-lymphocytes, facilitating their migration to lymphoid organs expressing the ligand CCL20. In contrast, the receptor for CCL11/eotaxin, CCR3, is expressed on a variety of cells involved in allergic responses, such as eosinophils, basophils and mast cells. T-helper lymphocytes of either the Th1 or Th2 subset also exhibit differential chemokine receptor expression profiles. Th1 lymphocytes selectively express CCR1, CCR5 and CXCR3, whereas CCR3, CCR4 and CCR8 are found on Th2 lymphocytes [8]. Such dynamic expression is thought to enable the T cells to respond to a variety of chemokines, undoubtedly providing flexibility in the adaptive immune response, and explains many of the earlier observations, such as CCL3 and CCL4 attracting distinct populations of lymphocytes.

Chemokine receptors bind chemokine ligands with high affinity to activate the heterotrimeric $G\alpha_i\beta_\gamma$ and a series of downstream effectors. A number of adaptor proteins dynamically interact with chemokine GPCRs to facilitate internalization and signal transduction, and form a dynamic "chemosynapse" that is vital for chemotaxis [9,10]. Overall, active $G\alpha_i$, $G\beta_\gamma$ and the "chemosynapse" facilitate directional sensing, polarization of the cell with the pseudopodium in the front, and the formation of a uropod at the back. Small GTPases Rac, Cdc42 and phosphatidylinositol-3-kinase (PI3K) accumulate at the leading edge with resultant actin polymerization and F-actin formation, leading to stabilization of the lamellipodium. Phosphatase and tensin homolog (PTEN) tyrosine/PIP3 phosphatase resides along the sides and at the posterior end of the migrating cells, Rho GTPase and its effectors accumulate at the trailing edge facilitating actomyosin contraction and tail retraction that result in migration of the cell [11].

1. Chemokines in Human Health

1.1. Chemokines in Human Health

Briefly, chemokine-receptor signaling enables human trophoblasts to migrate into the epithelial region of the endometrium. The human trophoblasts express the chemokine receptors CX3CR1, CCR1, CCR3, and their respective ligands are present in the endometrial microenvironment. The trophoblast also induces the expression of the chemokines CXCL1, CXCL2, CCL8 and the chemokine receptor CXCR4 in the maternal decidualized stromal cells. The bidirectional chemokine signaling between the trophoblast and the maternal endometrium may enable the successful implantation of the embryo [12]. Chemokines play a functional role in embryogenesis and development of the central nervous system (CNS) [13]. The chemokine receptors CXCR4 and CXCR7 and their ligand stromal cell derived factor-1α (SDF-1α) influence CNS development through homing of the neuronal precursor cells to their respective target areas of the developing brain. The chemokine CXCL12 is constitutively expressed in both neurons and astrocytes, and is involved in axon elongation and path finding [14]. The chemokine receptors CXCR4 and CXCR7 and their ligand CXCL12 also mediate and shape the migratory pattern of primordial stem cells [15]. Interference with the chemokines or their receptors mis-target the migratory patterns of progenitor cells, resulting in developmental abnormalities. The CXCL12/CXCR4 axis is also involved in myelopoiesis of the bone marrow, hematopoietic stem cell homing to sites of vascular expansion, and cardiac development [16,17].

1.2. Chemokines in Innate and Adaptive Immunity

Distinct pathogen-associated molecular patterns (PAMP) formed by the constitutive components and conserved products of microbial metabolism are recognized by receptors of the innate immune system, pattern-recognition receptors (PRRs), such as NOD1, NOD2 and Toll-like receptors (TLR) [18]. The TLR signaling triggers distinct biological responses through recruitment of different adaptors. During tissue damage, non-infectious endogenous molecules are released that cause sterile inflammation following activation of TLRs through damage-associated molecular patterns. Adenosine triphosphate (ATP) released from the necrotic cells was shown to activate the Nrlp3 inflammasome. This led to generation of an intravascular chemokine gradient in order to guide the neutrophils to the necrotic foci. Additionally, formyl-peptide signals released from the necrotic cells also navigated the neutrophils to the sites of sterile inflammation [19]. The cytokines generated through TLR signaling amplify the chemokine release, resulting in a polarized cytokine profile. The adaptive immune response is mediated by both homeostatic and inflammatory chemokines. The chemokines CCL19 and CCL21 activate the chemokine receptor CCR7 present on naive T cells, B cells, mature DC/Langerhans cells (LC) and CD56^{bright}NK cells, and induce their migration towards the T cell zone of secondary lymphoid organs (SLO) [20]. The B cells express the chemokine receptor CXCR5 that directs their homing to follicles of the lymph node (LN) that express the chemokine CXCL13, the ligand for CXCR5.

Immature DCs express CXCR1, CCR1, CCR2 and CCR6. The ligands for these receptors are inflammatory chemokines that recruit these immature DCs to the site of inflammation [21]. Infiltrated leukocytes at the sites of inflammation can promote oxidative stress.

1.3. Chemokines in Wound Healing

Wound healing is a well regulated process with three phases, namely inflammation, proliferation and remodeling. Chemokines are involved in each of these steps. The proliferative phase includes neo-angiogenesis, formation of granulation tissue and ECM, and re-epithelialization. The remodeling phase involves activation of matrix metalloproteinases through the mammalian target of the rapamycin (mTOR) pathway. The chemokines CXCL1 and CXCL8, as well as their receptor, CXCR2, are expressed in the human epidermis and are highly involved in wound healing [22]. Devalaraja et al. found that CXCR2 (-/-) mice exhibit impaired recruitment of neutrophils, reduction in migration and proliferation of keratinocytes during epithelialization, and also a significant delay in neovascularization and wound closure [23]. Expression of the angiostatic chemokine CXCL10 in transgenic mice was demonstrated to have an intense inflammatory phase coupled with an extended and disorganized granulation phase.

This resulted in impaired neovascularization of the wound bed and delayed wound healing [24]. Ishida et al. have shown that the chemokine CX3CL1 and its receptor CX3CR1 are upregulated at wound sites. The role of CX3CR1 in wound healing is further strengthened by the observation that there is reduced macrophage infiltration into the excisional wounds and reduced collagen deposition and neovascularization in CX3CR1 (-/-) mice [25]. The CCL27 chemokine accelerates wound healing through recruitment of bone marrow-derived keratinocyte precursor cells [26].

1.4. Chemokines in Angiogenesis

ELR^+ CXC-chemokines CXCL1, CXCL2, CXCL3, CXCL5, CXCL6, CXCL7, and CXCL8 are pro-angiogenic, and EL^- CXC-chemokines CXCL4,

CXCL4L1, CXCL9, CXCL10, CXCL11 and CXCL14 have angiostatic activity. The ELR^+ CXC-chemokines directly act on the chemokine receptors CXCR2 and CXCR1 in humans. These receptors are present on endothelial cells and induce migration, invasion and proliferation of the endothelial cells, resulting in angiogenesis. CXCR2 is the key receptor involved in neovascularization as all ELR^+ CXC-chemokines bind to CXCR2, but only CXCL6 and CXCL8 can bind to CXCR1. Additionally, only CXCR2 mediates chemotaxis of the endothelial cells, and blocking the function of CXCR2 blocks CXCL8 mediated angiogenesis, even though CXCL8 is capable of binding to CXCR1 [27]. ELR^+ CXC-chemokines also serve as a signal for the homing of endothelial progenitor cells from the bone marrow to the sites of angiogenesis. Neutralizing antibodies to CXCR2 blocks endothelial cell migration, and genetic ablation of CXCR2 results in profound inhibition of angiogenesis. Thus, CXCR2 exhibits a major role in wound healing. CC chemokines also have a role in angiogenesis. CCL2 mediates neovascularization by acting on CCR2 present in endothelial cells. CCL2, which acts on endothelial CCR2, also induces expression of VEGF-A and MCP-1 induced protein [28]. In addition, CCL11 and CCL16 regulate angiogenesis. Small molecule agents that either stimulate or inhibit chemokine mediated angiogenesis are vital for wound healing.

2. Chemokines in Human Disease

2.1. Human Autoimmune Diseases

Extensive knowledge of chemokines and chemokine receptors has had an impact on the many diseases where inflammation is a component. The differential distribution of chemokine receptors on different leukocyte types provides a template for determining the role of individual chemokines in particular diseases. Leukocyte recruitment represents a fundamental episode in inflammatory disorders, such as atherosclerosis, as well as in autoimmune diseases, such as in allergic asthma, rheumatoid arthritis, and chronic lung disease. The activated leukocyte at the chronic inflammatory site extrude their DNA decorated with antimicrobial peptides. This neutrophil extracellular trap (NET) is increased in chronic inflammation. The NET formation by neutrophils is mediated by CXCR2 and is independent of NADPH oxidase activity, but dependent on the Src family of kinases [29]. Excessive NET formation has been implicated in several autoimmune diseases, like preeclampsia, septic shock and autoimmune vasculitis, and also in cystic fibrosis.

2.1.1. Allergic asthma

Allergic asthma has an early phase of bronchoconstriction resulting from activation of mast cells and Th2 lymphocytes, and a marked infiltration of granulocytes, characteristically eosinophils as the predominant cells. Eosinophils are able to release cationic proteins and reactive oxygen species that are involved in host defence. It is thought that allergy is an aberration of this host defence system and that these eosinophil-derived products are also important in lung tissue damage and airway hyper-responsiveness, typical features of asthma. Several animal models of allergic airway disease, involving IL-5 gene deletion or neutralizing antibodies to IL-5, have supported the view that the eosinophil is important (IL-5 being necessary for eosinophil differentiation and proliferation in bone marrow). There has been considerable interest in the endogenous chemoattractants responsible for recruiting eosinophils into the asthmatic lung and into other tissues during allergic reactions. In the early 1990s, when we became interested in this area, there were no known selective eosinophil chemoattractant agents. This led to experiments on eotaxin, the last chemokine to be found by traditional biochemical techniques, and the only chemokine to be discovered using in vivo bioassay [30]. The generating system was also unique in being an in vivo system. Sensitized guinea pigs were challenged with an aerosol of ovalbumin, and bronchoalveolar lavage (BAL) fluid collected at intervals. The fluid was then injected intradermally into bioassay guinea-pigs previously injected intravenously with ^{111}In-labelled eosinophils. BAL fluid samples taken 1–3 h after challenge were found to induce the accumulation of radiolabelled eosinophils in the skin. The activity was purified by HPLC using the skin assay throughout to locate the fractions containing chemoattractant. Subsequent peptide sequencing revealed a novel CC chemokine that we named eotaxin. Research in other laboratories revealed that eotaxin signaled through a unique receptor, CCR3. Further work revealed eotaxin orthologues in several species, including man, and two other human eotaxins were discovered with low sequence similarity, but the same selective agonist activity on CCR3. The eotaxins are now known as eotaxin-1/CCL11, eotaxin- 2/CCL24 and eotaxin-3/CCL26. Eotaxins and CCR3 have been the subject of intensive investigation, because knowledge of their biology provides the opportunity to block eosinophil trafficking and its sequelae selectively [31]. Eotaxin-1 was found to be expressed in the lungs in allergen-challenged animals in early studies.

Eotaxin-1 was also found in the lungs of asthmatic patients, and its expression correlated with lung dysfunction. Several companies who successfully developed small-molecule antagonists suspended work in this area because of the lack of efficacy of IL-5 antibodies on lung function in asthmatic patients, implying that the eosinophil is not important in this respect. However, follow-up work in these trials showed that the antibody depleted eosinophils from the blood, but not effectively from the lungs. Interestingly, in support of work in mouse models, the antibodies were found to suppress indices of airway remodelling, suggesting that eosinophils are important in this respect. CCR3 is highly expressed in eosinophils, but also in other cells involved in allergic reactions, such as basophils, mast cells and a subpopulation of Th2 lymphocytes. This may have a bearing on the potential clinical activity of CCR3 antagonists. As mentioned above, allergic reactions have common features with host defence reactions to worm parasites, and the eosinophil has probably evolved as an effector cell specialized for this role. Accordingly, local eotaxin-1 generation has been observed in several parasite infection models. Interestingly, the hookworm Necator americanus produces a protease that selectively cleaves and inactivates eotaxin-1, another example of subversion of host-defence reaction by pathogenic organisms [32].

2.1.2. Rheumatoid arthritis

A marked infiltration of T- and B-lymphocytes, macrophages and neutrophils within the synovial lining is a hallmark of the inflammation clinically observed in RA, with Th1-associated chemokines thought to play a pivotal role [33]. In rheumatoid patients, CCL3/MIP-1a is readily detectable in synovial fluid, and increased levels have been reported to correlate with the severity of RA. In rodent models of RA, the murine orthologue of CCL3 has been found within arthritic joints during the evolution of both collagen-induced arthritis and adjuvant-induced arthritis, and neutralizing antibodies observed to provide a significant degree of protection, supportive of a role for the CCR1 axis in RA pathogenesis. Likewise, CCL2/MCP-1, a potent chemoattractant for monocytes, has been shown to be produced by both synovial cells and infiltrated cells in RA, with its specific receptor, CCR2, also attracting interest as a target for anti-inflammatory therapeutics. The murine orthologue of CCR2 is also upregulated in the joints from arthritic mice, and the use of a mutated form of the CCL2 molecule with antagonist properties led to a marked reduction in both the symptoms and histopathological damage observed. In contrast to peripheral blood T-cells, of which 40% are CXCR3 positive, almost all T-cells located within the inflammatory synovium of RA patients express CXCR3, which can be dissected further into increased numbers of memory CD3t/CD4t and CD3t/CD8t lymphocytes. This has led to the hypothesis that CXCR3 is important for T-lymphocyte homing to the RA joint, where the cells may orchestrate subsequent inflammatory events. Supportive of this, increased levels of CXCR3 agonists in the synovial tissue of RA patients have been reported in clinical studies. Thus, with respect to future RA treatments, at least three chemokine receptors present themselves as possible targets for the development of specific antagonists.

2.1.3. Chronic obstructive pulmonary disease (COPD)

Chronic obstructive pulmonary disease is due to the chronic inflammatory response of the respiratory system to smoking and/or repeated exposure to noxious pollutants in the air. COPD is characterized by lung infiltration of inflammatory cells of both the innate and adaptive immune systems, and both CC- and CXC-chemokines are involved in this process [34-36]. In COPD patients several CC chemokines are upregulated, including CCL2, CCL3, CCL5 and CCL20. In fact, in CCR5 (-/-) and CCR6 (-/-) mice exposed to nicotine there is reduced accumulation of macrophages, dendritic cells, neutrophils, and CD8+ T-cells in the lung. Furthermore, there is partial protection from developing COPD in these knockout mice [37]. The cytokine IL-17 upregulates several CXC chemokines and neutrophil survival factors in the airway epithelium, including chemokine ligands for CXCR2. This suggests that CXCR2 antagonists might be beneficial in the treatment of COPD [38]. The neutrophils chemoattractant proline-glycine-proline (PGP) is a biomarker of COPD. PGP is normally degraded by leukotriene A4 hydrolase (LTA4H), leading to resolution of the inflammation. Cigarette smoke selectively inhibited LTA4H, which led to an increased accumulation of PGP in the lungs. Elevated levels of PGP led to the prolonged presence of neutrophils in the lung, leading to the pathogenesis of COPD [39].

2.2. Human Cancer

Unresolved chronic inflammation leads to tumorigenesis. The tumor progression is influenced by the production of chemokines by tumor and stromal cells. Chemokines and their receptors provide directional cues for leukocytes or tumor cells for migration and metastasis, entry into the circulation, and homing and colonization to specific tissues. Many cancer types have upregulated expression of chemokines and chemokine receptors, leading to aberrant chemokine receptor signaling and expression. The amount and type of chemokines released determines the type and extent of leukocyte infiltration into the tumor microenvironment [40]. The literature demonstrates that most solid tumors are composed of malignant and stromal cells important for the growth and metastasis of the tumor. The stromal cells found in tumors are lymphocytes, macrophages, endothelial cells, fibroblasts, eosinophils, granulocytes, B cells, and natural killer cells. It has been clearly shown that the tumor microenvironment is composed of an extensive mix of both CC and CXC chemokine network that is involved in controlling the leukocyte infiltrate into the tumor. The chemokine mediated recruitment of different leukocytes into the chronic inflammatory and tumor microenvironment. CXCL12/CXCR4 and CCL21/CCR7 axes have been recognized as major factors in invasion and metastasis in several cancer types. The most frequently over-expressed chemokine receptor on tumor cells is CXC chemokine receptor 4 (CXCR4). CXCR4 selectively binds the CXC ligand 12 (CXCL12), also known as stromal cell-derived factor-1α (SDF-1α). The interaction between CXCR4 and CXCL12 is a normal process in cell migration. Muller et al. demonstrated high levels of CXCR4 expression on breast cancer cells that direct chemotaxis and invasive responses, enabling breast cancer cells to metastasize to other organs [41]. CXCL12 is expressed by stromal fibroblasts within tissues involved in the spread of breast cancer cells such as the lymph nodes, lung, liver, and bone marrow [41,42]. Inhibition of CXCR4 in MDA-MB-231 cells reduced breast cancer metastases to the lung and lymph node after tail vein or mammary fat pad injections [43,44]. CXCR4 is upregulated in more than 20 different tumor types. Other models in which CXCR4 has been suggested to play a role in metastasis are ovarian, prostate, and lung cancers. The CXCL12-CXCR4 axis has been implicated in the migratory behavior of glioma cells. Expression of CXCR1-CXCR5 has been reported in glioblastoma, with CXCR4 being the most frequently expressed and associated with cell invasion and poor patient prognosis. [45,46]. Human epidermal growth factor receptor 2 (HER2) upregulates the expression of the chemokine receptor CXCR4. This occurs through enhanced translation and attenuated degradation of CXCR4 upon activation of HER2. The upregulation of CXCR4 is required for the breast cancer cells to metastasize to the lungs. Elevated expression of CXCR4 and HER2 in human mammary carcinoma tissues correlates with poor overall survival [47]. CXCL12 activation of CXCR4 has also been shown to transactivate HER2 in lipid rafts in prostate cancer cells, and this HER2 activation promotes metastasis to the bone [48]. This was also shown to occur in breast cancer cells through src kinase activation [49]. The chemokine

CXCL12 also binds to CXCR7, further complicating the role of CXCL12-CXCR4 axis in regulating biological processes. It has been shown that the expression of CXCR7 is higher on transformed cells as compared to non-transformed cells [50]. The expression of CXCR7 has recently been shown in mammary carcinoma [50], prostate carcinoma [51] and non-small cell lung carcinoma [52]. Many cancer cells such as melanomas express a number of chemokines, including CXCL1 and CXCL8 and its receptors CXCR1 and CXCR2. Over-expression of CXCR1 and CXCR2 in melanoma cells leads to aggressive phenotype coupled with increased migration and tumor growth in mice, which can be reversed with antagonists, or neutralization of these receptors [53]. Other chemokines involved in melanoma are CXCL1-3, CCL5, and CCL2, which have been implicated in tumor growth and progression. Recent studies have demonstrated organ-specific patterns of melanoma metastasis that correlate with their expression of specific chemokine receptors, including CXCR4, CCR7, CCR9 and CCR10. The expression of CCR9 is associated with intestinal melanoma metastasis [54,55]. CCR9 is expressed on melanoma cells and melanoma cell lines established from small intestinal metastases. These cells were responsive to the CCR9 ligand CCL25, which is selectively expressed in the small intestine and thymus, suggesting an association between expression of CCR9 and organ specific metastasis [56]. In cutaneous melanocytic lesions, CCR10 was found expressed in both benign and malignant lesions. Evaluation of sentinel lymph nodes showed positive lymph nodes had increased CCR10 expression as compared to cases of negative lymph nodes, suggesting a role for CCR10 in human melanoma and its dissemination to lymph nodes [57]. CCR7 expression has been correlated with lymph node metastasis in gastric carcinoma [58], esophageal squamous cell carcinoma [59], non-small cell lung cancer [60], and colorectal carcinoma [61] and breast cancer [62]. A recent study demonstrated that CCR7 mediates leukemic T cell infiltration into the central nervous system in T-cell acute lymphoblastic leukemia (T-ALL). This study showed that CCR7 expression is regulated in leukemic cells by the T-ALL oncogene Notch1, and silencing of CCR7 or its ligand CCL19 in an animal model of T-ALL inhibits CNS infiltration [63]. Ghadjar et al. have shown a correlation of CCR6 expression with liver metastasis of colorectal cancer. Evaluation of CCR6 expression by immunohistochemistry verified the expression of CCR6 in all 64 primary tumor specimens: 24 of the 64 patients presented with liver metastasis [64]. In colorectal cancer, the predominant organ for metastasis is the liver. CCR6 has one known binding ligand, CCL20, which was originally identified in the liver, and is also expressed in mucosa and lymphoid tissues. The expression of

CCL20 in the liver generated interest in the role of CCR6 and its ligand in colorectal cancer and metastatic spread to the liver. Another study demonstrated that patients with colorectal cancer and liver metastasis had higher amounts of CCL20 in their liver compared to controls without metastases. Since CCR6 has been shown to play a role in tumor progression and metastasis, and CCL20 is expressed in the normal prostate, Ghadjar and colleagues used immunohistochemistry to show that CCR6 expression was associated with advanced and aggressive prostate cancer [65]. Rubie and colleagues have shown significant mRNA upregulation of CCL20 and CCR6 in samples from patients with pancreatic cancer. Additionally, CCL20 expression was strongly associated with advanced T-category (clinical stage) in patients with prostate cancer as compared to normal pancreatic tissue. Expression of CXCR3 has been detected in malignant epithelium from early stage breast cancer patients, and is associated with the promotion of lung metastasis in a murine mammary tumor model [66]. Silencing of CXCR3 in mammary tumor cell lines inhibited spontaneous lung metastasis and lung colonization from tumor cells implanted in the mammary gland in a murine model. Increased levels of CCL2 expression in breast cancer have been associated with increased levels of tumor- associated macrophages, correlating with invasive phenotype and poor prognosis. Myeloid monocytic cells like TAM, MDSC, DC and Tie-2 macrophages are recruited to the tumor through CCL2. Such CCL2 recruited immune cells produce VEGF, PDGF, TGFβ, CXCL8, MMP-2 and MMP-8. These proteins promote tumor angiogenesis and progression, and hence poor prognosis. Thus, the chemokine and its cognate receptors play a vital role in the migration of tumor cells from their primary site, via the circulation, to the preferential sites of metastases. Decoy receptors sequester chemokines without activating signaling pathways, and generally function as tumor suppressors [67]. In addition, studies have shown that decreased expression of the Duffy antigen receptor for chemokines (DARC) and D6 decoy receptors in breast cancer inversely correlate with lymph node metastases and increased survival rates [68]. The DARC receptor has also been shown to support enhanced chemokine-induced extravasation of leukocytes, and thus support the pro-migratory activity of chemokines [69].

3. THERAPEUTICS

Since chemokines and their receptors are involved in many diseases, there is hope that good antagonists will increase the means to treat them. The different interactions between chemokines can be used to find new targets against human diseases by blocking heterophilic interactions, multiple chemokine axes, direct blocking of chemokine receptors, as well as the blocking of glycosaminoglycan (GAGs)-binding sites. Several possibilities for therapeutically targeting chemokines are illustrated in the following part.

3.1. Modified Chemokines

Modifying the target chemokine is one option for creating antagonists since changing the molecular structure leads to a different binding pattern and receptor response. The N-terminal part is especially crucial for receptor signaling, and thus a change in this domain can lead to alteration or loss of receptor activation. Variants of chemokines with an extended or modified N-terminal part are, for example, N-methylated CCL5 (Met-RANTES) or amino-oxypentane-RANTES, which block the CCL5 receptors CCR1, CCR3, and CCR5 [70]. In liver fibrosis and atherosclerosis it was shown that inhibiting CCL5 receptors through Met-RANTES was sufficient to reduce inflammation in mice [71]. This phenomenon was observed in vivo and in vitro. Another example is R6H-CXCL8, a variant of CXCL8 with substitutions on the conserved ELR-triad and CXC-motif, which exclusively activates CXCR1 without effecting CXCR2. This is based on a distinct CXCL8 binding mechanism: the CXCR2 activation is mediated by the N-terminal ELR- and CXC-motif, whereas the N-loop of CXCL8 is essential for CXCR1 activation [72]. Furthermore, the CXCL8 variant displays anti-inflammatory properties since it activates CXCR1 by desensitization of the CXCR2 response in human neutrophils. In fact, this agonist could help to clarify the biological and physiological function, especially of CXCR1, in inflammatory diseases. Thus, R6H-CXCL8 is a potential candidate as a therapeutic molecule.

3.2. Glycosaminoglycan-Binding Affinity

Most chemokines have the ability to bind GAGs located on the cell surface. The enhancement or reduction of this property can diminish the Guanosine-binding protein coupled receptor (GPCR) signaling by an indirect blockade of chemokine binding to its receptor, since GAGs are co-factors for GPCR activation. It is assumed that chemokines first bind the GAG co-receptor followed by GPCR activation. A variety of chemokines was

previously designed with altered GAG-binding affinities, resulting in a loss of GPCR activation [73]. The activity of the pro-inflammatory chemokine CCL5 depends on the binding to GAGs. The substitution of positively charged residues into alanine in the 40s loop ([44AANA47]-CCL5mutant) results in defective heparin binding and loss of the ability to recruit monocytes. The heteromerization of both CCL5 variants leads to non-functional heteromers, with a lack of GAG-binding efficiency [74]. Another study additionally confirms [44AANA47]-CCL5 as a potential therapeutic agent against atherosclerosis [75]. But in contrast to Met-RANTES, [44AANA47]-CCL5 does not directly abolish GPCR activation. Thus, variations of CCL5 mutants acting in different ways can lead to anti-inflammatory properties by a direct blockade of the GPCR or by indirect inhibition through prevention of chemokine binding to GAGs on the cell surface. Another way to block chemokine activity using the affinity for GAGs is to design dominant-negative mutants with a higher GAG-binding affinity compared to the wild type chemokine[76]. H23K-RANTES showed attenuation of autoimmune uveitis in rats based on displacement of wild type CCL5 from its proteoglycan-co-receptor. Mutants with increased GAG-binding potential were designed for CCL2, as well. The PA508 mutant of CCL2 exhibits no ability for CCR2 activation but a fourfold higher GAG affinity compared to the wild type CCL2 [77]. H23K-RANTES showed attenuation of autoimmune uveitis in rats based on displacement of wild type CCL5 from its proteoglycan-co-receptor. Mutants with increased GAG-binding potential were designed for CCL2, as well. The PA508 mutant of CCL2 exhibits no ability for CCR2 activation but a fourfold higher GAG affinity compared to the wild type CCL2 [78]. In a recent study in mice, PA508-CCL2 showed prevention of neointima formation and reduction of tissue damage after myocardial infarction without notable side effects. Therefore, it could be a candidate for a therapeutic agent in reducing restenosis in stents. Additionally, a mutant of CXCL12 with a deficiency in heparan sulfate binding can still transduce signals through CXCR4 but is not able to promote transendothelial migration in vitro. In vivo experiments could further show that this mutant efficiently downregulates the CXCR4 expression and desensitizes the chemotactic response toward CXCL12. Hence, this modified chemokine might work in anti-inflammatory therapies [79].

3.3. Small Molecules and Antibodies

The development of small molecules blocking GPCR activation is a powerful tool for the treatment of inflammatory diseases. Major efforts have been made to find drugs for blocking HIV infection. Maraviroc was established as a functional anti-HIV drug by blocking CCR5 as an important entry receptor. In inflammatory diseases, like atherosclerosis, TAK779 and nbI-74330 antagonists for CCR5 and CXCR9, respectively, represent suitable therapeutic agents [80]. Antagonizing CXCR4 by TAK779 additionally blocks leukocyte trafficking induced by CXCL12 [81]. Recently, DF 2156A was introduced as a novel dual inhibitor of CXCL8 receptors CXCR1 and CXCR2 [82]. This dual function is based on a non-competitive inhibition resulting in a stabilized binding between DF 2156A and the two CXCL8 receptors due to the formation of specific ionic bonds in the allosteric binding site. Some CXCR2- and CCR2- specific antagonists (i.e. reparixin and MLN1202) have already been tested as therapeutic drugs in clinical trials, e.g., MLN1202, which is a CCR2-blocking monoclonal antibody shown to reduce high-sensitivity CRP as a surrogate parameter for atherosclerosis [83].

3.4. Chemokine Heteromerization

As mentioned before some chemokines inherently exhibit a syner- gistic function toward other chemokines which mostly depends on heteromerization. Disruption or changing these critical heterophilic interactions might entail a decrease in the physiological response which has an impact on the degree of inflammation. A prominent example is the heterophilic interaction betweenCXCL4 and CCL5 which results in a synergistic enhancement of CCL5-induced signaling accompanied by increased monocyte recruitment to the inflamed endothelium [84]. Interruption of the chemokine heteromerization by cyclic peptides was shown to eliminate the synergistic effect in vitro and in vivo. Recently, it was shown that small peptide antagonists, disrupting CXCL4–CCL5 heteromer formation in mouse models of acute lung injury, result in improved lung edema, less neutrophil infiltration, and reduced tissue damage [85]. Thus, targeting heterophilic chemokine interactions can act as a therapeutic approach by attenuating inflammatory disease in a mild way.

CONCLUSION

Chemokines play a paramount role in physiology and homeostasis, as well as in the pathogenesis of tumors and their metastasis. A multi-pronged therapeutic approach would be beneficial. This includes reduction or elimination of inflammation, oncogene inactivation, and polarization of immune cells to the Th1 type. A variety of approaches is involved in the reduction or removal of the inflammation. It may be the lowering of stress levels, chronic low dose employment of anti-inflammatory drugs, or removal or blocking the accumulation of the chemokines. If we can reduce chronic inflammation, this could have a huge, positive effect on cancer therapy.

Grant Sponsor: the National Natural Sciences Foundation of China (Grant No. 81000445), and State Key Specialist Construction Projects of China.

REFERENCES

[1] Hisayuki, N., Kunio, H., Naoki, O., Yoko, K. U., Kaori, O. O., Sumio, T., Toshiaki, I., Akio, Y., Yutaka, K., Sumio, T., Retsu, M., Jun, K., Miki, N. & Osamu, Y. (2008) Extensive expansion and diversification of the chemokine gene family in zebrafish: Identification of a novel chemokine subfamily CX. BMC. *Genomics*, *9*, 222.

[2] Nomiyama, H., Osada, N. & Yoshie, O. (2010). The evolution of mammalian chemokine genes. *Cytokine. Growth. Factor. Rev.*, *21*, 253–262.

[3] Mantovani, A., Bonecchi, R. & Locati, M. (2006). Tuning inflammation and immunity by chemokine sequestration: decoys and more. *Nat. Rev. Immunol. 6*, 907–918.

[4] Nomiyama, H., Osada, N. & Yoshie, O. (2011). A family tree of vertebrate chemokine receptors for a unified nomenclature. *Dev. Comp. Immunol.*, *35*, 705–715.

[5] Loetscher, P., Pellegrino, A., Gong, J. H., Mattioli, I., Loetscher, M., Bardi, G., Baggiolini, M. & Clark, L. I. (2001). The ligands of CXC chemokine receptor 3, I-TAC, Mig and IP10, are natural antagonists for CCR3. *J. Biol. Chem.*, *276*, 2986–2991.

[6] Lu, Z. H., Wang, Z. X., Horuk, R., Hesselgesser, J., Lou, Y. C., Hadley, T. J. & Peiper, S. C. (1995). The promiscuous chemokine binding profile

of the Duffy antigen/receptor for chemokines is primarily localized to sequences in the amino-terminal domain. *J. Biol. Chem., 270*, 26239–26245.

[7] Bleul, C. C., Farzan, M., Choe, H., Parolin, C., Clark, L. I., Sodroski, J. & Springer, T. A. (1996.) The lymphocyte chemoattractant SDF-1 is a ligand for LESTR/fusin and blocks HIV-1 entry. *Nature*, *382*, 829–833.

[8] Bonecchi, R., Bianchi, G., Bordignon, P. P., Ambrosio, D., Lang, R., Borsatti, A., Sozzani, S., Allavena, P., Gray, P. A., Mantovani, A. & Sinigaglia, F. (1998). Differential expression of chemokine receptors and chemotactic responsiveness of type 1 T helper cells (Th1 s) and Th2s. *J. Exp. Med.*, *187*, 129–134.

[9] Barzik, M., Raman, D., Sobolik, D. T., Sai, J., Ham, A. J., Mernaugh, R. L., Gertler, F. B. & Richmond, A. (2009). VASP is a CXCR2-interacting protein that regulates CXCR2-mediated polarization and chemotaxis. *J. Cell. Sci.*, *122*, 1882–94.

[10] Raman, D., Sai, J., Chew, C. S. & Richmond, A. (2010). LIM and SH3 protein-1 modulates CXCR2- mediated cell migration. *PLoS. One*, *5*, e10050.

[11] Ridley, A. J., Schwartz, M. A., Burridge, K., Firtel, R. A., Ginsberg, M. H., Borisy, G., Parsons, J. T. & Horwitz, A. R. (2003). Cell migration: integrating signals from front to back. *Science*, *302*, 1704–9.

[12] Hannan, N. J. (2007). Role of chemokines in the endometrium and in embryo implantation. *Curr. Opin. Obstet. Gynecol.*, *19*, 266–72.

[13] Boldajipour, B., Mahabaleshwar, H., Kardash, E., Reichman, F. M., Blaser, H., Minina, S., Wilson, D., Xu, Q. & Raz, E. (2008). Control of chemokine-guided cell migration by ligand sequestration. *Cell*, *132*, 463–73.

[14] Ohshima, Y., Kubo, T., Koyama, R., Ueno, M., Nakagawa, M. & Yamashita, T. (2008). Regulation of axonal elongation and pathfinding from the entorhinal cortex to the dentate gyrus in the hippocampus by the chemokine stromal cell-derived factor 1 alpha. *J. Neurosci.*, *28*, 8344–53.

[15] Sanchez, A. V., Camp, E., Leal, T. A., Atkinson, S. P., Armstrong, L., Diaz, L. M. & Mullor, J. L. (2010). Nanog regulates primordial germ cell migration through Cxcr4b. *Stem. Cells*, *28*, 1457–64.

[16] Ceradini, D. J., Kulkarni, A. R., Callaghan, M. J., Tepper, O. M., Bastidas, N., Kleinman, M. E., Capla, J. M., Galiano, R. D., Levine, J. P. & Gurtner, G. C. (2004). Progenitor cell trafficking is regulated by

hypoxic gradients through HIF-1 induction of SDF-1. *Nat. Med., 10*, 858–64.
[17] McGrath, K. E., Koniski, A. D., Maltby, K. M., McGann, J. K. & Palis, J. (1999). Embryonic expression and function of the chemokine SDF-1 and its receptor, CXCR4. *Dev. Biol., 213*, 442–56.
[18] Akira, S., Uematsu, S. & Takeuchi, O. (2006). Pathogen recognition and innate immunity. *Cell, 124*, 783– 801.
[19] McDonald, B., Pittman, K., Menezes, G. B., Hirota, S. A., Slaba, I., Waterhouse, C. C., Muruve, D. A. & Kubes, P. (2010). Intravascular danger signals guide neutrophils to sites of sterile inflammation. *Science, 330*, 362–6.
[20] Esche, C. & Stellato, C. (2005). Chemokines: key players in innate and adaptive immunity. *J. Invest. Dermatol., 125*, 615–28.
[21] Mueller, S. N. & Germain, R. N. (2009). Stromal cell contributions to the homeostasis and functionality of the immune system. *Nat. Rev. Immunol., 9*, 618–29.
[22] Nanney, L. B., Mueller, S. G., Bueno, R., Peiper, S. C. & Richmond, A. (1995). Distributions of melanoma growth stimulatory activity of growth-regulated gene and the interleukin-8 receptor B in human wound repair. *Am. J. Pathol., 147*, 1248–60.
[23] Devalaraja, R. M., Nanney, L. B., Du, J., Qian, Q., Yu, Y. & Richmond, A. (2000). Delayed wound healing in CXCR2 knockout mice. *J. Invest. Dermatol., 115*, 234–44.
[24] Luster, A. D., MacLean, J. A., Crowe, K. & Granstein, R. D. (1998). Delayed wound healing and disorganized neovascularization in transgenic mice expressing the IP-10 chemokine. *Proc. Assoc. Am. Physicians., 110*, 183–96.
[25] Ishida, Y., Gao, J. L. & Murphy, P. M. (2008). Chemokine receptor CX3CR1 mediates skin wound healing by promoting macrophage and fibroblast accumulation and function. *J. Immunol., 180*, 569–79.
[26] Inokuma, D., Abe, R., Fujita, Y., Sasaki, M., Shibaki, A., Nakamura, H., McMillan, J. R., Shimizu, T. & Shimizu, H. (2006). CTACK/CCL27 accelerates skin regeneration via accumulation of bone marrow derived keratinocytes. *Stem. Cells, 24*, 2810–6.
[27] Heidemann, J., Ogawa, H., Dwinell, M. B., Rafiee, P., Maaser, C., Gockel, H. R., Otterson, M. F., Ota, D. M., Lugering, N., Domschke, W. & Binion, D. G. (2003). Angiogenic effects of interleukin 8 (CXCL8) in human intestinal microvascular endothelial cells are mediated by CXCR2. *J. Biol. Chem., 278*, 8508–15.

[28] Keeley, E. C., Mehrad, B. & Strieter, R. M. (2008). Chemokines as mediators of neovascularization. *Arterioscler. Thromb. Vasc. Biol., 28*, 1928–36.

[29] Marcos, V., Zhou, Z., Yildirim, A. O., Bohla, A., Hector, A., Vitkov, L., Wiedenbauer, E. M., Krautgartner, W. D., Stoibe, W., Belohradsky, B. H., Rieber, N., Kormann, M., Koller, B., Roscher, A., Roos, D., Griese, M., Eickelberg, O. & Doring, G. (2010). Mall MA and Hartl D. CXCR2 mediates NADPH oxidase-independent neutrophil extracellular trap formation in cystic fibrosis airway inflammation. *Nat. Med., 16*, 1018–23.

[30] Jose, P. J., Collins, P. D., Walsh, D. T., Moqbel, R., Totty, N. F., Truong, O., Hsuan, J. J. & Williams, T. J. (1994). Eotaxin: a potent eosinophil chemoattractant cytokine detected in a guinea-pig model of allergic airways inflammation. *J. Exp. Med., 179*, 881–887.

[31] Pease, J. E. & Williams, T. J. (2001). Eotaxin and asthma. *Curr.Opin. Pharmacol.*, 1, 248–253.

[32] Culley, F. J., Brown, A., Conroy, D. M., Sabroe, I., Pritchard, D. I. & Williams, T. J. (2000). Eotaxin is specifically cleaved by hookworm metalloproteases preventing its action in vitro and in vivo. *J. Immunol., 165*, 6447–6453.

[33] Koch, A. E. (2005). Chemokines and their receptors in rheumatoid arthritis: future targets? *Arthritis. Rheum., 52*, 710–721.

[34] Barnes, P. J. (2009. The cytokine network in chronic obstructive pulmonary disease. *Am. J. Respir. Cell. Mol. Biol., 41*, 631-8.

[35] Larsson, K. (2008). Inflammatory markers in COPD. *Clin. Respir. J., 21*, 84–7.

[36] Traves, S. L. & Donnelly, L. E. (2008). Th17 cells in airway diseases. *Curr. Mol. Med., 8*, 416–26.

[37] Bracke, K. R., Demedts, I. K., Joos, G. F. & Brusselle, G. G. (2007). CC-chemokine receptors in chronic obstructive pulmonary disease. Inflamm. *Allergy. Drug Targets, 6*, 75–9.

[38] Barnes, P. J. (2008). Frontrunners in novel pharmacotherapy of COPD. *Curr. Opin. Pharmacol., 8*, 300– 7.

[39] Snelgrove, R. J., Jackson, P. L., Hardison, M. T., Noerager, B. D., Kinloch, A., Gaggar, A., Shastry, S., Rowe, S. M., Shim, Y. M., Hussell, T. & Blalock, J. E. (2010). A critical role for LTA4H in limiting chronic pulmonary neutrophilic inflammation. *Science, 330*, 90–4.

[40] Lazennec, G. & Richmond, A. (2010). Chemokines and chemokine receptors: new insights into cancer-related inflammation. *Trends. Mol. Med.*, *16*, 133–44.

[41] Muller, A., Homey, B., Soto, H., Ge, N., Catron, D., Buchanan, M. E., McClanahan, T., Murphy, E., Yuan, W., Wagner, S. N., Barrera, J. L., Mohar, A., Verastegui, E. & Zlotnik, A. (2001). Involvement of chemokine receptors in breast cancer metastasis. *Nature*, *410*, 50–6.

[42] Allinen, M., Beroukhim, R., Cai, L., Brennan, C., Lahti, D. J., Huang, H., Porter, D., Hu, M., Chin, L., Richardson, A., Schnitt, S., Sellers, W. R. & Polyak, K. (2004). Molecular characterization of the tumor microenvironment in breast cancer. *Cancer. Cell*, *6*, 17–32.

[43] Liang, Z., Wu, T., Lou, H., Yu, X., Taichman, R. S., Lau, S. K., Nie, S., Umbreit, J. & Shim, H. (2004). Inhibition of breast cancer metastasis by selective synthetic polypeptide against CXCR4. *Cancer. Res.*, *64*, 4302–8.

[44] Liang, Z., Yoon, Y., Votaw, J., Goodman, M. M., Williams, L. & Shim, H. (2005). Silencing of CXCR4 blocks breast cancer metastasis. *Cancer. Res.*, *65*, 967–71.

[45] Ehtesham, M., Mapara, K. Y., Stevenson, C. B. & Thompson, R. C. (2009). CXCR4 mediates the proliferation of glioblastoma progenitor cells. *Cancer. Lett.*, *274*, 305–12.

[46] Zhou, Y., Larsen, P. H., Hao, C. & Yong, V. W. (2002). CXCR4 is a major chemokine receptor on glioma cells and mediates their survival. *J. Biol. Chem.*, *277*, 49481–7.

[47] Li, Y. M., Pan, Y., Wei, Y., Cheng, X., Zhou, B. P., Tan, M., Zhou, X., Xia, W., Hortobagyi, G. N., Yu, D. & Hung, M. C. (2004). Upregulation of CXCR4 is essential for HER2-mediated tumor metastasis. *Cancer. Cell*, 6, 459–69.

[48] Chinni, S. R., Yamamoto, H., Dong, Z., Sabbota, A., Bonfil, R. D. & Cher, M. L. (2008). CXCL12/CXCR4 transactivates HER lipid rafts of prostate cancer cells and promotes growth of metastatic deposits in bone. *Mol. Cancer. Res.*, *6*, 446–57.

[49] Cabioglu, N., Summy, J., Miller, C., Parikh, N. U., Sahin, A. A., Tuzlali, S., Pumiglia, K., Gallick, G. E. & Price, J. E. (2005). CXCL-12/stromal cell-derived factor-1alpha transactivates HER2-neu in breast cancer cells by a novel pathway involving Src kinase activation. *Cancer. Res.*, *65*, 6493–7.

[50] Miao, Z., Luker, K. E., Summers, B. C., Berahovich, R., Bhojani, M. S., Rehemtulla, A., Kleer, C. G., Essner, J. J., Nasevicius, A., Luker, G. D.,

Howard, M. C. & Schall, T. J. (2007). CXCR7 (RDC1) promotes breast and lung tumor growth in vivo and is expressed on tumor-associated vasculature. *Proc. Natl. Acad. Sci. U.S.A.*, *104*, 15735–40.

[51] Wang, J., Shiozawa, Y., Wang, J., Wang, Y., Jung, Y., Pienta, K. J., Mehra, R., Loberg, R. & Taichman, R. S. (2008). The role of CXCR7/RDC1 as a chemokine receptor for CXCL12/SDF prostate cancer. *J. Biol. Chem.*, *283*, 4283–94.

[52] Iwakiri, S., Mino, N., Takahashi, T., Sonobe, M., Nagai, S., Okubo, K., Wada, H., Date, H. & Miyahara, R. (2009). Higher expression of chemokine receptor CXCR7 is linked to early and metastatic recurrence in pathological stage I nonsmall cell lung cancer. *Cancer*, *115*, 2580–93.

[53] Singh, S., Nannuru, K. C., Sadanandam, A., Varney, M. L. & Singh, R. K. (2009). CXCR1 and CXCR2 enhances human melanoma tumourigenesis, growth and invasion. *Br. J. Cancer*, *100*, 1638–46.

[54] Amersi, F. F., Terando, A. M., Goto, Y., Scolyer, R. A., Thompson, J. F., Tran, A. N., Morton, D. L. & Hoon, D. S. (2008). Activation of CCR9/CCL cutaneous melanoma mediates preferential metastasis to the small intestine. *Clin. Cancer. Res.*, *14*, 638–45.

[55] Richmond, A. (2008). CCR9 homes metastatic melanoma cells to the small bowel. *Clin. Cancer. Res.*, *14*, 621–3.

[56] Letsch, A., Keilholz, U., Schadendorf, D., Assfalg, G., Asemissen, A. M., Thiel, E. & Scheibenbogen, C. (2004). Functional CCR9 expression is associated with small intestinal metastasis. *J. Invest. Dermatol.*, *122*, 685–90.

[57] Simonetti, O., Goteri, G., Lucarini, G., Filosa, A., Pieramici, T., Rubini, C., Biagini, G. & Offidani, A. (2006). Potential role of CCL27 and CCR10 expression in melanoma progression and immune escape. *Eur. J. Cancer.*, *42*, 1181–7.

[58] Mashino, K., Sadanaga, N., Yamaguchi, H., Tanaka, F., Ohta, M., Shibuta, K., Inoue, H. & Mori, M. (2002). Expression of chemokine receptor CCR7 is associated with lymph node metastasis of gastric carcinoma. *Cancer. Res.*, *62*, 2937–41.

[59] Ding, Y., Shimada, Y., Maeda, M., Kawabe, A., Kaganoi, J., Komoto, I., Hashimoto, Y., Miyake, M., Hashida, H. & Imamura, M. (2003). Association of CC chemokine receptor 7 with lymph node metastasis of esophageal squamous cell carcinoma. *Clin. Cancer*, *Res. 9*, 3406–12.

[60] Takanami, I. (2003). Overexpression of CCR7 mRNA in nonsmall cell lung cancer: correlation with lymph node metastasis. *Int. J. Cancer*, *105*, 186–9.

[61] Gunther, K., Leier, J., Henning, G., Dimmler, A., Weissbach, R., Hohenberger, W. & Forster, R. (2005). Prediction of lymph node metastasis in colorectal carcinoma by expressionof chemokine receptor CCR7. *Int. J. Cancer*, *116*, 726–33.

[62] Cabioglu, N., Yazici, M. S., Arun, B., Broglio, K. R., Hortobagyi, G. N., Price, J. E. & Sahin, A. (2005). CCR7 and CXCR4 as novel biomarkers predicting axillary lymph node metastasis in T1 breast cancer. *Clin. Cancer Res.*, *11*, 5686–93.

[63] Buonamici, S., Trimarchi, T., Ruocco, M. G., Reavie, L., Cathelin, S., Mar, B. G., Klinakis, A., Lukyanov, Y., Tseng, J. C., Sen, F., Gehrie, E., Li, M., Newcomb, E., Zavadil, J., Meruelo, D., Lipp, M., Ibrahim, S., Efstratiadis, A., Zagzag, D., Bromberg, J. S., Dustin, M. L. & Aifantis, I. (2009). CCR7 signalling as an essential regulator of CNS infiltration in T-cell leukaemia. *Nature*, *459*, 1000-4.

[64] Ghadjar, P., Coupland, S. E., Na, I. K., Noutsias, M., Letsch, A., Stroux, A., Bauer, S., Buhr, H. J., Thiel, E., Scheibenbogen, C. & Keilholz, U. (2006. Chemokine receptor CCR6 expression level and liver metastases in colorectal cancer. *J. Clin. Oncol.*, *24*, 1910–6.

[65] Ghadjar, P., Loddenkemper, C., Coupland, S. E., Stroux, A., Noutsias, M., Thiel, E., Christoph, F., Miller, K., Scheibenbogen, C. & Keilholz, U. (2008). Chemokine receptor CCR6 expression level and aggressiveness of prostate cancer. *J. Cancer. Res. Clin. Oncol.*, *134*, 1181–9.

[66] Ma, X., Norsworthy, K., Kundu, N., Rodgers, W. H., Gimotty, P. A., Goloubeva, O., Lipsky, M., Li, Y., Holt, D. & Fulton, A. (2009). CXCR3 expression is associated with poor survival in breast cancer and promotes metastasis in a murine model. *Mol. Cancer Ther.*, *8*, 490-8

[67] Hansell, C. A., Simpson, C. V. & Nibbs, R. J. (2006). Chemokine sequestration by atypical chemokine receptors. *Biochem. Soc. Trans.*, *34*, 1009–13.

[68] Wu, F. Y., Ou, Z. L., Feng, L. Y., Luo, J. M., Wang, L. P., Shen, Z. Z. & Shao, Z. M. (2008). Chemokine decoy receptor d6 plays a negative role in human breast cancer. *Mol. Cancer Res.*, *6*, 1276–88.

[69] Pruenster, M., Mudde, L., Bombosi, P., Dimitrova, S., Zsak, M., Middleton, J., Richmond, A., Graham, G. J., Segerer, S., Nibbs, R. J. & Rot, A. (2009. The Duffy antigen receptor for chemokines transports chemokines and supports their promigratory activity. *Nat. Immunol.*, *10*, 101–8.

[70] Veillard, N. R., Kwak, B., Pelli, G., Mulhaupt, F., James, R. W., Proudfoot, A. E. & Mach, F. (2004). Antagonism of RANTES receptors reduces atherosclerotic plaque formation in mice. *Circ. Res., 94*, 253–261.

[71] Berres, M. L., Koenen, R. R., Rueland, A., Zaldivar, M. M., Heinrichs, D., Sahin, H., Schmitz, P., Streetz, K. L., Berg, T., Gassler, N., Weiskirchen, R., Proud, A., Weber, C., Trautwein, C. & Wasmuth, H. E. (2010). Antagonism of the chemokine Ccl5 ameliorates experimental liver fibrosis in mice. *J. Clin. Invest., 120*, 4129–4140.

[72] Sarmiento, J., Shumate, C., Suetomi, K., Ravindran, A., Villegas, L., Rajarathnam K., & Navarro, J. (2011). Diverging mechanisms of activation of chemokine receptors revealed by novel chemokine agonists. *PLoS. One, 6*, e27967.

[73] Rek, A., Krenn, E. & Kungl, A. J. (2009). Therapeutically targeting protein-glycan interactions. *Br. J. Pharmacol., 157*, 686–694.

[74] Koenen, R. R. & Weber, C. (2011). Chemokines: established and novel targets in atherosclerosis. *EMBO. Mol. Med., 3*, 713–725.

[75] Braunersreuther, V., Steffens, S., Arnaud, C., Pelli, G., Burger, F., Proudfoot, A. & Mach, F. (2008). A novel RANTES antagonist prevents progression of established atherosclerotic lesions in mice. *Arterioscler. Thromb. Vasc. Biol., 28*, 1090–1096.

[76] Brandner, B., Rek, A., Diedrichs, M. M., Wildner, G. & Kungl, A. J. (2009). Engineering the glycosaminoglycan-binding affinity, kinetics and oligomerization behavior of RANTES: a tool for generating chemokine-based glycosamino glycan antagonists. *Protein. Eng. Des. Sel., 22*, 367–373.

[77] Piccinini, A. M., Knebl, K., Rek, A., Wildner, G., Diedrichs, M. M. & Kungl, A. J. (2010). Rationally evolving MCP-1/CCL2 into a decoy protein with potent antiinflammatory activity in vivo. *J. Biol. Chem., 285*, 8782–8792.

[78] Ruth, J. H., Haas, C. S., Park, C. C., Amin, M. A., Haines, G. K., Shahrara, S. & Koch, A. E. (2006). CXCL16-mediated cell recruitment to rheumatoid arthritis synovial tissue and murine lymph nodes is dependent upon the MAPK pathway. *Arthritis. Rheum., 54*, 765–78.

[79] O'Boyle, G., Mellor, P., Kirby, J. A. & Ali, S. (2009). Antiinflammatory therapy by intravenous delivery of nonheparan sulfatebinding CXCL12. *FASEB J., 23*, 3906–3916.

[80] Koenen, R. R. & Weber, C. (2010). Therapeutic targeting of chemokine interactions in atherosclerosis. *Nat. Rev. Drug. Discov., 9*, 141–153.

[81] Sohy, D., Yano, H., Nadai, P., Urizar, E., Guillabert, A., Javitch, J. A., Parmentier, M. & Springael, J. Y. (2009). Hetero-oligomerization of CCR2, CCR5, and CXCR4 and the protean effects of "selective" antagonists. *J. Biol. Chem.*, *284*, 31270–31279.

[82] Bertini, R., Barcelos, L., Beccari, A., Cavalieri, B., Moriconi, A., Bizzarri, C., Benedetto, P., Giacinto, C., Gloaguen, I., Galliera, E., Corsi, M., Russo, R., Andrade, S., Cesta, M., Nano, G., Aramini, A., Cutrin, J., Locati, M., Allegretti, M. & Teixeira, M. (2012). Receptor binding mode and pharmacological characterization of a potent and selective dual CXCR1/CXCR2 noncompetitive allosteric inhibitor. *Br. J. Pharmacol.*, *16*, 436–454.

[83] Gilbert, J., Lekstrom, H. J., Don, D., Lee, Y., Hu, M., Xu, J., Wyant, T. & Davidson, M. (2011). Effect of CC chemokine receptor 2 CCR2 blockade on serum C-reactive protein in individuals at atherosclerotic risk and with a single nucleotide polymorphism of the monocyte chemoattractant protein-1 promoter region. *Am. J. Cardiol.*, *107*, 906–911.

[84] Koenen, R. R. & Weber, C. (2011). Manipulating the chemokine system: therapeutic perspectives for atherosclerosis. *Curr. Opin. Investig. Drugs*, *11*, 265–272.

[85] Grommes, J., Alard, J. E., Drech, M., Wantha, S., Morgelin, M., Kuebler, W. M., Jacobs. M., Von Hundelshausen, P., Markart, P., Wygrecka, M., Preissne, rK. T., Hackeng, T. M., Koenen, R. R., Weber, C. & Soehnlein, O. (2012). Disruption of platelet-derived chemokine heteromers prevents neutrophil extravasation in acute lung injury. *Am. J. Respir. Crit. Care. Med.*, *185*, 628–636.

In: Ligands
Editors: Paula Gawryszewska et al.
ISBN: 978-1-63117-143-7

Chapter 3

Calix[4]arene Methylene Bisphosphonic Acids as Effectors of Biochemical Processes

S. V. Komisarenko,[1] S. O. Kosterin,[1] E. V. Lugovskoy,[1] S. O. Cherenok,[2] V. Yu. Tanchuk,[3] A. I. Vovk[3] and V. I. Kalchenko[2,*]

[1] Palladin Institute of Biochemistry, NASU, Kyiv, Ukraine
[2] Institute of Organic Chemistry, NASU, Kyiv, Ukraine
[3] Institute of Bioorganic Chemistry and Petrochemistry, NASU, Kyiv, Ukraine

Abstract

Calix[4]arenes bearing methylene bisphosphonic acid moieties at the upper rim were synthesized with high yields by the reaction of the easily available formyl calixarenes with sodium salt of dialkylphosphites, and consecutive treatment of the bisphosphonates formed with trimethylbromosilane and methanol.

Calix[4]arene methylene bisphosphonic acids were shown to effectively inhibit ATPase activity of pig myometrium myosin subfragment-1 (inhibition coefficient I0.5 = 83 ± 7 μM). At the same time, these substances cause a significant increase of myosin

* Murmanska, 5 str. 02660, Kyiv-94, Ukraine; Email: vik@ioch.kiev.ua.

subfragment-1 hydrodynamic diameter, which may indicate the formation of an intermolecular complex between the calixarene and myosin head. Computer simulation methods (docking and molecular dynamics with the addition of grid technologies) enabled the elucidation of the grounds of intermolecular interactions between the calix[4]arene and myometrium myosin subfragment-1 that involve hydrophobic, electrostatic and π-π-stacking interactions, some of which are close to the ATPase active centre.

Calixarene tetrakis-methylene bisphosphonic acid is a powerful inhibitor of fibrin polymerization. The fibrin polymerization in the fibrinogen + thrombin reaction is decreased by 50% at calixarene concentrations of 0.52·10-6 M (IC50). At this concentration the molar ratio of the compound to fibrinogen was 1.7 : 1. For the case of desAB fibrin polymerization, the IC50 was 1.26·10-6 M at a molar ratio of calixarene to the fibrin monomer of 4 : 1. The calixarene blocks fibrin formation by combining with polymerization site ‘A’ (Aa17–19), which ordinarily initiates protofibril formation in a ‘knob-hole’ manner. The calixarene also doubled both the prothrombin time and the activated partial thromboplastin time in normal human blood plasma at concentrations of 7.13·10-5 M and 1.10·10-5M, respectively.

The inhibition of alkaline phosphatases by calix[4]arenes functionalysed at the macrocyclic upper rim by one or two methylenebisphosphonic acid fragments has been investigated. The compounds were found to display strong inhibition of alkaline phosphatase from bovine intestinal mucosa with inhibition constants in the micromolar range. The inhibiting activity of bis-substituted calix[4]arene was found to be considerably lower than that of monosubstituted derivative. At the same time the functionalysed macrocycles showed lower and almost similar levels of inhibitory activities on bovine kidney alkaline phosphatase or E. coli alkaline phosphatase. The tested compounds were docked computationally to the active site of the E. coli alkaline phosphatase. On the basis of results obtained the possible binding modes of inhibitors were analyzed.

Keywords: Calixarenes, methylene bisphosphonic acids, ATPases, computer simulation methods, fibrin polymerization, alkaline phosphatases

INTRODUCTION

One of most important methodological approaches applied in molecular and cellular technologies in biochemistry, molecular biology and biophysics, which is focused on the study of the structural and functional properties of biomolecules, supramolecular complexes [1] and subcellular and membrane structures, is based on the use of natural and artificial effectors *E* (activators and inhibitors) for modifying the activities of enzymatic, transport, regulator and signaling proteins. [2,3] Therefore, studying the physico-chemical and biochemical basis and mechanisms involving the interactions of small effector molecules with biomacromolecules and biomembranes has become one of the basic current problems of so- called "contact biology". It would be no exaggeration to say that "contact" events play a crucial role in the realization of practically all fundamental biological phenomena that are studied by transdisciplinary chemico-biological sciences.

It is obvious that an artificial effector *E* pretending to form a "molecular platform" able to effectively regulate the functional activity of an intracellular protein must satisfy, as a minimum, the following conditions: 1) be nontoxic and non-immunogenic for the organism; 2) to penetrate well through the plasmatic cellular membrane; 3) to function as a reversible mechanism; 4) to show a specific (selective) effect exactly on a specific protein; 5) to be characterized by a relatively high affinity towards a target protein (effector constant $K_e < 10^{-5}$ M).

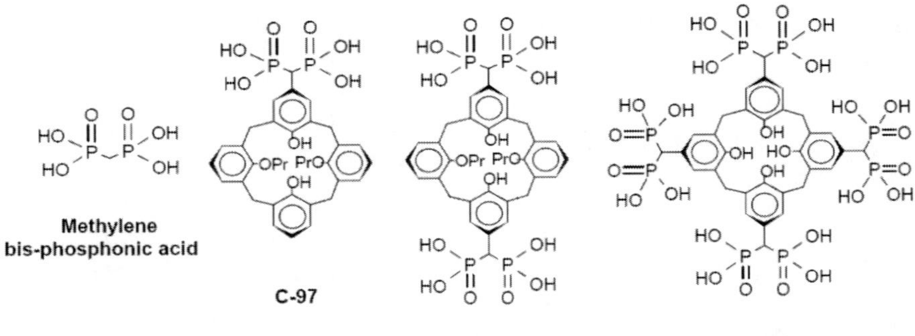

Methylene bis-phosphonic acid

C-97

C-98

C-192

Figure 1. Molecular structures of methylene bisphosphonic acid, calixarene methylene bisphosphonic acid C-97, calixarene di-methylene bisphosphonic acid C-98 and calixarene tetrakis-methylene bisphosphonic acid C-192.

In the context of the above-mentioned, much attention of researchers is currently devoted to phenol cyclic oligomers, calixarenes [4], as perspective artificial effectors for different biochemical processes. [5] This paper describes the results of a project focused on the synthesis, elucidation of biological activity, and biopharmaceutical application of the compounds C-97 C-98 and C-192 possessing one, two or four biophoric methylene bisphosphonic acid

(MBPA) groups at the upper rim of the calix[4]arene molecular platform (Figure 1).

1. Calixarenes of Bio-Medical Potential

Calixarenes [4] are vase-shaped macrocyclic compounds produced synthetically by precise cyclocondensation of *p*-substituted phenols and formaldehyde, which possess intramolecular lipophilic cavities formed by aromatic rings of the macrocyclic skeleton (Figure 2). Calixarenes may be easily functionalized at the upper or at the lower rims of the macrocyclic skeleton.

Due to their ability to form a variety of interactions – multi-site hydrogen bonding, specific stacking, and generalized electrostatic interactions – calixarene derivatives can be applied as valid recognition motifs when developing synthetic supramolecular systems. They have the unique ability to recognize and bind to the Host-Guest supramolecular complexes' cations, anions and neutral molecules, and biomolecules of appropriate size and architecture [6,7,8] These properties open wide perspectives for the practical application of calixarenes in different branches of chemistry, physics and biology. [9,10]

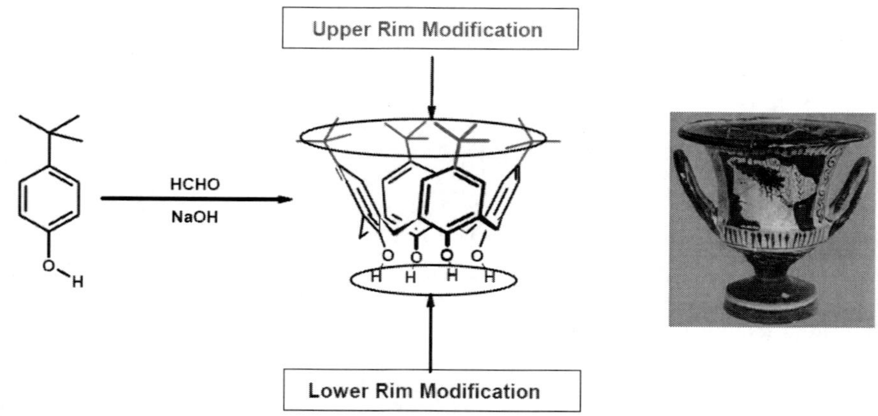

Figure 2. Synthesis and spatial structure of calix[4]arene.

During the last 10 years calixarene derivatives, especially water soluble and amphiphilic ones, have been the subject of growing interest in the development of biosensors in the field of biotechnology and drug discovery. [11,12] The highly diverse biomedical applications of these molecules now include antibacterial, anticancer, antiviral, antithrombotic, and membranotropic activities, and selective enzyme blocking and mimicking due to their ability of protein complexation [13,14] Calixarenes, substituted at the upper and/or lower macrocyclic rim with biologically active functional groups are considered to be promising compounds for treating many diseases. [15,16]

We were interested in modifying the calixarene platform by bio-relevant phosphorus [17,18,19], phosphonic [20,21], hydroxyphosphonic [22,23], aminophosphonic [24] and especially methylene bisphosphonic acid [25] groups. Methylene bisphosphonic acid, being a structural analog of natural pyrophosphate, demonstrates versatile bioactivities. [26,27]

2. Methylene Bisphosphonic Acids As Drugs

The first synthesis of methylene bisphosphonic acid was published by Menschutkin in1865, [28] and after that by von Baeyer and Hofmann in 1897. [29] The development of bisphosphonates (BPs) for medical use began with the realization by Fleisch and his colleagues that inorganic pyrophosphate (PPi), a naturally occurring compound in the body, and present also in the serum and urine, can prevent calcification by binding to newly forming crystals of hydroxyapatite. [30,31] Consequently, it was postulated that PPi might be the agent that normally prevents calcification of soft tissues, and regulates bone mineralization, and that pathological disorders might be linked to disturbances in PPi metabolism. Attempts to exploit these concepts by using pyrophosphate and polyphosphates to inhibit ectopic calcification in blood vessels, skin and kidneys in laboratory animals were successful only when the compounds were injected. Orally administered pyrophosphate and polyphosphates were inactive due to the hydrolysis of PPi in the gastrointestinal tract, probably by mucosal brush-border phosphatases. [32]

The bioactivities of BPs were investigated in detail by Komisarenko. The antitumor and immunomodulating activities of methylene bisphosphonic acid were revealed and carefully examined (Figure 1). [33,34] Phosphonates and bisphosphonates are structural analogs of inorganic pyrophosphate and phosphate, accordingly; they were chosen for the study *a priori* in consideration of the fact that PP_i takes part as a product or substrate in a large number of the most important enzymatic reactions. It may also be found in many metabolites. The presence of P–C bonds in phosphonates and P–C–P bonds in bisphosphonates, which usually cannot be hydrolyzed by enzymes, and the strong complex-forming abilities of phosphonates suggest that these substances may be used for the perturbation of cell metabolic pathways; although, according to some literature data, bisphosphonates do not influence the activity of a large quantity of enzymes, in particular pyrophosphatase. Contrary to this opinion, it was shown that bisphosphonates were effective inhibitors of inorganic pyrophosphatase and some other enzymes in reactions involving inorganic pyrophosphate. [35,36,37,38,39,40] The antitumor effect of MBPA was also shown – a property which needed further analysis and preclinical investigation. In particular, the mechanism of bisphosphonate effect and dependence of enzyme activities on the structure of bisphosphonates were investigated. MBPA was shown to possess immunomodulatory activity. Thus, when introduced in animals, MBPA caused inhibition of both biosynthesis of antibodies to T-dependent antigens and cell immune response reactions. It also inhibited biosynthesis of IgM-, IgE- and, especially, IgG-class antibodies, but did not influence the composition of the lymphocyte subpopulation. Other examined bisphosphonates (oxy- and aminoderivatives of MBPA) were strong complex-forming substances but did not possess an immunomodulatory effect. MBPA and its analogs – oxyethylidene bisphosphonic acid, aminobisphosphonic acid and phosphonoacetic acid – showed no toxicity and did not inhibit lymphocyte proliferation in culture when stimulated by mitogens. MBPA did not essentially influence the synthesis of specific interleukins by lymphocytes. It could not go through the placental barrier and showed no embryo-toxic effect. The use of ^{14}C-MBPA made it possible to reveal MBPA tropism towards lymphoid cells and to determine the kinetic and thermodynamic parameters of MBPA transport into cells. The data obtained suggest that MBPA could penetrate into lymphocytes with a transporter through the concentration gradient by means of the facilitated diffusion mechanism.

Such transport may be competitively inhibited by PP_i and oxyethylidene bisphosphonic acid (but no P_i), and does not depend on the intensity of ATP synthesis in the cell. The types of structural complexes that contain bisphosphonates inside cells and analyzed complexes with biologically important metals which may interact with cell enzymatic systems were calculated and determined. [41]

Substances which differed in the number of phosphoryl groups, type of bonds (P–C, P–O or P–N) or charge and molecular size, were used to study the mechanism by which bisphosphonates and phosphonates may influence the activity of a number of key enzymes responsible for the transformation of PP_i

or molecules including PP_i. It was concluded that tropism for lymphocytes is the basis of the immunomodulatory effects of phosphonates (first of all, MBPA). Transport and accumulation of MBPA in lymphocytes results in the inhibition of inorganic pyrophosphatase activity and an increase of local PP_i concentration. Later, different ligand complexes of bisphosphonates with bivalent metal ions and with PP_i are formed, and some enzymes change their activity, in particular, DNA-dependent RNA-polymerase II, enzymes of purine metabolism, and so on. [42]

On the basis of bisphosphonates, several prototypes of medicinal substances were developed. [43,44,45] Thus, a polyurethane composition that might serve as an immobilized immunomodulator with local anti-inflammatory and immunosuppressive effects was synthesized. It was shown that disodium salt of methylene bisphosphonic acid had antitumor properties, which resulted in the development of a new antitumor drug, MEBIFON, that was tested clinically and is currently produced by FARMAK in Kyiv.

The medical uses of bisphosphonates are summarized by Breuer. [46] Many BPs are currently in clinical use (Medronate, Oxidronate, Etidronate, Clodronate, Pamidronate, Alendronate, Tiludronate, Ibandronate, Incadronate, Risedronate, Zoledronate, Neridronate). BPs technetium complexes are used as bone markers in nuclear medicine for the diagnosis of bone metastases and other bone lesions. [47]

The most important clinical application of BPs is as inhibitors of osteolytic bone diseases, including osteoporosis, tumor-related bone destruction, and Paget's disease. BPs are widely used for the inhibition of abnormal calcification. [48]

The BPs have recently received considerable attention in the search of new vehicles for delivering drugs to bone [49], antiviral drugs [50], potential drugs for bipolar disorders [51], antihypercholesterolemic agents [52], and antiparasitic drugs. [53] The BPs demonstrate anti-inflammatory, antiarthritic and cardiovascular effects. [54] These bisphosphonate drugs have the common P-C-P pharmacophore function. The only variation among the different members of the family is in one substituent linked to the pharmacophore. However, the potency of drugs in the family of close structural and pharmalogical analogues could be improved ten thousandfold, simply by changing one substituent in the molecule.

3. Synthesis and Molecular Structure of Calix[4]arene Methylene Bisphosphonic Acids

The main method for the synthesis of metylene bisphosphonic acids consists in using the Arbuzov reaction of geminal dihalogenoalkanes with trialkyl phosphites, followed by the hydrolysis of the esters formed. [55] However, in the synthesis of the calixarene methylene bisphosphonic acids this method is limited by the availability of appropriate dihalogenoalkyl calixarenes.

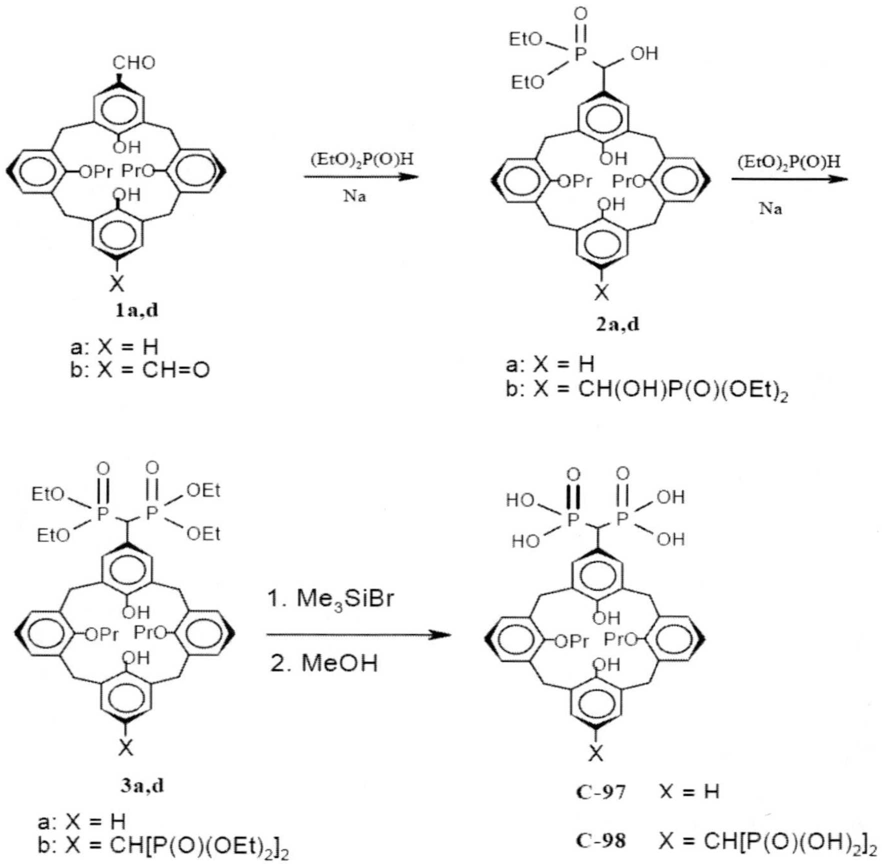

Scheme 1. Synthesis of calixarenes C-97, C-98.

Calixarene methylene bisphosphonic acids C-97, C-98 [56] are synthesized from formyl calixarenes **1** (Scheme 1). In the first stage diethylphosphite sodium salt is added to the C=O bond of formyl calixarene **1**, according to the classical scheme of the Abramov reaction, with the formation of alpha-hydroxyphosphonates **2**. Then, in the presence of sodium diethyl phosphite, the hydroxyphosphonates **2** cleave a water molecule to form a phosphorylated quinonmethyde as an intermediate (omitted in the scheme). The quinonmethydes further react with diethyl phosphite to form calixarene methylene bisphosphonates **3**. Acids C-97 and C-98 are obtained by the consecutive treatment of the esters **3** with trimetylbromosilane and methanol.

Calixarene tetrakis-methylene bisphosphonic acid C-192 is synthesized by the reaction of tetraformylcalixarene **4** with sodium salt of diisopropyl phosphite with the formation of tetrakis-bisphosphonate **5**, which, after subsequent dealkylation by treatment with trimetylbromosilane and methanol, gives C-192 (Scheme 2). [57,58]

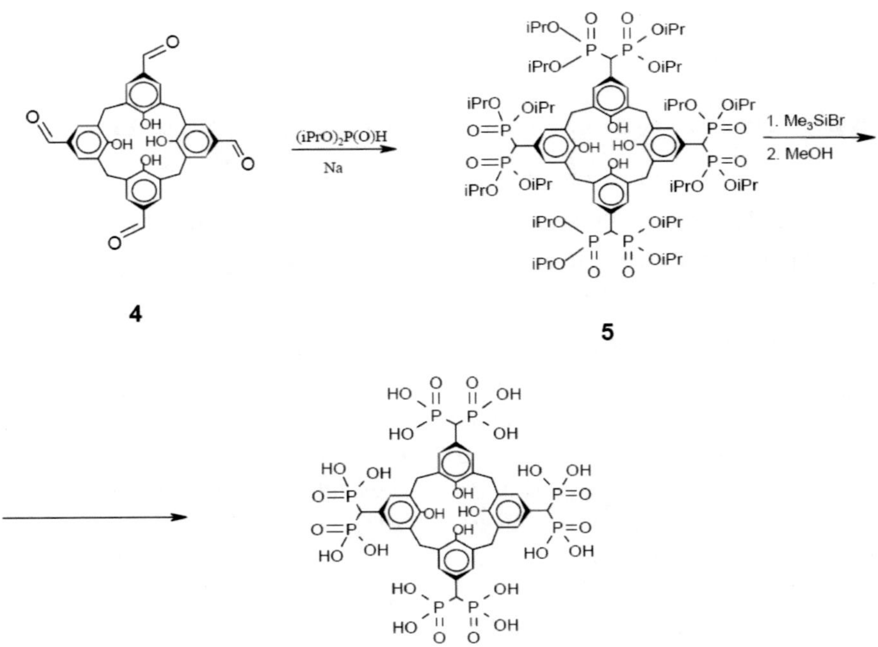

C-192

Scheme 2. Synthesis of calixarene C-192.

According to the ^{1}H NMR spectra and molecular modeling data, the acids C-97, C-98 and C-192 adopt *cone* conformations. The conformations are stabilized by intramolecular hydrogen bonds OH···O at the lower rim of the macrocycle. The *cone* conformation of C-192 is additionally stabilized by the intramolecular hydrogen bonds between the distal dihydroxyphosphoryl groups of the upper rim (Figure 3). The bonding at the upper rim makes the *cone* C-192 more regular compared with C-97 and C-98 ones.

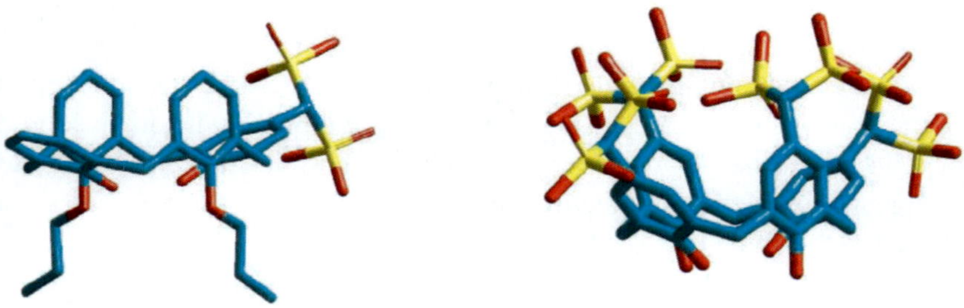

C-97 **C-192**

Figure 3. Energy minimized conformations of the calixarenes C-97 and C-192.

4. Structure-Functional Basis of the Intermolecular Interactions of Calix[4]Arene Methylene Bisphosphonic Acid C-97 with Myometrium Myosin Subfragment-1

It is of importance to understand the molecular and membrane mechanisms that control the contractile function of smooth muscle, in particular myometrium, in the normal state and in pathological states with different disorders; therefore, investigation of these questions is valid. The main function of the uterus in the female organism is baby carrying and delivery. Disturbance of uterine smooth muscle motility in females often leads to various pathologies – weak labour of childbirth, premature childbirth, miscarriage, etc. In this connection there is a need for the development of effective methods of correcting the disturbed contractile activity of the myometrium and, particularly, for investigations aimed at developing and examining new, highly efficient pharmacological substances capable of normalizing uterine motility.

It should be noted that at present a number of interesting experimental results concerning the influence of calixarenes on the transport and enzymatic activity of membrane-associated energy-dependent cation-transporting systems of the plasmatic membrane, sarcoplasmic reticulum and the mitochondria of uterine myocytes have been accumulated. It was shown that some calix[4]arenes may serve as rather selective and high-affinity inhibitors of Mg^{2+}, ATP- dependent sodium pump ($I_{0.5} < 100$ nM) and calcium pump ($I_{0.5} = 20$ - 30 μM) in the plasma membrane with Ca^{2+}-accumulation in the mitochondria.

Such membrane-associated energy-dependent transport systems are directly involved in the control of intracellular cation homeostasis (first of all, Ca^{2+} homeostasis). Nevertheless, in the case of muscles, in particular smooth ones, the functioning of both cation transporting ATP-dependent systems and ATP-hydrolase systems of contractile proteins is of great importance for maintaining the processes of electro- and pharmacomechanical coupling. Different pathological states of the female reproductive system (including those caused by hormonal disbalance) may lead to disfunction of the uterine contractile complex, which is revealed in abnormal super precipitation (*in vitro*) and changes of actomyosin ATPase activity. In some pathologies (e.g., hypertrophy) with disordered contractile function of smooth muscles, including the myometrium, some changes may take place in the expression of myosin isoforms and the kinetics of actin-myosin interaction. Accordingly, screening of reversible affine effectors of actin-myosin interaction (inhibitors, activators) is important for developing new pharmacological preparations capable of normalizing uterine contractile function in the pathological state of the myometrium.

Myosin ATPase transforms chemical energy, stored in ATP macroergic bonds, into mechanical work that is accompanied by directed movement; that is why this enzyme is also called a molecular motor. [59] Myosin subfragment-1 (head, S1) is the N-terminal part of the myosin heavy chain and consists of two domains: the N-terminal globular motor (catalytic) domain, which contains an active centre of ATPase, and a site of actin binding, and the regulatory domain or lever-arm, which is responsible for the movement of myosin relative to actin.

The following functional sub-domains were identified in the structure of myosin S1: 7-stranded β-sheet (amino acid residues 116–127, 170–180, 248–257, 265–271, 458–468 and 671–678), switch 2 helix (amino acid residues 469–509), converter (amino acid residues 716–772) and SH2/SH1 hinge (amino acid residues 688-715); these sub-domains undergo substantial conformational changes during energy transfer from the ATP-hydrolase centre to the actin binding site. [60]

The ATP-binding site contains the sequence GLY-GLU-SER-GLY-ALA-GLY-LYS-THR, similar to sequences in the active centres of other ATPases. [61] The ATPase active site is formed in part by the P-loop, switch 1 and switch 2 polypeptides that are linked to the 7-stranded β-sheet; conformations of these structures are sensitive to the position and coordination of γ-phosphate. The ATPase centre has been identified as a pocket, from which a gap is stretched to the actin binding site. When ATP binds to the myosin S1, the pocket becomes closed while the gap expands and disturbs the binding of S1 with actin; as a result, the myosin subfragment-1 dissociates from actin. When ATP is hydrolyzed by myosin ATPase to ADP and P_i, actin can again associate with S1, which facilitates the release of ATP hydrolysis products from the active centre. When ATP is hydrolyzed, the presence of associated actin is important for releasing inorganic phosphate from the nucleotide-binding pocket, since in its absence the salt bridge between Arg_{238} (switch 1) and Glu_{459} (switch 2) inhibits phosphate release. [62] The switch 2 α-helix transmits linear force (arising at ATP hydrolysis) from the active site to the converter domain; the last converts linear force into a torque that rotates the lever-arm and enables the shift of myosin relative to actin.

According to the available literature, subfragment-1 is a self-sufficient functional part of the myosin molecule; it preserves the properties of native myosin even in an isolated state – that is, it displays ATPase activity and the ability to interact with actin. [63,64] Earlier we showed the inhibiting effect of calix[4]arene C-97 on the ATPase activity of myometrium actomyosin (inhibition coefficient $I_{0.5}$ = 84±2 μM). [65] In this connection, a question arises concerning the mechanisms of calixarene C-97 effect on ATP-hydrolase activity of (exactly) myosin subfragment-1. Subfragment-1 is a convenient model for studying calixarene effects due to its high solubility in aqueous solutions with low ionic strength (unlike myosin) and because its specific ATPase activity is close to the myosin activity. Below we provide some data obtained in the study of calixarene C-97 effect on the ATP-ase activity of myometrium smooth muscle myosin subfragment-1, and consider possible mechanisms of complex formation between this calixarene and myosin subfragment-1 (with the use of computer simulation techniques – docking analysis and molecular dynamics). First, we will give some information concerning the methodological approaches used in the study.

Obtaining Actomyosin and Myometrium Myosin Subfragment-1

Smooth muscle actomyosin was purified from muscle tissue of pig uterus according to Barany [66] and Weber [67], with some modifications. Myometrium smooth muscle myosin subfragment-1 was obtained through actomyosin splitting by α-chymotrypsin according to the method of Weeds and Taylor [68], with some modifications. Separation and purification of subfragment-1 was performed by ion exchange chromatography using a column packed with DEAE-Sepharose CL-6B and chromatographic system BioRad (USA). Protein concentration was determined by the Bradford method [69]; also, concentrational dependence of light absorption at 280 nm was used with the curve standardized by serum albumin. Purity of smooth muscle myosin subfragment-1 was controlled by electrophoresis in polyacrylamide gel with sodium dodecyl sulphate, according to the Laemmli electrophoretic technique. [70] In accordance with electrophoretic data, the myosin head had $M_r \approx 100$ kDa, which corresponded to literature results. [71] Subfragment-1 was identified as a fraction with relatively high ATPase activity – 68 ± 9 μmol P_i / min per 1mg of protein (M±m, n=8).

Determination of the Atpase Activity of Myometrium Myosin Subfragment-1

The ATPase activity of myosin subfragment-1 was measured at 37 °C in an incubation medium (total volume 1 ml) of the following composition (in mM): tris-HCl buffer (pH 7.2) – 20, KCl – 100, $CaCl_2$ – 0.01, $MgCl_2$ – 5, ATP – 3 (standard conditions). The quantity of inorganic phosphate P_i removed from ATP during the hydrolase reaction was determined by Chen's method. [72]

Study of Calixarene C-97 Effect on ATP-Ase Activity of Myometrium Myosin Subfragment-1. Kinetic Measurements

To study calixarene C-97 effect on the ATP-ase activity of isolated myosin subfragment-1, aliquots of the calixarene solution in 50 mM tris-HCl buffer (pH 7.2) with an initial concentration of 10 mM were added to the standard incubation medium (see above). The final calixarene concentration in the sample was 10 – 100 μM. The value of ATP hydrolase activity, determined in the absence of the calixarene in the incubation medium, was taken as 100% ("zero point"). In the studied concentration the dependence of calixarene C-97 effect on the enzymatic activity of myosin subfragment-1 ATPase, the values of the apparent inhibition coefficient $I_{0.5}$ and Hill coefficient n_H were calculated with the use of linearized Hill graphs [73] according to the empirical equation

$$\lg[(V_0 - V)/V] = - n_H \cdot \lg I_{0.5} + n_H \cdot \lg I,$$

where V is specific enzymatic activity, V_0 is specific enzymatic activity in the absence of the inhibitor in the incubation medium, and I is the inhibitor concentration in the incubation medium. The typical value of the correlation coefficient r was 0.97–0.99.

Photon-Correlation Spectroscopy Experiments

Photon-correlation spectroscopy (PCS) experiments were performed on the ZetaSizer-3 Malvern Instrument (Great Britain), equipped with the Multi computing correlator type 7032 ce. Samples were irradiated by the helium-neon laser LGN-111 (λ=633 nm, power 25 mw) and laser radiation scattered at an angle of 90° was registered. The computer program PCS-Size mode v.1.61 [74] was used to analyze the experimentally measured autocorrelation function and to calculate the function of distribution by the hydrodynamic diameter (HDD) for microparticles in the volume of the measured cell. PCS was used to determine average HDD and the function of distribution by dimensions for particles of myometrium myosin subfragment-1, as well as for particles that are formed at the interaction of myosin subfragment-1 with calixarene C-97 (20, 60 та 100 μM).

Computer Simulation of the Ligand-Receptor Interaction in the System "Calixarene C-97 – Myosin Subfragment-1"

Computer simulation of the ligand-receptor interaction for the system "calixarene C-97 – myosin subfragment-1" was performed using the software MVD 4.1.0. This program allowed the realization of several algorithms of searching for the optimal ligand position in the active site of the protein (a variant of the evolutionary algorithm was used). [75] Actually, docking of the ligand (calixarene C-97) into the ligand-binding site (LBS) of myosin subfragment-1 was realized in conditions of the entirely ionized methylene bisphosphonic fragment. Standard AMBER force field parameters were used for the metal ion (Mg^{2+}). [76] When modeling interactions between calixarene C-97 and miometrium myosin subfragment-1 we realized so called "semiflexible" docking (only ligand motility was taken into account) with the selection of a series of complexes having the lowest total energy of the ligand binding. Calculation of the optimal geometry of the complexes formed, and determination of the energetically most advantageous spatial arrangement of the calixarene in the area of the myosin molecule were performed taking into account van der Waals, electrostatic and hydrophobic interactions, hydrogen bonds, as well as the contribution of desolvation energy. In the investigation, the three dimensional structure of myosin subfragment-1 with the identifier 1B7T in the RSCB Protein Data Bank was used. [77] The molecular dynamics of calixarene C-97 interaction with LBS of myosin subfragment-1 were investigated with the use of the Gromos96 force field. [78] The protein model was "placed" into a virtual cell in the form of a cut octahedron in such a manner that the spacing from the protein to the cell walls did not exceed 1.5 nm. The cell was "packed" with water molecule models (solvent) – SPC (Single Point Charge). The water molecules were partially substituted for Na and Cl ions to neutralize the system charge and simulate the physiological ionic strength (0.1 M). "Counterbalancing" of the solvent molecules was carried out during 0.5 ns with the protein atoms "attached" to their initial coordinates.

The temperature and the pressure were 298 K and 1 atm, respectively (Berendsen's method was used). [79] Following the solvent molecule counterbalancing, additional minimization of the system energy was performed. Actually, the MD simulation method was realized using the same parameters as at the solvent "counterbalancing", except for limiting the protein atom motility.

The coordinates were recorded in the initial trajectory file every 10 picoseconds. Motility of the calixarene C-97 molecule in the course of molecular dynamics was visualized with the use of the program UCSF Chimera 1.5.3. [80]

Investigation by Photon Correlation Spectroscopy of Calixarene C-97 Effect on the Effective Hydrodynamic Diameter (HDD) of Isolated Myosin Subfragment-1

The PCS method gives the distribution of different molecules and supramolecular structures in solution by their sizes. To determine HDD by the PCS method, the Stokes- Einstein formula was used, which is valid only for spherical particles. [81,82] Therefore, molecular diameters determined by this method will be close to the real ones only for globular proteins. When a protein is far from spherical form, the calculated HDD is a conventional value instead. The myosin head, the main functional unit of myosin [83], is rounded or pear-shaped; therefore, the preparation of isolated myometrium myosin subfragment-1 fits rather well for studying by the PCS method the effect of the calixarene on contractile proteins. The average subfragment-1 HDD and the function of particle size distribution were determined. Isolated myosin subfragment-1 had narrow a spectrum of particles in the interval from approximately 10 to 50 nm (Figure 4). The size of the most frequently registered particles (≈ 50%) was close to 20 nm. Particles with diameters close to 70 nm and particles less than 20 nm were also present in small quantities. It was shown that the average (most probable) HDD of myosin subfragment-1 was 22 ± 3 nm ($M \pm m$; n = 15). This value is in agreement with the literature data. [84,85] The concentration dependence of calixarene C-97 effect on the hydrodynamic indices of myosin subfragment-1 particles showed that the calixarene concentrations 20 and 60 μM did not cause significant changes of myosin head HDD vs. control – only a trend in HDD growth was observed. Only at the calixarene concentration 100 μM did myosin head HDD reliably increase versus the control: from 22 ± 3 nm (control) to 33 ± 3 nm (experiment) (Figure 5). The control examination of calixarene C-97 solution in 50 mM tris-HCl (pH 7.2) failed to discover any light-scattering microparticles. Thus, calixarene C-97 at a concentration of 100 μM caused a significant (vs. control) increase of myosin subfragment-1 HDD, which indicated the formation of a complex between the calixarene and myosin head. Of interest is the fact that using photon correlation spectroscopy with another supramolecular effector (activator) of myosin subfragment-1 ATP-hydrolase (calixarene C-107) we could observe time-dependent changes of myosin subfragment-1 HDD (data are not given). During the first 5 min after calixarene C-107 addition (60 μM), the values of HDD increased from 25 nm (control) to 70 – 80 nm, but in the next 10 to 15 min the size of the myosin head was stabilized at the HDD values of 45 нм (which is undoubtedly more vs. control).

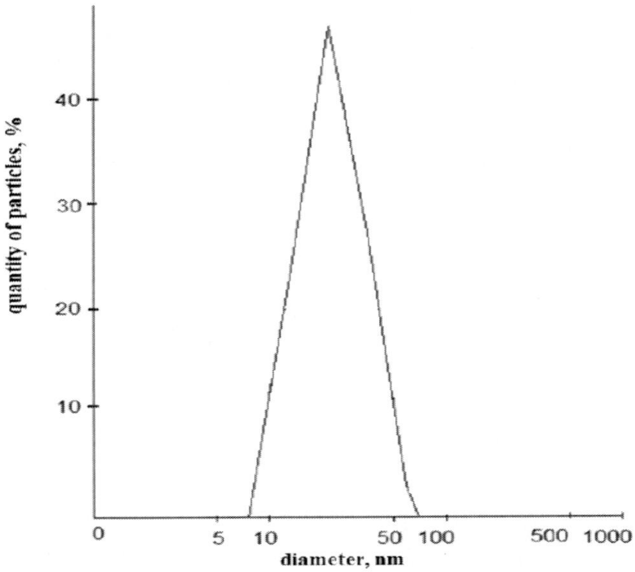

Figure 4. The hydrodynamic diameter (HDD) distribution for myosin subfragment-1 particles. Photon-correlation spectroscopy method. The quantity of particles, equivalent HDD, which corresponded to the values in the interval from HDD_{min} to HDD_{max} with the accuracy 0.1% (according to the characteristics of the laser-correlation spectrophotometer) was taken for 100%.

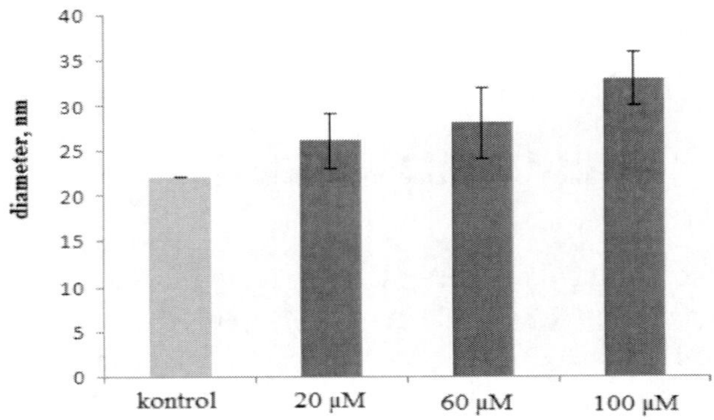

Figure 5. Calixarene C-97 increases the myosin subfragment-1 hydrodynamic diameter (HDD) (M±m; n=6). Photon-correlation spectroscopy method.

Examination of Calixarene C-97 Effect on Aтpase Activity of Myometrium Myosin Subfragment-1

The study of calixarene C-97 effect on myometrium myosin subfragment-1 ATPase showed that in a concentration of 10 μM it almost did not influence ATPase activity. At a further increase of this calixarene concentration, it inhibited myosin head ATPase in dose-dependent mode, and at 100 μM concentration the inhibiting effect vs. control was 60 ± 5% (*M* ± *m; n* =5) (Figure 6). According to the linearized graph in Hill coordinates (Figure 6,7), inhibition coefficient $I_{0.5}$=83 ± 7 μM (*M* ± *m; n* = 5).

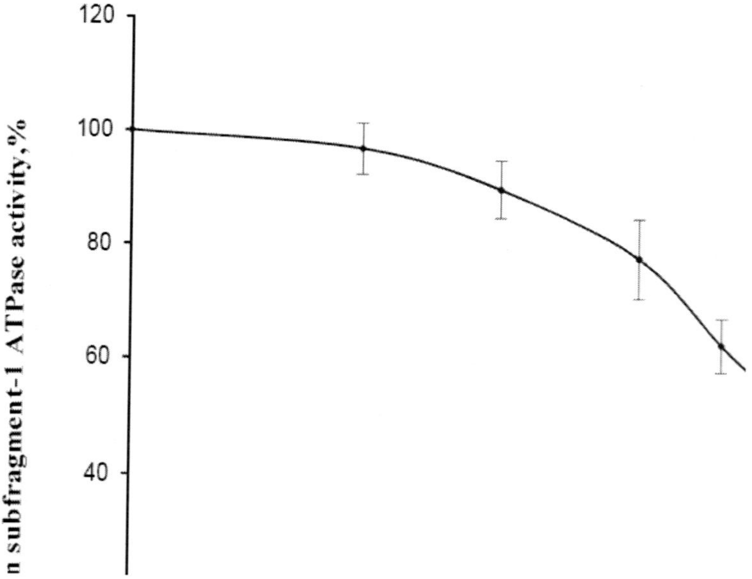

Figure 6. Catalytic titration of myosin subfragment-1 ATPase activity with calixarene C-97 (M ± m; n=5).

When comparing calixarene C-97 inhibition of ATPase activity for actomyosin complex ($I_{0.5}$ = 84 ± 2 μM) [86] and myometrium myosin subfragment-1 ($I_{0.5}$ = 83 ± 7 μM) one can see that the myosin head and actomyosin complex display practically identical affinity towards this calixarene. Of interest is the fact that Na^+, K^+-ATPase of plasmatic membranes has a several orders of magnitude higher affinity for calixarene C-97 ($I_{0.5}$ = 98 ± 8 nM) as compared with actomyosin and myosin subfragment-1 ATPase.

This observation should be taken into account in the analysis of calixarene C-97 effect on the contractile and electrical activity of smooth muscle cells.

Using linearized graphs in the analysis of the concentration dependence of subfragment-1 ATPase activity on calixarene C-97, we also calculated the Hill coefficient n_H, which amounts to 1.3 ± 0.5 (Figure 7). Therefore, this value of the Hill coefficient most probably suggests that the myosin head may bind only one molecule of calixarene C-97.

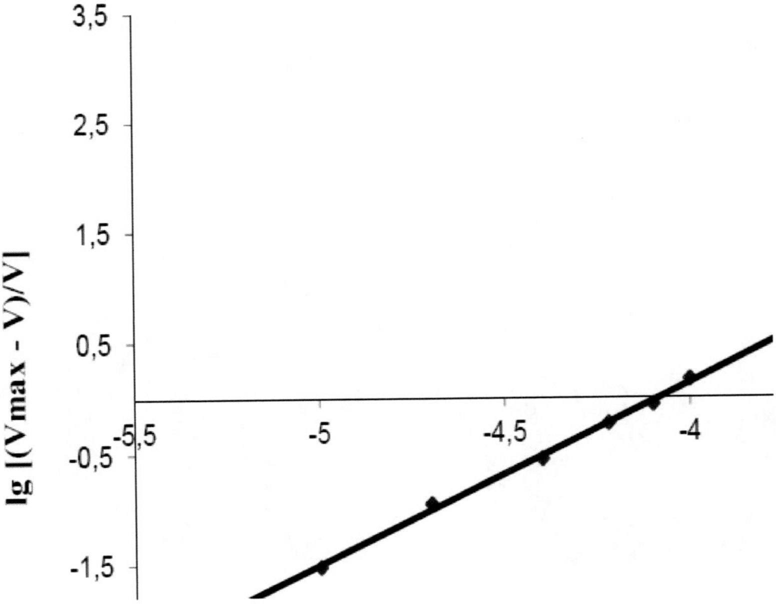

Figure 7. Averaged curve linearized in Hill coordinates for catalytic titration of myosin subfragment-1 ATPase activity with calixarene C-97.

Therefore, calixarene C-97 effect on ATPase of the contractile complex may follow from its ability to bind with myosin subfragment-1; consequently, the myosin head is assumed to be one of the targets by means of which calixarene C-97 may influence the contractile complex (based on the above value of the Hill coefficient n_H, the stoichiometry of binding is 1 : 1).

Using the Computer Simulation Method for Examining the Structural Basis of Molecular Interactions between Calixarene C-97 and Myometrium Myosin Subfragment-1

To understand the molecular mechanism of calixarene C-97 inhibiting effect on myosin subfragment-1 ATPase, it is necessary to have information on the three dimensional structure of the enzyme in complexes with a substrate or/and inhibitor. With this purpose we carried out the computer simulation for the interaction of calixarene C-97 (as a separate ligand or coupled with ATP) with myometrium myosin subfragment-1 in the presence of Mg^{2+} as a cofactor (Figure 8, A, B). Mg cation takes part in the binding of ATP to the active site, and in the process of its hydrolysis.

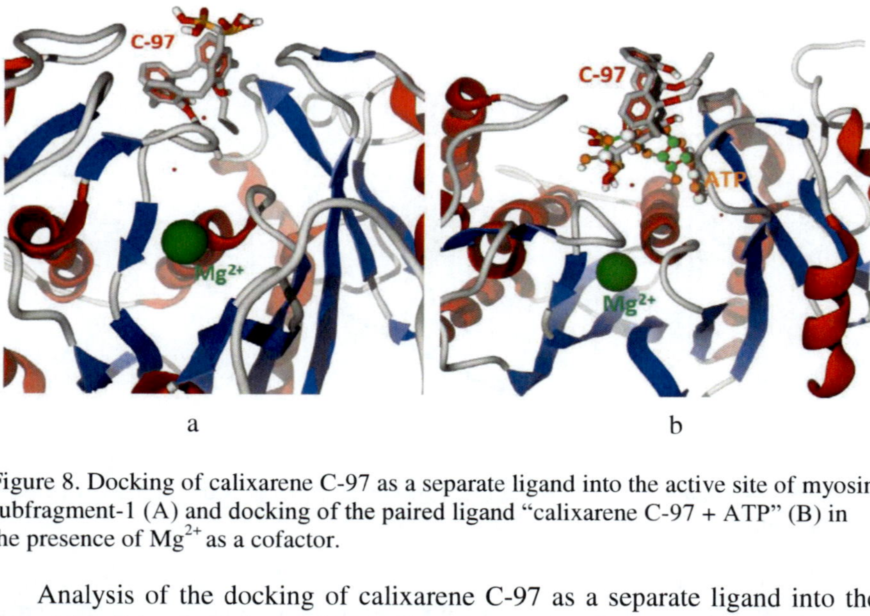

a b

Figure 8. Docking of calixarene C-97 as a separate ligand into the active site of myosin subfragment-1 (A) and docking of the paired ligand "calixarene C-97 + ATP" (B) in the presence of Mg^{2+} as a cofactor.

Analysis of the docking of calixarene C-97 as a separate ligand into the ligand binding site of myosin subfragment-1 suggests that residues Asp_{320}, Asn_{321}, Leu_{676} and Gln_{678} may be involved in the fixation of the calixarene phosphonate groups (the data are not shown in the Figure). Calixarene fragments are oriented in the space formed by residues Ile_{322}, Asn_{238} and Lys_{677} on one side, and by Tyr_{126}, Arg_{127} and Arg_{128} on the other. Besides, a positively charged nitrogen atom of Lys_{677} residue is located next to oxygen atoms of the methylene bisphosphonate fragment (the distance between nitrogen atom of Lys_{677} residue and ionized oxygen atoms of the phosphonate group is 0.31 nm). Bonds taking part in the calixarene binding to a site close to the active centre of myosin S-1 were indentified; they are: hydrogen bonds, π-π-stacking interactions between aromatic fragments of the calixarene bowl and Tyr_{126} of the benzene ring in the myosin head, as well as electrostatic interactions with the involvement of Arg_{127}, Arg_{128}, Asp_{320} and Lys_{677}.

Simultaneous docking of the paired ligand "calixarene C-97 + ATP" into the ligand binding centre of myosin subfragment-1 in the presence of Mg^{2+} revealed the possible formation of a complex between calixarene C-97 and ATP, which binds with the involvement of amino acid residues to a site of the myosin head close to the active centre; this is somewhat different from the results obtained with the docking of calixarene C-97 alone. In addition, the spatial orientation of calixarene C-97 in the ligand binding centre of the myosin head differs: the methylene bisphosphonate fragment of the calixarene in its interaction with the ATP molecule becomes oriented towards the active centre. Binding of the complex "calixarene C-97 + ATP" to a site close to the myosin S1 active centre is realized with the participation of hydrogen bonds, π-π-stacking interactions between the aromatic rings of the calixarene bowl and the benzene rings of Tyr_{126} and Tyr_{132} in the myosin head, as well as electrostatic interactions with the participation of Arg_{127}. Residues Asn_{238} and Asn_{321}, which are quite close to oxygen atoms of phosphonate groups, may form hydrogen bonds in interaction with the ligand phosphonate groups. It is also obvious that the hydrophobic platform of calixarene C-97 may contribute to the binding due to the interaction with hydrophobic sites in the centre of substrate binding.

The Table 1 shows amino acid residues of the active centre of myometrium myosin S1, which are involved in interactions with calixarene C-97 and the complex "calixarene C-97 + ATP". According to the literature, residues Tyr_{126}, Arg_{127}, Leu_{676} and Lys_{677}, enabled in fixation of calixarene C-97 phosphonate groups, are incorporated into the 7-stranded β-sheet structure of myosin subfragment-1, which is joined with polypeptide sites switch 1 and switch 2. These polypeptides participate directly in the ATPase active centre formation [14]. Therefore, it is obvious that the interaction of calixarene C-97 with myosin subfragment-1 in the site located near the ATPase centre of myosin will influence this centre's conformation and, accordingly, the process of ATP hydrolysis catalyzed by myosin subfragment-1.

Table 1. Amino acid residues of the substrate binding domain of myosin subfragment-1, which take part in the binding of calixarene C-97 in the case of its docking as a separate ligand, and when docking is performed for a couple of ligands "C-97 + ATP" in the presence of Mg^{2+} as cofactor

C-97 + Mg^{2+}	C-97 + ATP+Mg^{2+}
Tyr_{126}	Asn_{124}
Arg_{127}	Pro_{125}
Arg_{128}	Tyr_{126}
Asn_{238}	Arg_{127}
Asp_{320}	Tyr_{132}
Asn_{321}	Glu_{184}
Ile_{322}	Asn_{185}
Leu_{676}	Asn_{238}
Lys_{677}	Asn_{321}
Gln_{678}	Ile_{322}

Therefore, the results obtained by the method of molecular docking indicate that: 1) calixarene C-97 may bind with myosin subfragment-1 to the site close to the active centre, which leads to conformational changes in the latter; 2) this calixarene may form a complex with ATP prior to the interaction with the protein, which complicates access of nucleoside triphosphate to the ATP binding site. It may well be that both these variants of calixarene C-97 interaction with subfragment-1 may cause their inhibiting effect on ATPase activity of myometrium myosin subfragment-1. Nevertheless, it should be noted that the last variant cannot explain the considerable differences (almost three orders of magnitude) between affinity towards calixarene C-97 for Na^+,K^+-ATPase and myosin subfragment-1 ATPase (values of inhibition coefficient $I_{0.5}$ is equal 100 nM and 80 – 90 μM, respectively).

Note that the docking method fails to take into account the motility of myosin S1 and ligands, which may occur in real conditions. Therefore, to verify results obtained by the docking method we investigated the interaction of calixarene C-97 with the ligand binding centre of myosin subfragment-1 by the method of molecular dynamics, taking into consideration the motility of both the receptor (myosin subfragment-1) and the ligand.

Dynamic analysis of the calixarene C-97 molecule position in the ligand binding site of myosin subfragment-1 allowed the determination of time-dependent deflections of the active site atoms from their initial state; these deflections were characterized by changes in distances (in nm) between the calixarene centre of mass and the nearest amino acid residue (Asn_{321}).

Accordingly, the time dependence of the C-97 position in the ligand binding site of myosin subfragment-1 was plotted (at intervals of 4 ns). Analysis of the deflections showed that the initial calixarene location changed noticeably, starting from approximately 1 ns: the calixarene conformational mobility increased, and afterwards again decreased, which correlated with the change of distance between the calixarene and Asn_{321} centres of mass. Therefore, the interaction of calixarene C-97 with the myosin head is accompanied by time variation of the calixarene location. In the process of molecular dynamics, the total energy of the system "calix[4]arene C-97 – myosin subfragment-1" becomes somewhat lower; therefore, calixarene C-97 apparently takes up a more energy-advantageous position in the ligand binding site of myosin subfragment-1. It was shown that the calixarene C-97 molecule shifts relatively to amino acid residue Asn_{320} by about 1.81 Å, and that the arrangement of calixarene C-97 in the final binding location is energetically stable.

Analysis of hydrogen bonds during a 4 ns interval between calixarene C-97 and amino acid residues of the subfragment-1 ligand binding site (Figure 9) shows that the number of hydrogen bonds enabled in the interaction between calixarene C-97 and the myosin head remains, on average, unchanged. Consequently, the calixarene location changes in the course of molecular dynamics are most probably caused by the optimization of hydrophobic and electrostatic interactions.

The method of 4 ns molecular dynamics was also used to identify amino acid residues taking part in the calixarene binding to the ligand binding centre of myometrium myosin subfragment-1 (Figure 10). They include Tyr_{126}, Arg_{127}, Leu_{676} and Lys_{677}. These residues take part in the fixation of the calixarene phosphonate groups that are located next to the ATPase active centre. Overall, these results correlate well with the data obtained by the docking method (see above).

The computer simulation results broadened our notions about the structural basis of intermolecular interactions between calixarene C-97 and myometrium myosin subfragment-1. In particular, the role of hydrophobic, electrostatic and π-π-stacking interactions between the calixarene and amino acid residues of myosin subfragment-1, some of which are close to ATPase active centre, was elucidated.

Thus, calixarene C-97 effectively inhibits ATPase activity of myosin subfragment-1 ($I_{0.5}$ = 83 ± 7 μM). This substance causes a statistically significant (vs. control) increase of the myosin head's hydrodynamic diameter, which is indicative of a complex formation between calixarene and the myosin head. The use of the computer simulation method resulted in the identification of amino acid residues, and hydrophobic and electrostatic forces that take part in interactions between the calixarene and myometrium myosin subfragment-1. The results obtained suggest that the myosin head is one of the targets for the calixarene effect on the contractile complex of smooth muscle.

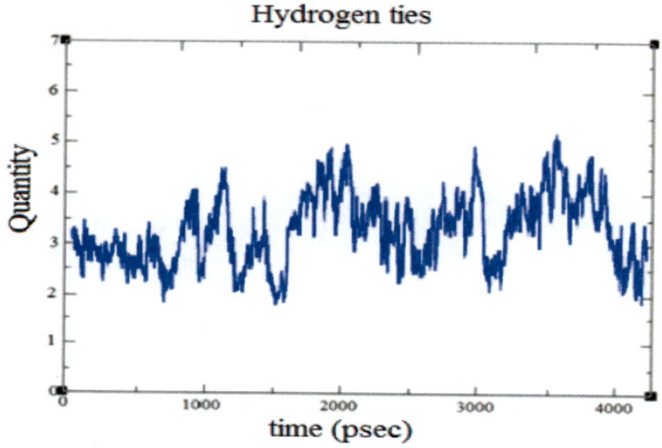

Figure 9. Dynamics of hydrogen bonds generated between calixarene C-97 and amino acid residues of the ligand-binding centre of myosin subfragment-1.

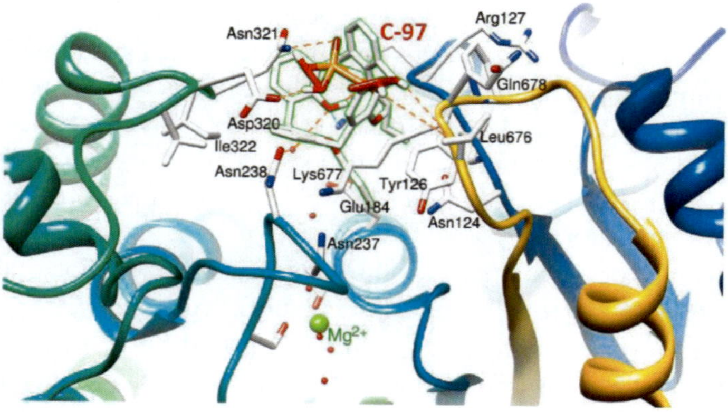

Figure 10. Binding of calixarene C-97 in the ligand-binding centre of myosin subfragment-1 determined by 4 ns molecular dynamics. The Figure shows amino acid residues taking part in epy calixarene binding to the LBC of myosin subfragment-1. Dotted lines show H-bonds between amino acid residues of subfragment-1 and the inhibitor. Interactions of C-97 with the myosin head also include π-π stacking interactions and electrostatic bonds.

5. Calix[4]arene Methylene Bisphosphonic Acids as Inhibitors of Fibrin Polymerization

The present study aimed to investigate the anticoagulant properties of phosphorus containing calixarenes in the last two steps of blood coagulation: thrombin+fibrinogen reaction and fibrin monomer polymerization, which lead to the formation of the fibrin network of the thrombus. Fibrinogen is a glycoprotein (MW ≈ 344 kDa) composed of two monomeric units connected by disulfide bonds. Each monomer consists of three non-identical polypeptide chains Aα, Bβ and γ, also connected by disulfide bridges. [87] The NH_2-terminal ends of all six polypeptide chains are situated in the central region of fibrinogen known as the E-domain. The two peripheral regions of the molecule are historically called the D-domains.

When the blood coagulation system is activated, thrombin is formed from prothrombin and attacks fibrinogen, splitting off two fibrinopeptides A (Aα1–16), and thereby exposing fibrin polymerization sites 'A' (Aα17–19). [88] Removal of fibinopeptides A leads to a form of fibrinogen called 'desA', which polymerizes spontaneously to form two-molecule-thick protofibrils. Removal of fibrinopeptides B ('desAB') encourages lateral associations that lead to fibrils. [89,90] The fibrils continue to associate, branching and forming a 3D network: the framework of the blood thrombus. [91] It is widely accepted that the initial step of fibrin polymerization (protofibril formation) is carried out by the intermolecular pairing of 'A' and 'a' polymerization sites of fibrin monomers. Site 'a' is a cavity ('hole') that includes amino acid res-idues γGln329, γAsp330, γHis340 and γAsp364, and is situated in the γC-domain of the fibrinogen/fibrin molecule. [92]

In particular, we have focused on compounds in which the molecular scaffold is decorated with methylene bis-phosphonic acid groups, mainly on calix[4]arene tetrakis-methylene bisphosphonic acid (C-192) possessing four such substituent groups (Figure 1).

Preparation of Fibrinogen, Fibrin Desab and Turbidity Analysis of Fibrin Polymerization

Human fibrinogen was prepared by sodium sulphate precipitation from human plasma. [93] DesAB fibrin monomer was prepared as described previously. [94] The effects of compounds on fibrin polymerization were studied spectrophotometrically at 350 nm, as described previously. [95] The curve of increasing turbidity during fibrin clotting shows the parameters: τ, lag time, which corresponds to the time of protofibril formation; Vmax, maximum rate of fibrin polymerization, which was defined by graphic calculation of the angle of the tangent to the tur-bidity increase curve at the point of maximum steepness; and Δh, final turbidity of fibrin clots. The polymerization of desAB fibrin was studied at its final concentration 0.1 $mg{\cdot}mL^{-1}$ in the polymerization medium containing 0.05 M ammonium acetate (pH 7.4) with 0.1 M NaCl and $1{\cdot}10^{-4}$ M $CaCl_2$. The polymerization of fibrin formed in the fibrinogen + thrombin reaction was investigated in a final concentration of fibrinogen of 0.1 $mg{\cdot}mL^{-1}$ and thrombin of 0.4 NIH $units{\cdot}mL^{-1}$ in the same polymerization medium.

Electron Microscopy

The samples of polymerizing fibrin produced in the thrombin–fibrinogen reaction in the absence or the presence of calixarene C-192 (10^{-5} M) were taken out of the reaction medium at various times, placed on a carbon-coated grid for 2 min and then stained with 1% (w/v) uranyl acetate for 1 min. Transmission electron microscopy was performed with an H-600 electron microscope (Hitachi, Tokyo, Japan) operated at 75 kV. Electron micrographs were obtained at ×50 000 magnification.

The Determination of Association Constants of C-192 Complexes with Gly-Pro-Arg-Pro

The HPLC consisted of a high-pressure pump (type HPP 4001) (Laboratorni Pristroje, Prague, Czech Republic) connected to a Rheodyne sample 7120 injector (Rheodyne, Berkeley, CA, USA) and an ultraviolet-visible detector (type LCD 2563) (Laboratorni Pristroje). The column (150·3.3 mm inner diameter) was packed with Separon SGX CN (Lachema, Prague, Czech Republic). The mobile phase was a mixture of methanol–water in the ratio 50 : 50 (v/v), with the calixarene C-192 additive at a concentration in the range $1.48\cdot10^{-4}$ to $5{\cdot}10^{-4}$ M. The flow rate was 0.6 $mL{\cdot}min^{-1}$. The concentration of the analytes in the solution for analysis was 10^{-5} M. The solvent was identical to the mobile phase composition. Each of the samples was analyzed five times. All chromatograms were obtained at 20 °C.

Association constants (Table 2) of the calixarene complexes with amino acids Gly, Pro, Arg and tetrapeptyde Gly-Pro-Arg-Pro (280–3395 M^{-1}) were calculated from the dependence of the 1/k' value versus the calixarene concentration [CA] in the mobile phase (Fig 11) by the equation

$$K_A = \frac{k'_0(1/k' - 1/k'_0)}{[CA]}$$

where k_0' and k ' are the capacity factors determined in the absence and presence of the calixarene in the mobile phase, and [CA] is the calixarene concentration in the mobile phase.

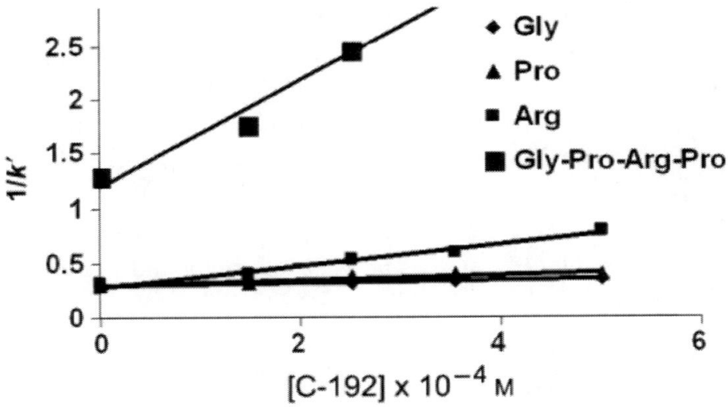

Figure 11. Dependence of 1/k' for Gly, Pro, Arg and Gly-Pro-Arg-Pro on the C-192 concentration in the mobile phase.

Table 2. Values 1/k' of the guests and association constants K_A (M^{-1}) for their complexes with calixarene C-192

Guest	1/k'					KA, M-1 (RSD, %)
	Calixarene concentration					
	0.00	1.48	2.52	3.54	5.00	
Gly	0.302	0.313	0.324	0.331	0.349	280 (10)
Pro	0.294	0.318	0.367	0.396	0.403	814 (26)
Arg	0.311	0.395	0.532	0.592	0.794	2576 (21)
Gly-Pro-Arg-Pro	1.287	1.754	2.453	3.015	3.693	3395 (19)

PT and APTT assays

The effects of calixarene C-192 on the coagulation of human blood plasma were studied using a coagulometer (CT 2410 'Solar', Minsk, Belarus). Reagents from Renam (Moscow, Russia) were used to calculate PT and APTT. PT and APTT ratios were calculated using the formula: t_c/t_o, where t_o is the time of clot formation in blood plasma without calixarene C-192, and t_c is the time of clot forma-tion in blood plasma with calixarene C-192.

Influence of Calix[4]Arene Methylene Bisphosphonic Acids on Fibrin Polymerization

Calixarene C-192 was studied with respect to its effects on fibrin polymerization in two kinds of assay. In the first assay the formation of fibrin was followed directly after the addition of thrombin. In the second assay previously prepared fibrin was dispersed and then allowed to repolymerize under appropriate conditions. In both cases fibrin formation was gauged by turbidity measurements. Turbidity analysis showed that the compound decreased the maximum rate of fibrin polymerization in the thrombin–fibrinogen reaction by 50% at a molar ratio of compound to starting fibrinogen of 1.7 : 1 (Figure 12).

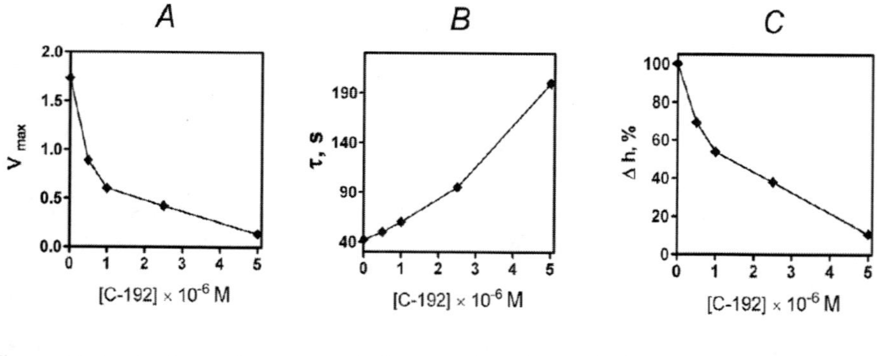

Figure 12. Turbidity analysis of the influence of C-192 on fibrin polymerization in the fibrinogen + thrombin reaction. The dependence on calixarene C-192 concentration is shown for (A) the maximum rate of fibrin polymerization Vmax, (B) the lag-time τ, and (C) the final turbidity of fibrin clots Δh.

The final turbidity of clots was decreased by 50% at a molar ratio of 4.3 : 1 (compound: starting fibrinogen) (Figure 12C). The lag-time was also increased strongly in the presence of C-192, indicating either an decrease of the rate of protofibril formation or, conceivably, an increase of protofibril critical length (Figure 12B). Similar results were obtained when calixarene C-192 inhibited the re-association of dispersed desAB fibrin (Figure 13A–C); in this case, IC_{50}= $1.26 \cdot 10^{-6}$ M.

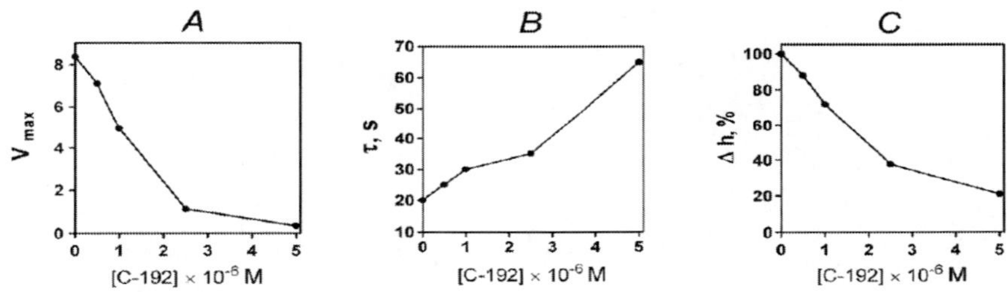

Figure 13. Turbidity analysis of the influence of C-192 on fibrin desAB polymerization. The dependence on calixarene C-192 concentration is shown for (A) the maximum rate of fibrin polymerization Vmax, (B) the lag-time τ, and (C) the final turbidity of fibrin clots Δh.

Such a strong and specific inhibition by calixarene C-192 of both the thrombin–fibrinogen reaction and the re-association of fibrin desAB indicates that calixarene retards clotting not as a result of the inhibition of thrombin, but entirely because of the blocking of fibrin polymerization sites.

We also performed electron microscopy studies to determine the stage of fibrin polymerization that was inhibited by C-192. Transmission electron microscopy of the thrombin + fibrinogen media showed that fibrin remained in monomer state in the presence of calixarene C-192 at its final concentration of 10^{-5} M, whereas, at the same time, mature fibrils were formed in the absence of C-192 (Figure 14).

The results of turbidity analysis and electron microscopy indicate that the inhibition by C-192 occurs at the first stage of fibrin polymerization, presumably by blocking one of the sites of protofibril formation.

We also investigated the inhibitory effects of two structural fragments of C-192: methylene-bis-phosphonic acid and *para*-hydroxyphenyl-methylene-bis-phosphonic acid (HPMBPA) (Figure 15). The results showed that these constituents inhibit fibrin polymerization with considerably less activity: the IC_{50} value was > $1.0 \cdot 10^{-4}$ M for 1 and $0.72 \cdot 10^{-4}$ M for 2.

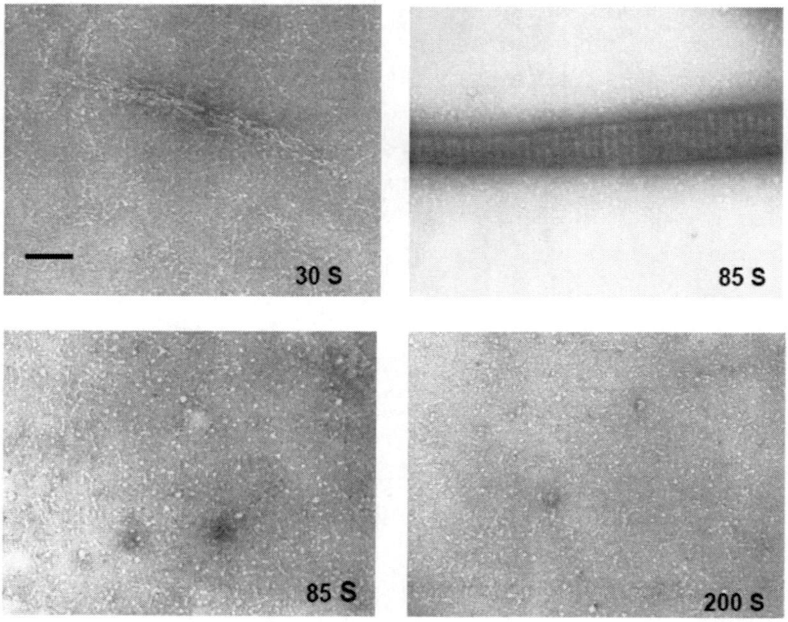

Figure 14. Electron micrographs of fibrinogen + thrombin reaction media in the absence of C-192 (A, B), as well as in its presence (C, D). Scale bar = 100 nm.

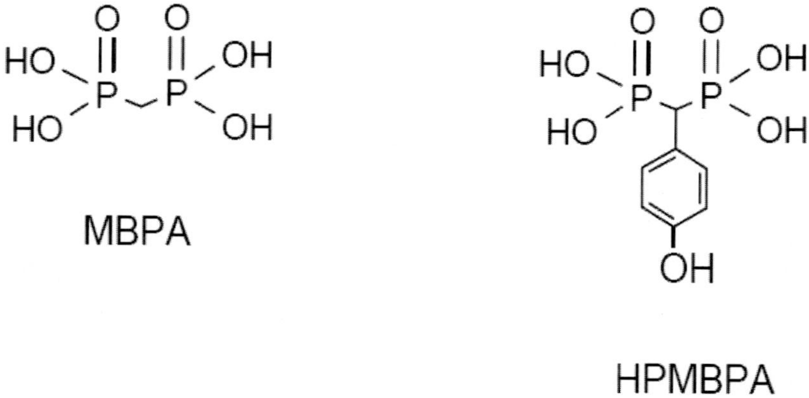

Figure 15. Two structural fragments of C-192: methylene-bis-phosphonic acid and *para*-hydroxyphenyl-methylene-bis-phosphonic acid.

It is of interest that the inhibitory activity of C-98, which has the two methylene bis-phosphonic acid substituents, is much less (Table 3) (IC_{50}=1.31·10^{-4} M, indicating that the calixarene scaffold and the number of phosphoryl groups in the molecule play a crucial role in the inhibitory effect.

Table 3. Concentration values of compounds that cause 50% inhibition of the maximum rate of the polymerization of fibrin produced in the fibrinogen + thrombin reaction

Compound	IC_{50}
C-192	1.26 × 10^{-6} M
C-98	1.31 × 10^{-4} M
para-Hydroxyphenyl-methylene bis-phosphonic acid	>1.00 × 10^{-4} M
Methylene bis-phosphonic acid	0.72 × 10^{-4} M

Furthermore, calixarene C-192 doubles both the prothrombin time (PT) and APTT in normal human blood plasma at concentrations of 7.13·10^{-5} M and 1.10·10^{-5} M, respectively (Figure 16). The molar ratios of C-192 to plasma fibrinogen were approximately 21 : 1 and 3.3 : 1 for the PT and APTT assays, respectively. The delays in clotting times in the blood plasma coagulation experiments are what would be expected by the inhibition of the fibrin polymerization from fibrinogen after the activation of the blood coagulation system. Electron microscopy confirmed that C-192 inhibits the first stage of fibrin polymerization (i.e., the formation of protofibrils).

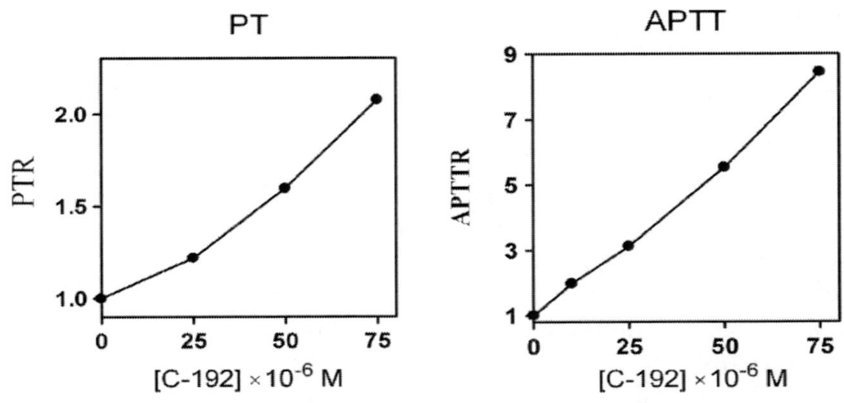

Figure 16. The dependence of the PT and APTT ratios on the calixarene C-192 concentration.

Because this stage is fulfilled through the intermolecular binding of the complementary sites 'A'–'a', it appeared to be certain that this inhibition is a result of the blocking of site 'A' (Aa17-19, GlyProArg) by the calixarene in a 'knob-hole' manner. To confirm that this was the case we employed HPLC to study the complex formation between C-192 and the synthetic peptide Gly-Pro-Arg-Pro, a synthetic analogue of the A knob; the free amino acids Gly, Pro and Arg were used as controls. The association constants of calixarene C-192 complexes with amino acids Gly, Pro, Arg and tetrapeptide Gly-Pro-Arg-Pro in a methanol–water mobile phase (50 : 50, v/v) were determined as previously described. [96] The addition of calixarene C-192 to the mobile phase decreased the capacity factor, k', of the guest molecules (Table 2). This confirms the formation of inclusion host–guest complexes. There is linear dependence of 1/k' versus the concentration of C-192 in the mobile phase (Figure 16); the correlation coefficient is 0.98–0.99, indicating a 1:1 ratio of calixarene to Gly-Pro-Arg-Pro in the complex. The association constants of the complexes are within 280-3395 M^{-1} (Table 2). The highest one, 3395 M^{-1}, is observed for tetrapeptide Gly-Pro-Arg-Pro.

In accordance with the molecular modelling data (Figure 17 A, B) there are two modes for complexation of C-192 with Gly-Pro-Arg-Pro. In the first mode (A) cooperative electrostatic interactions of two proximal negatively-charged phosphonyl groups with the Gly α-amino terminal group and the Arg guanidinium residue play the principal role in complex formation. Most of the tetrapeptide molecule is exposed outside the calixarene cavity. In the second mode (B) the hydrophobic part of the Gly-Pro-Arg-Pro molecule is deeply embedded into the calixarene cavity. The complex is stabilized by $P\text{-}O^{-}...H_3N^{+}$ electrostatic interactions of the phosphonyl group with the Gly amino acid residue, as well as by CH-π interactions of the CH_2-group in the Pro residue with the calixarene aromatic ring. Hydrophobic forces can additionally stabilize the complex in a water solution.

Thus, we have shown for the first time that C-192 is a powerful and specific inhibitor of the final step of blood coagulation, fibrin polymerization, and can be used as the basis for the design of a new class of anti-thrombotic agents. We found that the mechanism of C-192 inhibition involves its effect on the first step of fibrin polymerization, protofibril formation, which is carried out via intermolecular interactions of comple-mentary polymerization sites 'A' and 'a' of fibrin molecules.

We have also shown that C-192 forms complex with synthetic peptide Gly-Pro-Arg-Pro, which imitates polymerization site 'A' (Aa17 Gly-Pro-Arg), suggesting that the inhibitory effect of C-192 may be a result of the blocking of site 'A' by this calixarene.

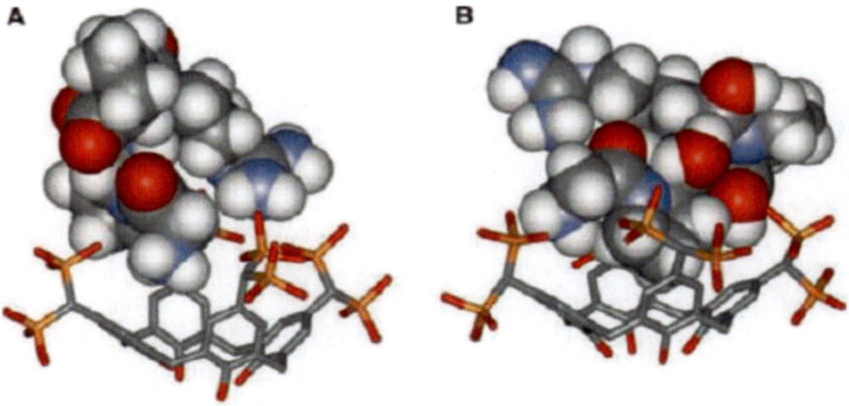

Figure 17. Two energy minimized structures of calixarene C-192 complexes with GlyProArgPro (A, B) formed through cooperative electrostatic interactions of two proximal negatively charged phosphonyl groups of the C-192 molecule with the Glya-amino terminal group and Arg guanidinium residue.

Previous scientific experiments with healthy adult mice have shown that calixarene C-192 is a median toxic compound (LD_{50} is 780 mg/kg, *per orally*).

These experiments demonstrate that calix[4]arene C-192 (calixarene tetrakis-methylene bisphosphonic acid) is a specific inhibitor of fibrin polymerization and blood coagulation, and can be used for the design of a new class of antithrombotic agents.

6. Calixarene Methylene Bisphosphonic Acids as Phosphatase Inhibitors

Calixarene methylene bisphosphonates were found to regulate the activities of alkaline phosphatases and protein tyrosine phosphatases *in vitro*. Nonspecific alkaline phosphatases catalyse phosphoryl transfers in the hydrolysis or the transphosphorylation of phosphate monoesters. [97] They are involved in a variety of biological phenomena, including their heightened expression in osteoblasts, [98] with possible consequences for osteoporotic patients. Protein tyrosine phosphatases regulate a number of biochemical processes which proceed through dephosphorylation of phosphotyrosine residues in proteins, including cell growth, differentiation, metabolism, and movement. [99,100,101] Because of the importance of the protein phosphorylation–dephosphorylation cycle in living cells, the activities of these enzymes are associated with diabetes, obesity, cancer, and other human diseases. [102,103,104]

In this study calixarenes C-97 and C-98 were analyzed as inhibitors of alkaline phosphatases from bovine intestinal mucosa, bovine kidney, and *E. coli.* The activities of these macrocyclic compounds were compared with MBPA and HPMBPA (Figure 15).

The effects of inhibitors on the rate of enzymatic reactions catalyzed by the alkaline phosphatases were studied in Tris-HCl buffer (pH 9), as described previously. [105] The incubation medium included 0.1 M buffer and *p*-nitrophenyl phosphate as the substrate in the range from 15 mM to 0.75 mM. In the study of the inhibition of alkaline phosphatase from bovine intestinal mucosa, the concentration of MBPA ranged from 0.1 mM to 0.4 mM; concentrations of HPMBPA and C-97 varied from 10 □M to 40 □M, C-98 - from 3 μM to 12 μM. In the case of alkaline phosphatase from bovine kidney, concentrations of inhibitors MBPA and HPMBPA varied from 0.2 mM to 0.8 mM, inhibitors C-97 and C-98 - from 20 μM to 80 μM. In the case of alkaline phosphatase from *E. coli,* concentration of HPMBPA changed from 2 mM to 8 mM, concentrations of inhibitors C-97 and C-98 - from 0.1 mM to 0.4 mM.

To prepare working solutions, MBPA, HPMBPA and C-98 as tetrasodium salts were dissolved in water; compound C-97 in dimethyl sulfoxide. The volume concentration of dimethyl sulfoxide in experiments with C-97 and in blank experiments without it was 2%. During kinetic studies the reaction mixture was incubated for 5 min at 25° C, and the reaction initiated by adding 50 ml of enzyme solution to the incubation medium. The total volume of the reaction mixture was 2 ml. The rate of hydrolysis of *p*-nitrophenyl phosphate was measured by the increase in the absorption of *p*-nitrophenol at 410 nm using a molar extinction coefficient of 18300 $M^{-1}cm^{-1}$.

The results of measurements of initial rates of hydrolysis of *p*-nitrophenyl phosphate at different concentrations of substrate and inhibitor were used to determine the type of inhibition and values of the kinetic parameters. Michaelis constant (K_m), maximum velocity of enzyme reaction (V_{max}) in the absence of inhibitor, and apparent values of K_m' and V_{max}' at different concentrations of C-97 and C-98 were calculated from the Lineweaver–Burk plot, which displayed a mixed type of inhibition for alkaline phosphatases from bovine intestinal mucosa, bovine kidney and *E. coli.*

The kinetics of mixed type inhibition at $[S]_o >> [E]$ [106] is described by the following equations:

$$V_{max}' = V/(1+[I]/K_i')$$

and

$$V_{max}'/K_m' = V_{max}/K_m(1+[I]/K_i),$$

where K_i is a dissociation constant of the enzyme-inhibitor complex, and K_i' is a dissociation constant of the enzyme-substrate-inhibitor complex. Inhibition constants K_i and K_i' calculated from the slopes and intercepts of double-reciprocal plots were the average values obtained at four different concentrations of inhibitors. Figure 18 shows a typical dependence of $1/V_0$ on $1/[S]_0$ for the enzymatic reaction, where alkaline phosphatase from bovine intestinal mucosa was inhibited by compound C-98.

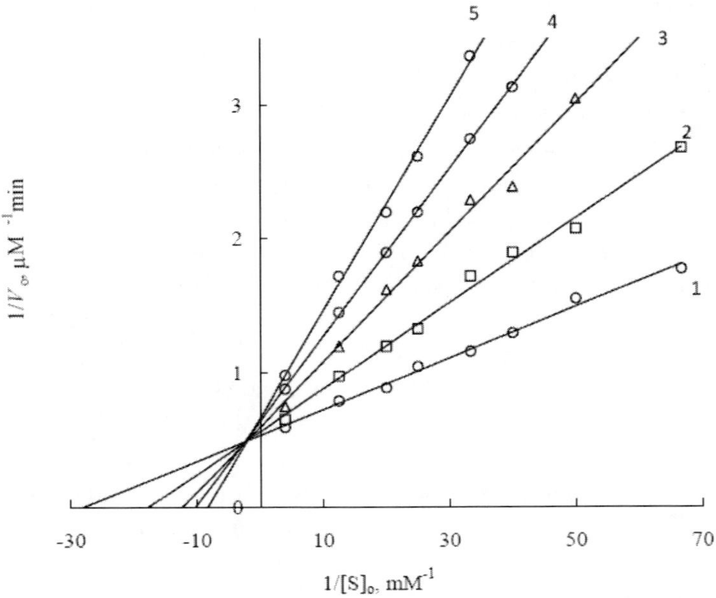

Figure 18. Lineweaver–Burk graphical representation of the inhibition of alkaline phosphatase from bovine intestinal mucosa by compound C-98. *p*-Nitrophenyl phosphate was used as a substrate. The inhibitor concentrations were *1* – 0; *2* - 0,3; *3* - 0,6 ; *4* - 0,9 ; *5* - 1,2 μM.

In this case the *y*-intercepts as well as the slopes of the straight lines are proportional to the concentration of inhibitor. The data obtained are in agreement with the formation of enzyme-inhibitor and enzyme-substrate-inhibitor complexes during the inhibition of the alkaline phosphatase by calix[4]arene C-98. The mixed type of inhibition was also observed for the methylene bisphosphonic acid and model compound HPMBPA. For all the inhibitors the values of K_i' are one order of magnitude or even more higher than the values of K_i. This means that the competitive component provides the main contribution to the inhibition of the enzymes. The greatest contribution of non-competitive binding among all compounds (HPMBPA, C-97, and C-98) was observed for the inhibition of alkaline phosphatase from bovine intestinal mucosa by macrocyclic inhibitor C-98.

The data presented in Table 4 show different affinities of the alkaline phosphatase from bovine intestinal mucosa to methylene bisphosphonic acid and its derivatives. The inhibitory effect of HPMBPA is higher than that of methylene bisphosphonic acid. Introducing the calix[4]arene platform into the structure of the inhibitor leads to compound C-97 being a slightly more effective inhibitor compared to HPMBPA. The introduction of the second methylene bisphosphonic acid residue (compound C-98) provides a further decrease in the K_i value by more than one order of magnitude. Compound C-98 is a potent inhibitor of alkaline phosphatase from bovine intestinal mucosa with K_i of 0.4 μM and K_i' of 5.2 μM. The affinity of the enzyme to the calix[4]arene C-98 is almost two orders of magnitude higher than the affinity of the enzyme to *p*-nitrophenyl phosphate, and about 300 times higher than the affinity of methylene bisphosphonic acid. Approximately the same inhibitory activity was observed for macrocyclic inhibitor C-98 against isoenzyme of calf intestine alkaline phosphatase. This compound, being an effective inhibitor of alkaline phosphatases, is suggested to have the additional fixation of phosphonate residues and better hydrophobic binding of the macrocyclic platform involved in the interaction with the enzyme.

Table 4. Inhibition of alkaline phosphatase from bovine intestinal mucosa by methylene bisphosphonic acid derivatives

No.	Inhibitor	K_i, □M	K_i', □M
1	MBPA	130±20	1800±500
2	HMBPA	17±2	820±250
3	C-97	6.3±0.6	160±60
4	C-98	0.40±0.04	5.2±0.8

In the case of alkaline phosphatase from bovine kidney, the inhibitory ability of compound C-97 is about 15 times higher compared to methylene bisphosphonic acid (Table 5). However, the use of bismethylene bisphosphonate C-98 instead of monomethylene bisphosphonate C-97 almost does not change the value of K_i.

Table 5. Inhibition of alkaline phosphatase from bovine kidney by methylene bisphosphonic acid derivatives

No.	Inhibitor	K_i, □M	K_i', □M
1	MBPA	150±45	1300±370
2	HMBPA	53±3	1500±150
3	C-97	8.8±1,9	480±90
4	C-98	12±3	330±130

The inhibition of alkaline phosphatase from *E. coli* by methylene bisphosphonic acid is not significant. [107] In addition, the rate of hydrolysis of the substrate by alkaline phosphatase from *E. coli* changed over time in the presence of high concentrations of methylene bisphosphonate, making it difficult to determine the inhibition constants. However, the same tendency in the activity of the compounds was observed even for the enzyme from *E. coli*. For example, methylene bisphosphonic acid, at a concentration of 0.1 mM, did not affect the activity of alkaline phosphatase from *E. coli* (substrate concentration of 0.25 mM), whereas macrocyclic inhibitors C-97 and C-98 decreased the initial rate of enzymatic hydrolysis of *p*-nitrophenyl phosphate (by 12% and 21 %, respectively) under similar conditions. Detailed studies showed that the inhibition of alkaline phosphatase from *E. coli* by model bisphosphonate HMBPA was competitive, while the inhibition by compounds C-97 and C-98 was of the mixed type. Data from Table 6 show that the macrocyclic inhibitors C-97 and C-98 are characterized by lower values of the inhibition constants as compared to HMBPA, forming more stable complexes with the enzyme. As in the case of alkaline phosphatase from bovine kidney, the effect of the second mehtylene bisphosphonic acid residue was not observed..

Computer simulations were carried out to elucidate possible ways of binding methylenebisphosphonic acid derivatives to alkaline phosphatases. Among the enzymes studied only alkaline phosphatase from *E. coli* was represented in the RSCB Protein Data Bank, and the crystal structure of this enzyme was employed for molecular docking calculations. AutoDock 4.2 was used as a docking package [108], and visualization was done using the Swiss PDB-Viewer. [109]

Table 6. Inhibition constants, experimental and calculated values of free energy of binding for the inhibition of alkaline phosphatase from *E. coli* by methylene bisphosphonic acid derivatives

No.	Inhibitor	K_i, mM	K_i', mM	ΔG_{exp}, kcal/mol*	ΔG_{calc}, kcal/mol
1	MBPA	-	-	-	-6.81
2	HMBPA	1.6±0,2	-	-3.82	-7.53
3	C-97	0.23±0.09	3.0±0.9	-4.96	-9.50
4	C-98	0.19±0.07	1.4±1.2	-5.07	-9.02

* ΔG_{exp}=-RTln(1/K_i) at 298.15 K.

Table 6 presents theoretical and experimental binding energies. It can be seen that the calculated values follow the same trends as the experimental ones. The greatest change of binding energy occurs between model compound HMBPA and macrocyclic inhibitor C-97. Complexes of the inhibitors with alkaline phosphatase from *E. coli* were obtained by docking the respective ligands into the active site of the enzyme. In this case competitive inhibition was simulated. Phosphonate groups were in monoanionic form; macrocycles were in *cone*-conformations. Effective metal charges were as low as +1.1, otherwise electrostatic interactions prevailed, giving very low binding energies, most of which consisted of electrostatic interactions. Four water molecules, three of which were coordinated around the Mg^{2+} ion, were left in the active centre of the enzyme. The binding mode of methylenebisphosphonic acids is similar to that of inorganic phosphate for all the inhibitors. One of the phosphonic acid groups interacts with both Zn ions, while the ion of Mg is shielded by several water molecules which interact with another phosphonic acid group. The second phosphonic acid group also interacts with a molecule of water coordinated by Asp327 and Lys328. Ser102 and Arg166 can also form hydrogen bonds with phophonic groups of the ligand. Calixarene parts of the inhibitors are fitted into the cavity formed by residues of Lys328, His 331, and Ser 409 from one side, and by the residues of Tyr169, Arg166, and Lys167 from the other side. Compound C-98 binds differently from compound C-97 (Figure 19). It is shifted towards Lys166, which forms hydrogen bonds with the second bisphosphonate group. In the case of phospatase from *E. coli* this does not give any advantages (the energy of binding becomes even worse), but in the case of alkaline phosphatase from bovine intestinal mucosa similar interactions may be favourable.

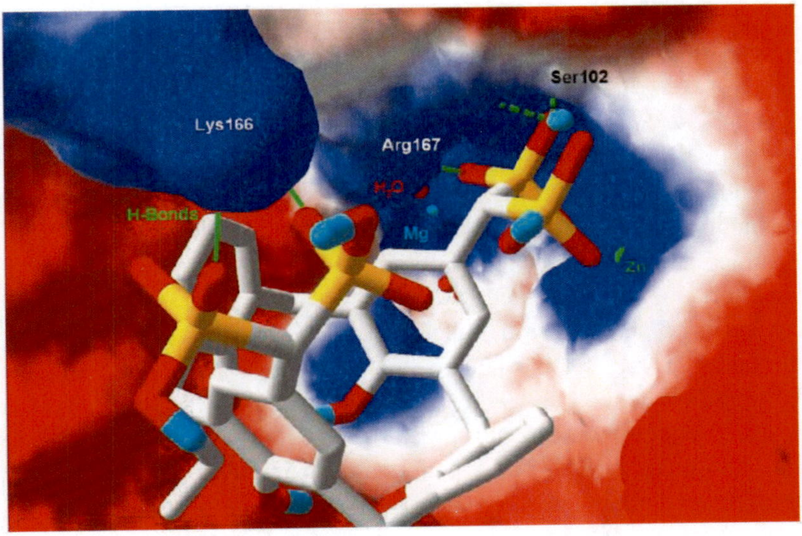

Figure 19. Possible binding mode of compound C-98 with alkaline phosphatase from *E. coli.*

The different affinity of inhibitor C-98 with respect to alkaline phosphatase from bovine intestinal mucosa and bovine kidney or *E. coli* is probably due to differences in the structure of the active site and its surroundings. This is consistent with other known properties of these enzymes. [110,111,112] The sequence of amino acids in the structure of alkaline phosphatase from mammalian tissues and *E. coli* are 25-30% similar. Amino acid residues of the active centres of these enzymes are preserved in evolution, and the active centre of the enzyme from mammalian tissues is organized about the same as that of alkaline phosphatase from *E. coli* [110,112]. Conservation of amino acid residues involved in direct binding of the substrate causes the same mechanism of catalysis [111]. However, Lys328 and Asp153 residues near the catalytic centre of the enzyme from *E. coli* are replaced by histidine residues in alkaline phosphatases from mammalian tissues [111]. The presence of an additional substrate binding site, more distant from the active site, is an important structural feature of alkaline phosphatases from mammalian tissues [110,112]. Data on alkaline phosphatases from calf and bovine intestine also indicate the presence of amino acid residues located near to the active centre that are not directly involved in the mechanisms of catalysis, but spatially hinder access to the substrate binding site. This amino acid residue at position 322, depending on its nature, can affect the position of the substrate and the hydrolysis of a covalent enzyme-substrate intermediate [112]. Therefore, we can assume that in the case of the isoform of alkaline phosphatase from bovine intestinal mucosa better binding of compound C-98 is provided by the hydrophobic contacts of the macrocycle, as well as by additional hydrogen bonds formed by the second bisphoshonate group.

Extending this approach [24] to other enzymes, calix[4]arenes bearing methylenebisphosphonic or hydroxymethylenebisphosphonic acid fragments were studied as inhibitors of protein tyrosine phosphatases [113, 25]. Some of the inhibitors of PTP1B showed IC_{50} values in the micromolar range and good selectivity in comparison with other phosphatases, such as TC-PTP, PTPb, LAR, and CD45. Calix[4]arene mono-, bis- and tetrakis(methylene bisphosphonic) acids, as well as calix[4]arene and thiacalix[4]arene tetrakis(methylphosphonic) acids were found to inhibit *Yersinia* protein tyrosine phosphatase [25]. The kinetic studies revealed that thiacalix[4]arene bearing four methylphosphonate fragments at the wide rim of the macrocycle has the highest affinity for the enzyme. It was also shown that thiacalix[4]arene-tetrakis-methylphosphonic acids displayed stronger inhibition properties towards alkaline phosphatases from bovine intestine mucosa, shrimp and human placenta [21], and inhibit the activity of human protein tyrosine phosphatase 1B. [114] All these results show that calix[4]arenes and thiacalix[4]arenes substituted by phosphonate or methylenebisphosphonate groups might be considered as potentially bioactive compounds which could be used for the further development of phosphatase inhibitors.

CONCLUSION

We believe that our data make a contribution to modern knowledge about the biochemical and biophysical properties of calixarenes and the molecular mechanism of their interactions with proteins. Taking into account the ability of calixarenes to penetrate into the cell, and their low toxicity, these data may be used as a basis for the further development of a new generation of supramolecular effectors for the regulation of smooth muscle contractile activity at the level of ATP dependent actin-myosin interaction, and could be used for the design of a new class of antithrombotic agents.

The results of this study demonstrate the applicability of the macrocyclic platform for the design of phosphatase inhibitors. The preorganization of bioisosteric phosphonic acid fragments on the calix[4]arene platform results in a significant increase in the inhibition of alkaline phosphatases and protein tyrosine phosphatases. The strength of the enzyme-inhibitor interactions is determined both by cooperative effects and the accuracy of the fit of the macrocyclic inhibitor to the surface of the enzyme at the inhibitor binding site.

ACKNOWLEDGMENT

This interdisciplinary project of the Palladin Institute of Biochemistry NASU, the Institute of Organic Chemistry NASU, and the Institute of Bioorganic Chemistry and Petrochemistry NASU was supported by the State Programme of Ukraine "Nanomaterials and Nanotechnologies", chart "Nanobiothechnologies."

REFERENCES

[1] Lehn, J.-M. *(1995).* Supramolecular chemistry. Concepts and perspectives. *VCH. Weinheim.*

[2] Huang, P., Xiao, A., Zhou, M., Zhu, Z., Lin, S., Zhang, B. (2011). Heritable gene targeting in zebrafish using customized TALENs. *Nature Biotechnology.* 29, 699-700.

[3] Deng, D., Yan, C., Pan, X., Mahfouz, M., Wang, J., Zhu, J.-K., Shi, Y., Yan, N. (2012). Structural Basis for Sequence-Specific Recognition of DNA by TAL Effectors. *Science.* 335, 720-723.

[4] Gutsche, C.D. (2008). Calixarenes: an introduction, monographs in supramolecular chemistry. *Royal Society of Chemistry.* Cambridge.

[5] Rodik, R.V., Boyko, V.I., Kalchenko, V.I. (2009). Calixarenes in bio-medical researches. *Curr. Med. Chem.* 16, 1630-1655.

[6] Asfari, M.-Z., Böhmer, V., Harrowfield, J., Vicens, J. (Eds.). (2001). Calixarenes 2001. *Kluwer Academic Publishers.* Dordrecht.

[7] Kim, H.J., Lee, J.H., Mutihac, L., Vicens, J., Kim, J.S. (2012). Recognition of amino acids by functionalized calixarenes. *Chem. Soc. Rev.* 41, 1173-1190.

[8] Mutihac, L., Lee, J.H., Kim, J.S., Vicens, J. (2011). Recognition of amino acids by functionalized calixarenes. *Chem. Soc. Rev.* 40, 2777-2796.

[9] Lumetta, G.J., Rogers, R.D., Gopalan, A.S. (eds). (2000). Calixarenes for Separations. *American Chemical Society.* Washington.

[10] Vicens, J., Harrowfield J. (Eds). (2007). Calixarenes in the Nanoworld. *Springer.* Dordrecht.

[11] Nimse, S.B., Kim, T. (2013). Biological applications of functionalized calixarenes. *Chem. Soc. Rev.* **42**, 366-386

[12] Rodik, R.V., Klymchenko, A.S., Miroshnichenko, S.I., Richert, L., Kalchenko, V.I., Mély, Y. (2011). Virus-sized DNA nanoparticles for gene delivery based on micelles of cationic calixarenes. *Chem. Eur. J.* 17, 5526-5538.

[13] Da Silva, E., Lazar, A.N., Coleman, A.W. (2004). Biopharmaceutical applications of calixarenes. *J. Drug. Sci. Tech.* 14, 3-20.

[14] Coleman, A.W., Perret, F., Moussa, M., Dupin, M., Guo, Y., Perron, H. (2007). Calix[n]arenes as protein sensors. *Top. Curr. Chem.* 277, 31-88.

[15] Sansone, F., Segura, M., Ungaro, R. (2001). Calixarenes in Bioorganic and Biomimetic Chemistry. In Calixarenes 2001; Asfari, M.-Z.; Bohmer, V.; Harrowfield, J.; Vicens, J. (Eds.). *Kluwer Academic Publishers: Dordrecht.*

[16] Ukhatskaya, E.V., Kurkov, S.V., Matthews, S.E., Loftsson, T. (2013). Encapsulation of drug molecules into calix[n]arene nanobaskets. Role of aminocalix[n]arenes in biopharmaceutical field. *J. Pharm. Sci.* 102, 3485-512.

[17] Lipkowski, J., Simonov, Yu., Kalchenko, V.I., Visotsky, M.A., Markovsky, L.N. (1998). Molecular and crystal structure of water-soluble 25,27-bis(dihydroxyphosphoryloxy)calix[4]arene. *Anales de Quimica.* 94, 328-331.

[18] Cherenok, S., Dutasta, J.-P., Kalchenko, V. (2006). Phosphorus-containing chiral macrocycles. *Current Organic Chemistry.* 10, 2307-2331.

[19] Tairov, M.A., Vysotsky, M.O., Kalchenko, O.I., Pirozhenko, V.V., Kalchenko, V.I. (2002). Symmetrical and inherently chiral water-soluble calix[4]arenes bearing dihydroxyphosphoryl groups. *J. Chem. Soc., Perkin Trans.* 1. 1405-1411.

[20] Houel, E., Lazar, A., Da Silva, E., Coleman, A.W., Solovyov, A.V., Cherenok, S., Kalchenko, V.I. (2002). Interfacial interaction of cations

with amphiphilic dihydroxyphosphonylcalix[4]arene mesosystems. *Langmuir.* 18, 1374-1379.

[21] Vovk, A.I., Kononets, L.A., Tanchuk, V.Yu., Drapailo, A.B., Kalchenko, V.I., Kukhar, V.P. (2010). Thiacalix[4]arene as Molecular Platform for Design of Alkaline Phosphatase Inhibitors. *J. Incl. Phenom. Macrocyclic Chem.* 66, 271–277.

[22] Solovyov, A.V., Cherenok, S., Tsymbal, I., Failla, S., Consiglio, G., Finocchiaro, P., Kalchenko, V.I. (2001). Calix[4]arenes bearing alpha-amino- or alpha-hydroxyphosphonic acid fragments at the upper rim. *Heteroatom Chemistry.* 12, 58-67.

[23] Cherenok, S., Yushchenko, O., Tanchuk, V., Mischenko, I., Samus, N., Ruban, O., Matvieiev, Y., Karpenko, J., Kukhar, V., Vovk, A., Kalchenko, V. (2012) Calix[4]arene-α-hydroxyphosphonic acids. Synthesis, stereochemistry, and inhibition of glutathione S-transferase. *Arcivok.* 278 – 298.

[24] Cherenok, S., Vovk, A., Muravyova, I., Shivanyuk, A., Kukhar, V., Lipkowski, J., Kalchenko, V. (2006). Calix[4]arene α-aminophosphonic acids: asymmetric synthesis and enantioselective inhibition of alkaline phosphatases. *Organic Letters.* 8, 549-552.

[25] Vovk, A.I., Kononets, L.A., Tanchuk, V.Yu, Cherenok, S.O., Drapailo, A.B., Kalchenko, V.I., Kukhar, V.I. (2010). Inhibition of Yersinia protein tyrosine phosphatase by calixarene phosphonous and methylenebisphosphonous acids. *Bioorg. Med. Chem. Lett.* 20, 483–487.

[26] Weinstein, R.S., Roberson, P.K., Manolagas, S.C. (2009). Giant osteoclast formation and long-term oral bisphosphonate therapy. *N. Engl. J. Med.* 360, 53–62.

[27] Gnant, M., Mlineritsch, B., Schippinger, W. (2009). Zoledronic acid in breast cancer: latest findings and interpretations. *N. Engl. J. Med.* 360, 679–691.

[28] Menschutkin N. (1865). Uber die Einwirkung des Chloroazetyls auf Phosphorige Saure. *Ann. Chem. Pharm.* 133, 317–320.

[29] Von Baeyer, H., Hofmann, K. A. (1897). Acetodiphosphorige Saure. *Ber. Dtsch. Chem. Ges.*, 30, 1973–1978.

[30] Fleisch, H., Bisaz, S. (1962). Isolation from urine of pyrophosphate, a calcification inhibitor. *Am. J. Physiol.* 203, 671–675.

[31] Fleisch, H., Russell, R.G.G., Straumann, F. (1966). Effect of pyrophosphate on hydroxyapatite and its implications in calcium homeostasis. *Nature.* 212, 901–903.

[32] Fleisch, H., Russell, R.G.G., Bisaz, S., Muhlbauer, R.C., Williams, D.A. (1970). The inhibitory effect of phosphonates on the formation of calcium phosphate crystals in vitro and on aortic and kidney calcification in vivo. *Eur. J. Clin. Invest.* 1, 12–18.

[33] Gulaya, N.M., Komisarenko, S.V. (1982). Inorganic pyrophosphate, its structural analogues and inorganic pyrophosphatase. *Uspekhi Biologicheskoy Khimii.* 22, 195-213.

[34] Komisarenko, S.V., Karlova, N. P., Kolesnikova, I.N., Gaivoronskaya, G.G., Popova, G.N., Borisevich, A.N., Lozinsky, M.O. (1985). The use of diphosphonates as imunomodulators. Chemistry and biology of immunoregulators: [Collected papers]. *Acad. Sci. Zinante. Latv.*

[35] Komisarenko, S.V., Gulaya, N.M., Borisenro, A.M., Veller, O.S. (1979). Diphosphonate analogs of pyrophosphate and inorganic mice pyrophosphatase. *Reports Acad. Sci. USSR.* 7, 563-566.

[36] Komisarenko, S.V., Kolesnikova, I.N., Fomovskaya, G. N. (1985). The influence of diphosphonic analogues of inorganic pyrophosphate on reactions catalyzed by DNA-dependent RNA-polymerase II. Ukr. *Biokhim. Zh.* 57, 62-66.

[37] Komisarenko, S.V., Fomovskaya, G. N., Kolesnikova, I.N., Tarusova, N.B., Borisevich, A.N., Lozinsky, M.O. (1985). The influence of diphosphonic analogues of inorganic pyrophosphate on activity of RNA-polymerases from the calf thymus. *Ukr. Biokhim. Zh.* 57, 56-61.

[38] Fomovskaya, G. N., Komisarenko, S.V. (1985). Kinetics of the interaction of methylene diphosphonic acid and inorganic pyrophosphate with DNA-dependent RNA-polymerase from calf thymus. *Biokhimia.* 50, 839-843.

[39] Gaivoronskaya, G.G., Komisarenko, S.V. (1983). Effect of diphosphonic acids on alkaline phosphatase activity. *Ukr. Biokhim. Zh.* 55, 403-407.

[40] Smirnova, I. N., Kudryavtseva, N. A., Komissarenko, S.V., Tarusova, N.B., Baykov A.A. (1988). Diphosphonates are potent inhibitors of mammalian inorganic pyrophosphatase. *Archives of Biochemistry and Biophysics.* 267, 280-284.

[41] Karlova, N. P., Komissarenko, S.V. (1981). Influence of methylene diphosphonic acid on lymphoid organs morphology. *Reports Acad. Sci. USSR.* 5, 68-70.

[42] Gulaya, N.M., Bogomolets, E. O., Karlova, N. P., Komissarenko, S.V. (1980). Influence of methylene diphosphonic acid on some processes in animal tissues. *Farmakology and Toxicology.* 43, 192-195.

[43] Gaivoronskaya, G.G., Strelchuk, S.I., Komissarenko, S.V. (1981). Mutagenic and antimutagenic effect of diphosphonic acids. *Cytology and Genetics.* 15, 41-45.

[44] Fomovskaya, G. N., Komissarenko, S.V. (1987). Pharmacokinetics and immunomodulatory effects of the anticancer drug methylene diphosphonic acid. *Chemistry of tumors in USSR.* 48, 14-16.

[45] Sharykina, N. I., Kudriavtseva, I. G., Komissarenko, S.V., Karlova, N. P. (1987) Anticancer effect methylene diphosphonic acid (M-1) in experiment. *Chemistry of tumors in USSR.* 48, 12-13.

[46] Breuer, E. (2006). The development of bisphosphonates as drugs. Analogue-based drug discovery. IUPAC. Fischer, J. and Ganellin, C.R. (Eds.) *WILEY-VCH Verlag GmbH & Co.* KGaA, Weinheim.

[47] Fogelman, I., Bessent, R.G., Turner, J.F., Citrin, D.L., Boyce, I.T., Greig, W.R. (1978). The use of whole-body retention of Tc-99m diphosphonate in the diagnosis of metabolic bone disease. *J. Nucl. Med.* 19, 270–275.

[48] Fleisch, H. (1997). Bisphosphonates in Bone Disease. From the Laboratory to the Patient. 3rd edn. *The Parthenon Publishing Group.* New York.

[49] Uludag, H. (2002). Bisphosphonates as a foundation of drug delivery to bone. *Current Pharm. Design.* 8, 99–110.

[50] Wagstaff, A.J., Bryson, H.M. (1994). *Phosphonoformate. Drugs.* 48, 199–226.

[51] Atack, J.R. (1997). Inositol monophosphatase inhibitors – Lithium mimetics? *Med. Res. Rev.* 17, 215–222.

[52] Ciosek, C.P., Jr., Magnin, D.R., Harrity, T.W., Logan, J.V.H., Dickson, J.K., Jr., Gordon, E.M., Hamilton, K.A., Jolibois, K.G., Kunselman, L.K., Lawrence. R.M., Mookhtiar, K.A., Rich, L.C., Slusarchyk, D.A., Sulsky, R.B., Biller, S.A. (1993). Lipophilic 1,1-bisphosphonates are potent squalene synthase inhibitors and orally active cholesterol-lowering agents in vivo. *J. Biol. Chem.* 268, 24832–24837.

[53] Martin, M.B., Grimley, J.S., Lewis, J.C., Heath, H.T., Bailey, B.N., Kendrick, H., Yardley, V., Caldera, A. Lira, R., Urbina, J.A., Moreno, S.N.J., Docampo, R., Croft, S.L., Oldfield, E. (2001). Bisphosphonates inhibit the growth of Trypanosoma brucei, Trypanosoma cruzi, Leishmania donovani, Toxoplasma gondii, and Plasmodium falciparum: a potential route to chemotherapy. *J. Med. Chem.* 44, 909–916.

[54] Danenberg, H.D., Fishbein, I., Gao, J., Monkkonen, J., Reich, R., Gati, I., Moerman, E., Golomb, G. (2002). Macrophage depletion by

clodronate-containing liposomes reduce neointimal formation following balloon injury in rats and rabbits. *Circulation.* 106. 599–605.

[55] Curry, J.D., Nicholson, D.A., Quimby, O.T. (1972). Oligophosphonates. *Topics in Phosphorus Chemistry.*7, 37-102.

[56] Vovk, A.I., Kalchenko, V.I., Cherenok, S.O., Kukhar, V.P., Muzychka, O.V., Lozynsky, M.O. (2004). Calix[4]arene methylenebisphopshonic acids as calf intestine phosphatase inhibiters. *Org. Biomol. Chem.* 2, 3162-3166.

[57] Gritsenko, P.G., Lugovskoy, E.V., Koshel, T.A., Cherenok, S.O., Yuschenko, O.A., Chernyshov, V.I., Kalchenko, V.I., Komisarenko, S.V. (2010). Influence of calixarene methylenebisphosphonic acids on fibrin polymerization. *Reports Acad. Sci. NASU.* 175-179.

[58] Lugovskoy, E.V., Gritsenko, P.G., Koshel, T.A., Koliesnik, I.O., Cherenok, S.O., Kalchenko, O.I., Kalchenko, V.I., Komisarenko, S.V. (2011). Calix[4]arene methylenebisphosphonic acids as inhibitors of fibrin polymerization. *FEBS J.* 278, 1244-1251.

[59] Khataee, H.R., Khataee, A.R. (2009). Application of molecular motors in intelligent nanosystems. *Digest J. Nanomaterials and Biostructures.* 4, 613–621.

[60] Burghardt, T.P., Neff, K.L., Wieben, E.D., Ajtai, K. (2010). Myosin individualized: single nucleotide polymorphisms in energy transduction. *BMC Genomics.* 11, 172.

[61] Bárány, M., Bárány, K. (1996). Biochemistry of Smooth Muscle Contraction. Bárány, M. (Ed.). Chicago. *Academic Press.*

[62] Kaliman, I.A., Grigorenko, B.L., Shadrina, M.S., Nemukhin, A.V. (2009). Opening the Arg-Glu salt bridge in myosin: computational study. *Phys. Chem. Chem. Phys.* 11, 4804–4807.

[63] Popov, E.M., Demin, V.V., Shibanova, E.D. (1996). Problem of protein. 2: Three-dimensional structure of protein. Sorokina, T.I. (Ed.). *Nauka.* Moskow.

[64] Levitsky, D.I. (2004). Actomyosin system of biological mobility. *Biochimia.* 69, 1447–1462.

[65] Bevza, A.A., Labintseva, R.D., Rodik, R.V., Cherenok, S.O., Kosterin, S.O., Kalchenko, V.I. (2009). Effect of calix[4]arenes on the activity of actomyosin ATPase and actomyosin subfragment-1 ATPase from the myometrium. *Ukr. Biokhim. Zh.* 81, 49–58.

[66] Bárány, M., Bárány, K., Gaetijens, M., Balin G. (1966). Chicken gizzard myosin. *Arch. Biochem. Biophys.* 113, 205–211.

[67] Weber, A. (1956). The ultracentrifugal separation of L-myosin and actin in an actomyosin sol under the influence of ATP. *Biochem. Biophys. Acta.* 19, 345–351

[68] Weeds, A.G., Taylor, R.S. (1975). Separation of subfragment-1 isoenzymes from rabbit skeletal muscle myosin. *Nature.* 257, 54–56.

[69] Bradford, M.M. (1976). A rapid and sensitive method for the quantitation of microgram quantities of protein utilizing the principle of protein-dye binding. *Anal. Biochem.* 72, 248–282.

[70] Laemmly, U.K. (1970). Cleavage of structural proteins during the assembly of the head of bacteriophage T4. *Nature* (London). 227, 680–685.

[71] Iwane, A.H., Kitamura, K., Tokunaga, M., Yanagida, T. (1997) Myosin subfragment-1 is fully equipped with factors essential for motor function. *Biochem. Biophys. Res. Commun.* 230, 76–80.

[72] Chen, P.S., Toribara Jr., T.Y., Warner, H. (1956). Microdetermination of phosphorus. *Analyt. Chem.* 28, 1756–1758.

[73] Burghardt, T.P., Neff, K.L., Wieben, E.D. Ajtai, K. (2010). Myosin individualized: single nucleotide polymorphisms in energy transduction. *BMC Genomics.* 11, 172.

[74] Merkus, H.G. (2009). Particle Size Measurements. Fundamentals, Practice, Quality. *Springer.*

[75] Storn, R.M., Price, K. (1997). Differential evolution – a simple and efficient heuristic for global optimization over continuous spaces. *J. Glob. Optim.* 341–359.

[76] Case, D.A., Pearlman, D.A., Caldwell, J.W. et al. (2002). AMBER 7. Users' Manual. University of California, San Francisco.

[77] Berman, H.M., Westbrook, J., Feng, Z., Gilliland G, Bhat TN, Weissig H, Shindyalov IN, Bourne PE. (2000). The Protein Data Bank. *Nucl. Acids Res.* 28, 235–242.

[78] Van Der Spoel, D., Lindahl, E., Hess, B., Groenhof, G., Mark, A.E., Berendsen, H.J. (2005). GROMACS: fast, flexible, and free. *J. Comput. Chem.* 26, 1701–1719.

[79] Allen, M.P., Tildesley, D.J. (2002). Computer Simulation of Liquids. Oxford. *Clarendon Press.*

[80] Li, L., Jose, J., Xiang, Y., Kuhn, R.J., Rossmann. M.G. (2010). Structural changes of envelope proteins during alphavirus fusion. *Nature*. 468, 645–646.

[81] Korostylev, P.P. (1964). Preparing solutions for chemico-analytical researches. *Publishers Acad. Sci. USSR.*

[82] Burgess, S.A., Yu, S., Walker, M.L., Chalovich, J.M., Knight, P.J. (2007). Structures of smooth muscle myosin and heavy meromyosin in the folded, shutdown state. *J. Mol. Biol.* 372, 1165–1178.

[83] Coluccio, L.M. (Ed). (2008). Proteins and Cell Regulation. Myosins. A Superfamily of Molecular Motors. *Springer.* Watertown. MA. USA.

[84] Baby, G. J., Warsaw, D. M., Periasamy, M. (2000). *Microscopy Research and Technique.* 50, 532–540.

[85] Burgess, S.A., Yu, S., Walker, M.L., Chalovich, J.M., Knight, P.J. (2007). Structures of smooth muscle myosin and heavy meromyosin in the folded, shutdown state. *J. Mol. Biol.* 372, 1165–1178.

[86] Veklich, T.O., Kosterin, S.A., Rodik, R.V., Cherenok SO, Boĭko VI, Kalchenko VI. (2006). Effect of calixarene-phosphonic acid on Na+, K+-ATPase activity in plasma membranes of the smooth-muscle cells. *Ukr. Biokhim. Zh.* 78, 70–78.

[87] Blomback, B. (1996) Fibrinogen and fibrin – proteins with complex roles in haemostasis and thrombosis. *Thromb. Res.* 83, 1–75.

[88] Laudano, A.P., Doolittle, R.F. (1978). Synthetic peptide derivatives that bind to fibrinogen and prevent the polymerization of fibrin monomers. *Proc. Natl. Acad. Sci. USA.* 75, 3085–3089.

[89] Fowler, W.E., Hantgan, R.R., Hermans, J., Erickson, H.P. (1981). Structure of the fibrin protofibril. *Proc. Natl. Acad. Sci. USA*. 78, 4872–4876.

[90] Weisel, J.W., Veklich, Y., Gorkun, O. (1993). The sequence of cleavage of fibrinopeptides from fibrinogen is important for protofibril formation and enhancement of lateral agregation in fibrin clots. *J. Mol. Biol.* 232, 285–297.

[91] Weisel, J.W., (1986). Fibrin assembly. Lateral aggregation and the role of the two pairs of fibrinopeptides. *Biophys. J.* 50, 1079–1093.

[92] Spraggon, G., Everse, S.J., Doolittle, R.F., (1997). Crystal structures of fragment D from human fibrinogen and its crosslinked counterpart from fibrin. *Nature.* 389, 455–462.

[93] Varetskaya, T.V. (1960). Microheterogeneity of fibrinogen. Cryofibrinogen. *Ukr. Biokhim. Zh.* 32, 13–24.

[94] Belitser, V.A., Varetskaja, T.V., Malneva, G.V., (1968). Fibrinogen-fibrin interaction. *Biochim. Biophys. Acta.* 154, 376.

[95] Lugovskoi, E.V., Makogonenko, E.M., Chudnovets, V.S., Derzskaya, S.G., Gogolinsikaja, G.K., Kolesnikova, I.N., Bukhanevich, A.M., Komisarenko, S.V. (1991). The study of fibrin polymerization with monoclonal antibodies. *Biomed. Sci.* 2, 249–296.

[96] Kalchenko, O.I., Lipkowski, J., Kalchenko, V.I., Vysotsky, M.A., Markovsky, L.N. (1998). Effect of octakis(diethoxyphosphoryloxy)-tert-butyl-calix[8]arene in mobile phase on the reversed-phase retention behavior of aromatic compounds: host-guest complex formation and stability constants determination. *J. Chrom. Sci.* 36, 269–273.

[97] Sun, L., Martin, D.C., Kantorowitz, E.R. [1999]. Rate-determining step of Escherichia coli alkaline phosphatase altered by the removal of a positive charge at the active centre. *Biochemistry-US.* 39, 2842-2848.

[98] Boyan, B.D., Schwartz, Z., Lohmann, C.H., Sylvia, V.L., Cochran, D.L., Dean, D.D., Puzas, J.E. (2003). Pretreatment of bone with osteoclasts affects phenotypic expression of osteoblast-like cells. *J. Orthop. Res.* 21, 638-647.

[99] Tonks, N.K. (2006). Protein tyrosine phosphatases: from genes, to function, to disease. *Nat. Rev. Mol. Cell Biol.* 7, 833-846.

[100] Vintonyak, V.V., Antonchick, A.P., Rauh, D., Waldmann, H. (2009). The therapeutic potential of phosphatase inhibitors. *Curr. Opin. Chem. Biol.* 13, 272-283.

[101] Tabernero, L., Aricescu, A.R., Jones, E.Y., Szedlacsek, S.E. (2008). Protein tyrosine phosphatases: structure-function relationships. *FEBS J.* 275, 867-882.

[102] Scott, L.M., Lawrence, H.R., Sebti, S.M., Lawrence, N.J., Wu, J. (2010). Targeting protein tyrosine phosphatases for anticancer drug discovery. *Curr. Pharm. Des.* 16, 1843-1862.

[103] Kasibhatla, B., Wos, J., Peters, K.G. (2007). Targeting protein tyrosine phosphatase to enhance insulin action for the potential treatment of diabetes. *Curr. Opin. Invest. Drugs.* 8, 805-813.

[104] Zhang, S., Zhang, Z.Y. (2007). PTP1B as a drug target: recent developments in PTP1B inhibitor discovery. *Drug Discovery Today.* 12, 373-381.

[105] Muzychka, O.V., Tanchuk, V. Yu., Vovk, A.I., Muravyova, I.V., Kalchenko, V.I., Cherenok, S.O., Kukhar, V.P. (2007). Calix[4]arene methylenebisphosphonic acids: the features of alkaline phosphatases inhibition. *Ukr. Biochem. J.* 79, 26-33.

[106] Cornish-Bowden, A. (2004). Fundamentals of enzyme kinetics. 3rd Edition; *Portland Press.* London.

[107] Myers, J.K., Antonelli ,S.M., Widlanski, T.S. (1997). Motifs for metallophosphatase inhibition. *J. Am. Chem. Soc.* 119, 3163-3164.

[108] Morris, G.M., Huey, R., Lindstrom, W., Sanner, M.F., Belew, R.K., Goodsell, D.S., Olson, A.J. (2009). Autodock4 and AutoDockTools4:

automated docking with selective receptor flexiblity. *J. Comput. Chem.* 30, 2785-2791.

[109] Guex N., Peitsch M.C. (1995). SWISS-MODEL and the Swiss-PdbViewer: An environment for comparative protein modeling. *Electrophoresis.* 18, 2714-2723.

[110] Le Du, M.H., Millan, J.L. (2002). Structural evidence of functional divergence in human alkaline phosphatases. *J. Biol. Chem.* 277, 49808-49814.

[111] Holtz, K.M., Stec, B., Kantrowitz, E.R. (1999). A model of the transition state in the alkaline phosphatase reaction. *J. Biol. Chem.* 274, 8351-8354.

[112] Manes, T., Hoylaerts, M.F., Muller, R., Lottspeich, F., Hölke, W., Millán, J.L. (1998). Genetic complexity, structure, and characterization of highly active bovine intestinal alkaline phosphatases. *J. Biol. Chem.* 273, 23353-23360.

[113] Trush, V.V., Cherenok, S.O., Tanchuk, V.Y., Kukhar, V.P., Kalchenko V.I., Vovk, A.I. (2013). Calix[4]arene methylenebisphosphonic acids as inhibitors of protein tyrosine phosphatase 1B. *Bioorg. Med. Chem. Lett.* 23, 5619-5623.

[114] Trush, V.V., Tanchuk, V.Yu., Kononets, L.A., Drapailo, A.B., Kalchenko, V. I., Vovk, A. I., Kukhar, V.P. (2012). Inhibition of protein tyrosine phosphatases by phosphonic acids on alix[4]arene and thiacalix[4]arene platform. *Reports Acad. Sci. NASU.* 145-151.

In: Ligands
Editors: Paula Gawryszewska et al.
ISBN: 978-1-63117-143-7

Chapter 4

SYNTHESIS AND BIOLOGICAL APPLICATIONS OF TRIS(PYRAZOL-1-YL)-METHANE AND BORATE METAL COMPLEXES

***Luísa M. D. R. S. Martins*[1,2*],**
***Elisabete C. B. A. Alegria*[1,2,†]**
***and Armando J. L. Pombeiro*[2]**
[1]Chemical Engineering Department,
Instituto Superior de Engenharia de Lisboa,
R. Conselheiro Emídio Navarro, Lisboa, Portugal
[2]Centro de Química Estrutural, Instituto Superior Técnico,
Universidade de Lisboa, Lisboa, Portugal

ABSTRACT

A survey of tris(pyrazol-1-yl)-borate and -metane homoscorpionate $RX(R'pz)_3$ (pz = pyrazol-1-yl, X = C or B) metal complexes with biological activity is presented. The main biological applications, enzyme modelling and antiproliferative and antimicrobial activity, of such scorpionate-type complexes are classified by ligand and also by transition metal.

---

* Corresponding Author address Email: lmartins@deq.isel.ipl.pt.
† Corresponding Author address Email: ebastos@deq.isel.ipl.pt.

Keywords: tris(pyrazol-1-yl)methane, tris(pyrazol-1-yl)borate, scorpionate, scorpionate metal complexes, enzyme modelling, cytotoxicity, antiproliferative activity, antitumor activity, antimicrobial activity, pharmacological activity

1. INTRODUCTION

Tris(pyrazol-1-yl)borates, Tp, where the central atom is boron, and related tris(pyrazol-1-yl)methanes, Tpm, formally derived by replacing the bridging $(BH)^-$ moiety with the CH group (Figure 1), are widely used ligands in chemistry [1-5]. The coordination chemistry of Tp has been extensively developed (Tp have been used to synthesize complexes of most metals of the periodic table) [2, 3] with particular interest arising out of the versatility of this class of ligands, namely allowing an easy modification of the steric and electronic environment of the metal centre by variation of the pyrazol-1-yl moiety or the substituents in the pyrazol-1-yl rings, to fine-tune the stability, inertness and reactivity of their metal complexes.

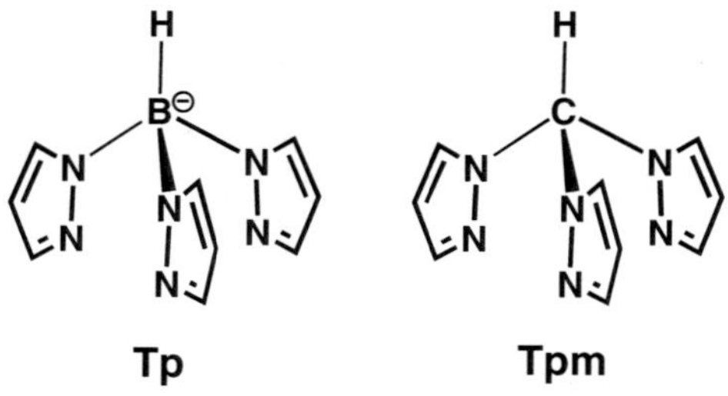

Figure 1. Hydrotris(pyrazol-1-yl)borate $HB(pz)_3^-$ (Tp), and hydrotris(pyrazol-1-yl)methane $HC(pz)_3$ (Tpm).

The intriguing designation of these versatile tripodal ligands, introduced by Trofimenko in 1967, [6] derives from their peculiar coordination mode: with the three nitrogen heteroatoms of the corresponding pyrazol-1-yl rings able to occupy three facially adjacent vacant positions of the coordination sphere of a metal centre. This configuration (i.e., *fac*-chelation) mimics the analogue position of a scorpion that uses the three stings (two pincers and one claw) to attack its prey (Figure 2).

A similar analogy has been described for another class of chelating ligand, the pincer ligands. The latter are, in fact, characterized by their peculiarity to bind two or three adjacent coplanar positions of a coordination sphere of a metal centre. However, differently, the scorpionate ligands extend their coordination ability out of a planar dimension, employing the third 'arm' to a facially capping coordination.

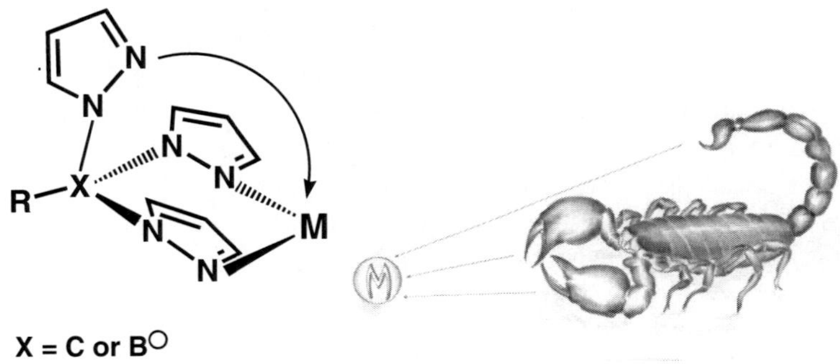

Figure 2. Illustration of a parallel between the tris(pyrazol-1-yl) -methane or -borate binding and the scorpion attack.

More precisely, the term scorpionate has been used to describe the interchange between bidentate and tridentate coordination modes of these ligands (Figure 2), which is at the core of the structural and chemical versatility of many metal complexes of this kind.

Moreover, it has been recently recognized that synthetic progress toward the functionalization of the central (carbon or boron) atom at the apical position is greatly advantageous, as such changes to the backbone can dramatically influence the properties of the complexes, extend the coordination properties of these ligands and opening to a large variety of applications.

One application of enormous significance is performed by complexes soluble and stable under physiological conditions, and several examples of the use of tris(pyrazol-1-yl)-borate or -methane complexes to mimic enzymatic systems, as well as biomedical applications as metallodrugs (in view of their anti-tumour properties or antibacterial activity), have been reported [2].

2. SCORPIONATE FEATURES

The main characteristics of the Tp ligands can be compared [7-10] with other face-capping ligands, in particular taking into account the parallel between tris(pyrazol-1-yl)borate (Tp) and cyclopendadienyl (Cp^-). Three important features are common: they are both mononegative, six electron donor and occupy three coordination positions (*i.e.* face-capping chelating ligand). The carbon analogue to Tp, hydrotris(pyrazol-1-yl)methane, Tpm, maintains the tripodal face capping aspect and the same electron donor count, but differs from the others on the charge (Tpm is neutral).

Tris(pyrazol-1-yl)-borate or -methane ligands are then potentially tridentate ligands (Figure 3) able to coordinate metal centres in a N_3-chelating mode; nevertheless it is also reported [11] that Tpm and Tp can afford different coordination modes, adapting its geometry to the demands of the metal centre, binding in a κ^2-mode (i.e., bidentate, Figure 3) [12-16] or acting as a bridging ligand [17, 18].

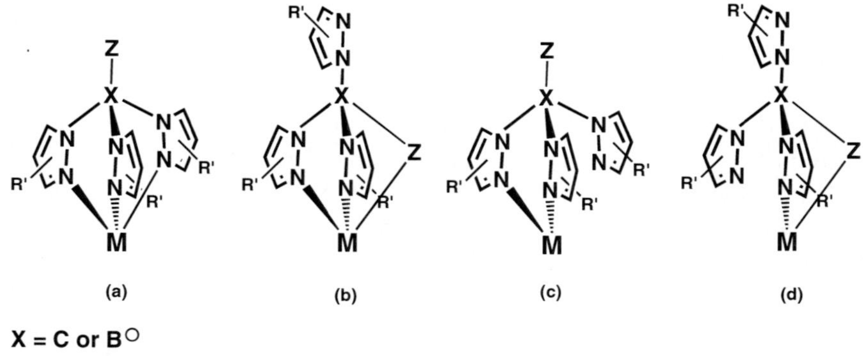

Figure 3. Selected coordination modes of carbon or boron scorpionates. (a) Tripodal N,N,N (κ^3-*N,N,N*), (b) Tripodal N,N,Z (κ^3-*N,N,Z*), (c) Bipodal N,N (κ^2-*N,N*) and (d) Bipodal N,Z (κ^2-*N,Z*); Z = donor atom.

Compared to the anionic species, the isoelectronic neutral analogues, tris(pyrazol-1-yl)methanes (Tpm), are considerably less well developed [3-5]. The underdevelopment of their chemistry results from the relatively small number of such ligands currently available, the difficulties associated with their synthesis and the usually low yields.

Taking into account the different preparative methods of Tp and Tpm (Scheme 1), it is simple to notice how the synthesis of tris(pyrazol-1-yl)borate and of its derivatives is undoubtedly easier and more versatile than those of Tpm. In fact, the Tp preparation proceeds with H_2 formation from reaction between the highly reactive potassium borohydride and pyrazole. [7] On the contrary, the synthesis of Tpm proceeds through the deprotonation of pyrazole to pyrazolate and the successive nucleophilic substitution at a chloroform molecule. This difference of the synthetic processes is even more evident in the case of preparation of 3-, 4- or 5-substituted tris(pyrazol-1-yl) species (see below), [2] where the synthesis of a tris(3,5-diphenylpyrazol-1-yl)borate (Tp^{Ph2}) proceeds regularly as described,[7] but the preparation of its carbon analogue does not occur.

Another difference is as follows. It has been reported [6] that a vigilant control of temperature in the Tp synthesis is required to avoid the formation of the 'over-substituted' product, tetrakis(pyrazol-1-yl)borate, whereas the synthesis of tetrakis(pyrazol-1-yl)methane, using carbon tetrachloride, does not proceeds easily.

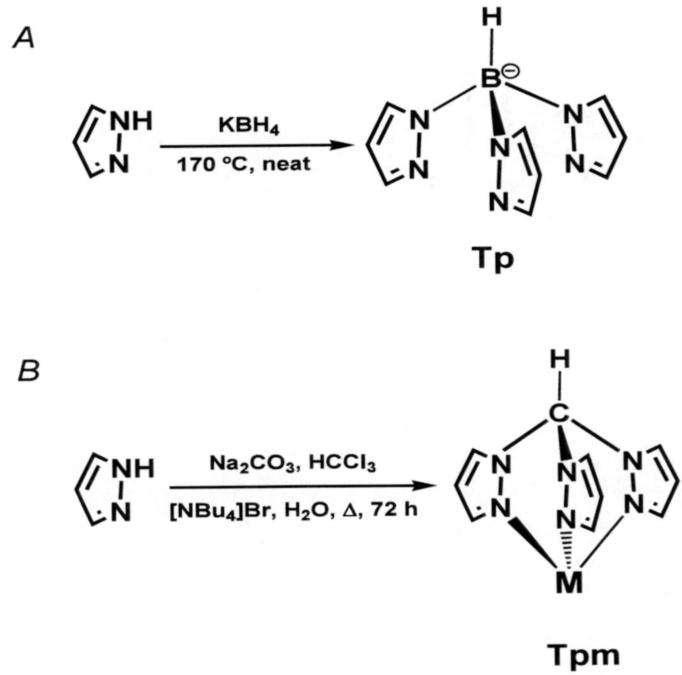

Scheme 1. Synthesis of hydrotris(pyrazol-1-yl)borate (Tp, reaction A, [7]) and of hydrotris(pyrazol-1-yl)methane (Tpm, reaction B, [19]).

Recent breakthroughs in the synthesis of tris(pyrazol-1-yl)methanes,[19, 20] namely microwave-assisted synthesis, have offered the opportunity for the development of this useful and promising class of ligands. [13-16, 18, 21-29].

2.1. Functionalization

The derivatization [2, 7] of the homoscorpionate ligands Tp or Tpm is an important topic and allows to direct the design of the derivatives toward highly specific applications.

Several positions of the homoscorpionate molecule are commonly recognized of significance for the functionalization and the two main sites are the apical position and the heterocyclic ring (Figure 4). Their modification could lead to unique effects on the coordination behaviour of the corresponding ligand.

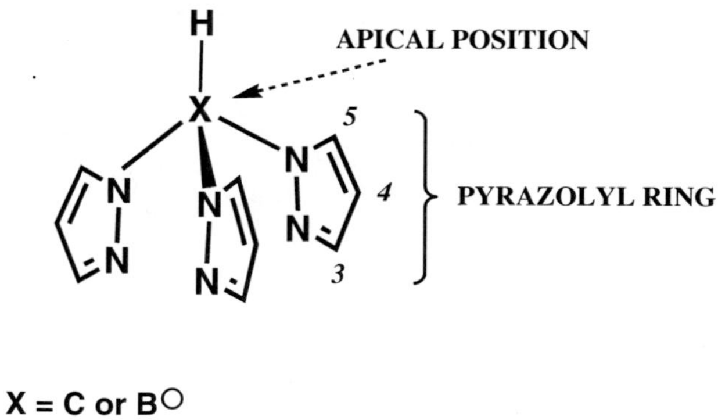

Figure 4. Schematic representation of Tp or Tpm with highlighted positions for functionalization.

The synthesis of $[RB(pz)_3]^-$ ligands is possible by a number of routes [1]. Some of them involve the reaction of the appropriate RBX_2 (where X is a halogen or a leaving group such as tosylate), boron hydride RBH_2 or boronic acid $RB(OH)_2$ [31, 32] with the pyrazolate ion plus excess pyrazole.

At Tpm, the substitution at the central methine carbon atom can be achieved by reacting a suitable electrophile with the carbanion $^-C(pz)_3$ formed by deprotonation of Tpm. [19, 33-41] However, this method resulted unsuccessful for the sterically hindered tris(pyrazol-1-yl)methane at 5-position of pyrazol-1-yl rings. The removal of the apical proton of Tpm could be achieved by different strong bases, such as butyl lithium, sodium hydride, sodium ethoxide or potassium metoxide, [41, 42] and the choice of the suitable base could be a crucial step for the success of the reaction.

It is possible to define separately the effect of the substituents at the pyrazol-1-yl ring (Figure 4):

Position 3: the major contribution of the substituents at this position is of steric nature. These groups, in fact, are facially oriented toward the coordination to metal centre and hold an important role for the modulation of coordination chemistry of the corresponding complex.[28, 43, 44] In the case of bulky substituents (i.e., ^{i}Pr, Ph and ^{t}Bu), the ligand is forced to engage a definite coordination geometry. In particular, for those metals with a high affinity for octahedral N_6-coordination mode (e.g., Fe^{II}), this steric limitation obliges them to abandon the typical full-sandwich coordination of two of such type of ligands: $[ML_2]X$ (L = Tp or Tpm derivative). Moreover, it has been reported [45-47] that substituents at 3-position could be directly involved in the coordination to the metal centre. Many relevant applications in bioinorganic chemistry are also known,[47-50] specifically due to this modulation effect on the coordination chemistry of metal centres with biological interest (e.g., Cu, Fe, Zn). Furthermore, it has been proved that the bulky substituents at the 3-position could also enhance the thermodynamic and hydrolytic stability of the corresponding complexes. [18]

Position 4: substituents in this position of the pyrazolyl ring are not common for Tp and Tpm derivatives, [51,52] and this type of functionalization has not been markedly studied probably because of its minor influence on the coordination properties of the ligand. In fact, substituents at 4-position are directed neither towards the metal centre nor towards the apical position.

This ‘meridian’ orientation is, indeed, more related to electronic effects on the heterocyclic ring and do not interfere with the coordination ability of the scorpionate ligand.

Position 5: the greater difficulty to prepare Tpm derivatives bearing bulky substituents in position 5 compared to Tp analogues is indicative of the steric interactions existing in this type of compounds. As mentioned previously, the low versatility of the Tpm synthesis has driven to a limited investigation of its 5-substituted derivatives. With the exception of methyl and *iso*-propyl substituents, [53-55] no bulkier groups have been introduced at position 5. Herein, the groups can modulate the ‘opposite face’ of the scorpionate ligand [56] and recently some works report important improvements in this field. [57-59]

3. Synthesis of Tris(pyrazol-1-yl) -Methanes or -Borates With Biological Interest

The homoscorpionate ligands used to prepare transition metal complexes with biological interest, as well as their applications, are summarized on Table 1. Beside Tp and Tpm, which were presented previously, the synthesis of the above ligands follows.

Table 1. Tris(pyrazol-1-yl)-methane or -borate ligands for biological applications

Ligand	Metal	Application	Ref.
Tpm	Mn, Ru, Rh, Cu	antitumor activity	[67, 73-75]
Tpm^{OH}	Co	antitumor activity	[23]
Tpm^{Py}			
Tpm^{OMs}			
Tpms	Ag	antitumor and antimicrobial activity	[76]
$Tpms^{iPr}$	Fe, Ni, Co, Cu, Zn	enzyme modeling	[62]
Tp	V, Ru	antitumor activity	[71, 72]
	Ru	antimicrobial activity	[78]
	Fe	enzyme modeling	[96, 100]
Tp^{Me2}	V	antitumor activity	[71]
	V, W, Mo, Ni, Cu, Zn	enzyme modeling	[48, 81, 82, 90, 91, 94, 95, 103]
Tp^{iPr}	Mo	enzyme modeling	[94]
Tp^{iPr2}	Mn, Fe, Zn	enzyme modeling	[83, 84, 98, 104]
Tp^{Ph}	Cu	enzyme modeling	[99]
$Tp^{tBu,Me}$	Zn	enzyme modeling	[49, 102]
$Tp^{Ph,Me}$			[103]
$Tp^{tBu,iPr}$		antitumor activity	[77]
$Tp^{(CF_3)_2}$	Ag	antimicrobial activity	[79, 80]

3.1. Carbon Scorpionates

3.1.1. Tris(pyrazol-1-yl)methanesulfonate, $SO_3C(pz)_3$ (Tpms) and Tris(3-iso-propylpyrazol-1-yl)methanesulfonate, $SO_3C(3\text{-}iPrpz)_3$ ($Tpms^{iPr}$)

The Tpm sulfonated analogue, tris(pyrazol-1-yl)methanesulfonate, $Tpms^{R}$ [R = H (Tpms) or *i*Pr ($Tpms^{iPr}$), Scheme 2], is a crucial carbon-functionalized derivative. Firstly reported by Kläui, [36] in 2000, it has rapidly gained a considerable significance on the scorpionate chemistry.[21,26,28, 60-64], The additional sulfonate group can be involved in the coordination to the metal, converting the N_3-type Tpm^R ligand in a more versatile Tpms^R able to coordinate as either a tripodal or a bipodal ligand (*i.e.* with N_3-, N_2O-, N_2- or NO-coordination modes, see Figure 5).[28, 60, 62,63]

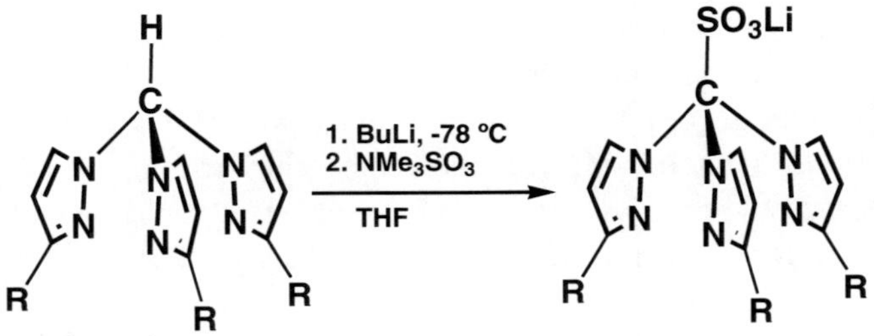

Scheme 2. Synthesis of tris(pyrazol-1-yl)- or tris(3-*iso*-propylpyrazol-1-yl)-methane sulfonate lithium salt (R = H or *i*Pr, respectively).[36].

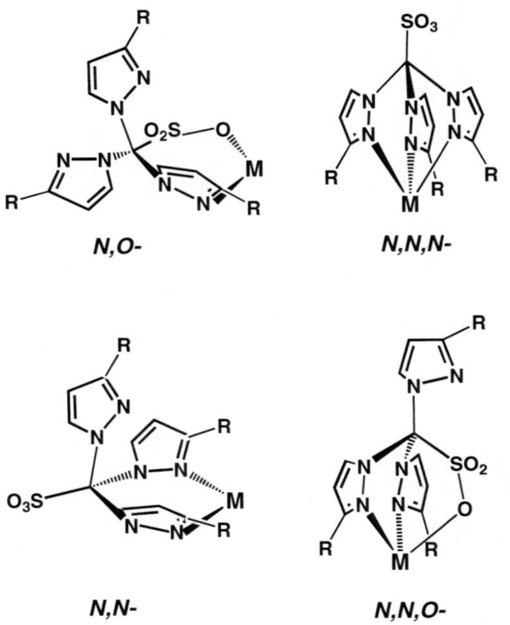

Figure 5. Coordination modes of Tpms^R ligands [R = H (Tpms) or *i*Pr (Tpms^{iPr}).

In contrast to the Tpm, the sulfonate derivatives are highly soluble in water and moderately soluble in alcohols. Moreover, Li[Tpms^R] (*i.e.* the corresponding lithium salts) are stable over a wide range of pH values in aqueous solution. This feature turned out to be decisive in many areas: the corresponding water soluble metal complexes represent a favourable class of compounds to be tested in catalysis and bioinorganic chemistry.

3.1.2. 2,2,2-Tris(pyrazol-1-yl)ethanol, $HOCH_2C(pz)_3$, Tpm^{OH}

An important derivative of Tpm is the 2,2,2-tris(pyrazol-1-yl)ethanol (Tpm^{OH}, Scheme 3). The latter compound is easily obtained from carbon functionalization of [65] Tpm and represents a sort of 'chemical scaffold' for further functionalizations of the molecule (see below). It is also easily predictable that the reactivity of alkoxide prepared upon deprotonation of the Tpm^{OH} is a more versatile scaffold than the carbanion derived from Tpm.

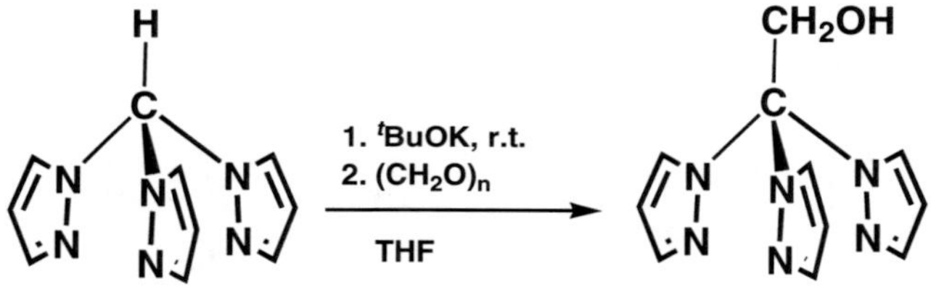

Scheme 3. Synthesis of 2,2,2-tris(pyrazol-1-yl)ethanol [65].

The preparation of Tpm^{OH} passes through a mild deprotonation of the apical methine carbon and the successive nucleophilic attack to paraformaldehyde to yield the alcohol derivative.

3.1.3. 2,2,2-Tris(pyrazol-1-yl)mesyloxymethyl, $MeSO_2OCH_2C(pz)_3$, Tpm^{OMs}

The mesylation of the –OH moiety of tris-2,2,2-(pyrazol-1-yl)ethanol (Tpm^{OH}) yields easily the desired product, Tpm^{OMs} (Scheme 4) [66] which exhibits a high stability.

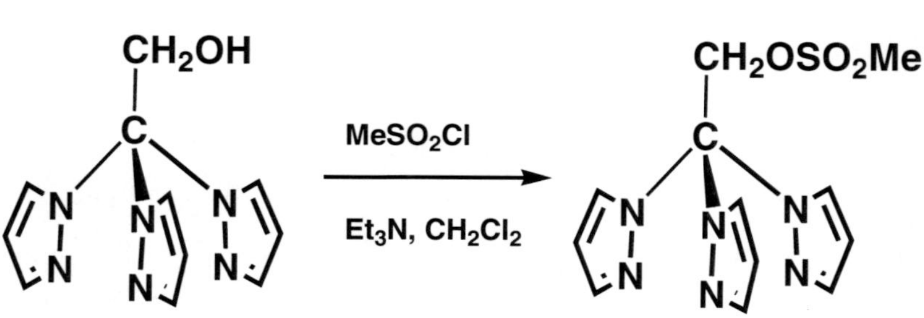

Scheme 4. Synthesis of 2,2,2-tris(pyrazol-1-yl)mesyloxymethyl [66].

This stability can be partially explained considering the steric hindrance of the three pyrazolyl rings around the methylene moiety, which can obstruct the access to substitution reactions.

3.1.4.4-((Tris-2,2,2-(pyrazol-1-yl)ethoxy)methyl)-pyridine, $PyCH_2OCH_2C(pz)_3$, Tpm^{Py}

The hydroxy group of Tpm^{OH} can be easily deprotonated with sodium hydride, in a classical procedure of deprotonation of alkyl alcohols. The sodium salt of tris-2,2,2-(1-pyrazol-1-yl)ethanol (*i.e.* $NaOCH_2C(pz)_3$), prepared *in situ*, is usually used without delay for the next step with a suitable electrophile to give the desired product, Tpm^{Py} (Scheme 5). [18]

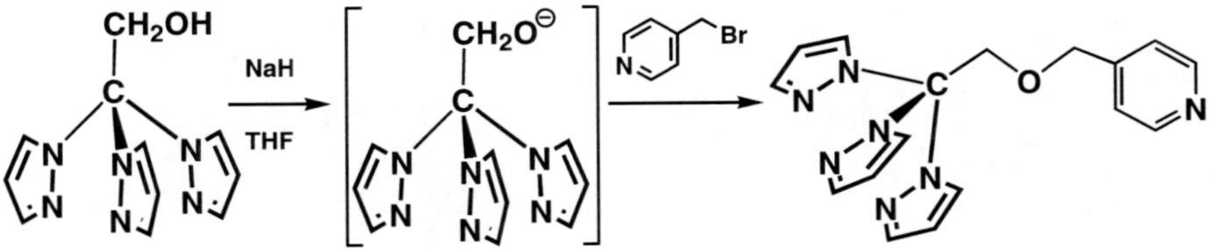

Scheme 5. Synthesis of 4-((tris-2,2,2-(pyrazol-1-yl)ethoxy)methyl)-pyridine [18].

3.2. Boron Scorpionates

As mentioned before, the synthesis of tris(pyrazol-1-yl)borate and of its derivatives proceeds with H_2 formation from reaction between the highly reactive borohydride (BH_4^-) salt in molten pyrazole (Scheme 6). A large variety of 1-H pyrazoles may be employed to synthesize the ligands by this route, with the exception of those containing functionalities incompatible with the borohydride ion.[1,2]

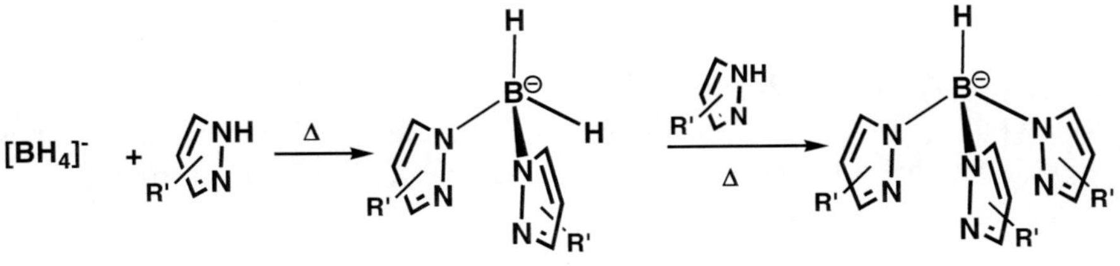

Scheme 6. Synthesis of hydrotris(pyrazol-1-yl)borates.

Hydrotris(3-phenylpyrazol-1-yl)borate, HB(3-Phpz)$_3$ (Tp^{Ph}) and Hydrotris(3-iso-propylpyrazol-1-yl)borate, HB(3-^{i}Prpz)$_3$ (Tp^{iPr})

The ligands Tp^{Ph} and Tp^{iPr} are prepared in standard fashion. Their synthesis involves thermolysis of a mixture of the pyrazole and the borohydride salt M[BH$_4$] (M = K or Na) in a 3.5:1 ratio, either in a melt or in a high-boiling solvent (Scheme 6). The reaction is monitored by hydrogen evolution. With these asymmetric 3-R-pyrazoles, the reaction proceeds with boron being bonded to the least hindered nitrogen atom. [2]

*Hydrotris(3,5-dimethylpyrazol-1-yl)borate, HB(3,5-Me$_2$pz)$_3$ (Tp^{Me2}), Hydrotris(3,5-diisopropylpyrazol-1-yl)borate, HB(3,5-^{i}Pr$_2$pz)$_3$ (Tp^{iPr2}), Hydrotris(3,5-di-trifluoromethylpyrazol-1-yl)borate, HB{3,5-(CF$_3$)$_2$pz}$_3$ (Tp$^{(CF3)2}$), Hydrotris(3-tert-butyl-5-isopropylpyrazol-1-yl)borate, HB(3-^{t}Bu-5-^{i}Prpz)$_3$ (Tp^{tBu,iPr}), Hydrotris(3-*tert*-butyl-5-methylpyrazol-1-yl)borate, HB(3-^{t}Bu-5-Mepz)$_3$ (Tp^{tBu,Me}) and Hydrotris(3-phenyl-5-methyl-pyrazol-1-yl)borate, HB(3-Ph-5-Mepz)$_3$ (Tp^{Ph,}Me)*

The ligands Tp^{Me2}, Tp^{iPr2}, Tp$^{(CF3)2}$, Tp^{tBu,iPr}, Tp^{tBu,Me} and Tp^{Ph,Me} are prepared by neat reaction of the appropriate substituted pyrazole (Hpz^{3R,5R} or Hpz$^{3R,5R'}$) and the borohydride ion in a 6-8:1 ratio, at 180–220 ºC (Scheme 6). With the 3,5-disubstituted pyrazoles (Hpz^{3R,5R} or Hpz$^{3R,5R'}$) no careful temperature control is needed to avoid the formation of tetrakis(pyrazol-1-yl)borates [2].

4. Biological Applications

In the design of a new potentially active metallodrug, beyond the nature of the metal, it is of paramount importance the choice of appropriate ligands, which affect the thermodynamic and kinetic stability, as well as solubility and lipophilicity of the complexes.

A number of tris(pyrazol-1-yl)-methane or -borate metal complexes have already proven to function as potent cytotoxic agents *in vitro*.[67] Moreover, these ligands have also become particularly important in the development of biomimetic coordination chemistry of metalloproteins, such as: Cu-proteins (e.g. , hemocyanin, ascorbate oxidase, superoxide dismutase), Fe-protein (hemerythrin) or Mn-protein (superoxide dismutase).

One particular advantage of the Tpm or Tp ligands is that the basic scaffold is readily modified, both on the pyrazolyl rings [4, 5, 68] and at the bridgehead carbon or boron atom, [69] so that the electronic, steric and coordination properties can be tailored as desired. In addition, simple modifications to the Tpm ligand can impart aqueous solubility on its complexes [60].

Tris(pyrazol-1-yl)borates have also been applied as biologic sensors. In particular, the determination of thiocyanate and azide in environmental and biological samples is of important practical significance.[70]

4.1. Antitumor Activity

Vanadium

The vanadium(IV) complexes [VO(Tp)(acac)] and [VO(Tp^{Me2})(acac)]·CH_3CN (acac = acetylacetonate) show a biphasic effect upon rat CBRH-7919 hepatoma cells proliferation: a slight stimulation of growth at low doses (1-10 μM) of both complexes and a significant cytotoxic effect at higher doses (100-1000 μM).

The bio-effects are different for both complexes, possibly due to the greater steric requirements and electronic effects of methyl groups in the 3,5-positions of pyrazole ring in Tp^{Me2}, showing that the structural modifications on the pyrazole moiety can have some effects on the toxicity of vanadium complexes [71].

Manganese

Hydrotris(pyrazol-1-yl)methane manganese tricarbonyl complex [Mn$(CO)_3$(Tpm)][PF_6] was explored as a potential PDT (photodynamic therapy) agent.

It undergoes photoinduced CO release upon irradiation with UV light in aqueous buffer, with two out of three ligands being lost under those conditions. This manganese complex shows photoinitiated cytotoxicity against the human colon cell line HT29, with a reduction in cell biomass comparable to that of established 5-fluorouracil agent 5-FU, but is inactive in the dark at concentrations up to 100 μM. [67]

Ruthenium and Rhodium

Tp-ruthenium(II) complexes containing the water-soluble phosphane 1,3,5-triaza-7-phosphaadamantane (PTA) or 1-methyl-3,5-diaza-1-azonia-7-phosphaadamantane (m-PTA) ligands [72], [RuCl(Tp)(PPh_3)(m-PTA)], [RuH(Tp)(PPh_3)(PTA)], [RuH(Tp)(PPh_3)(m-PTA)][CF_3SO_3] and [Ru(Tp)(NCMe)(PPh_3)(PTA)][CF_3SO_3] (PPh_3 = triphenylphosphine), show promising cytotoxic (IC_{50}, 10^{-6} M), and display differential effects on cancer and normal cell growth.

They interact with DNA, and have inhibitory effects against human tumour cell lines that are much stronger than those reported for other ruthenium complexes, and are close to the antitumor activity of anticancer drugs currently used in clinical practice, such as doxorubucin.

It was also found [73] that Tpm-ruthenium(II) complexes of the type $[Ru(Tpm)(L)(dppn)]^{n+}$ (dppn = benzo[*i*]dipyrido-[3,2-*a:2',3'-c*]phenazine, n = 1; L = Cl, n = 2; L = MeCN or pyridine), as well as the rodium(III) complex $[RhCl(dppz)(Tpm)]^{2+}$ (dppz = dipyrido[3,2-*a*:2′,3′-*c*]phenazine), exhibit intercalative binding into the DNA double helix.

Moreover, the cytotoxic potential of $[RhCl(pp)(Tpm)]^{2+}$ (pp = polypyridyl ligand: bpy = 2,2'-bipyridine, bpm = 2,2'-bipyrimidine, phen = 1,10-phenanthroline; tap = pyrazino[2,3-*f*]quinoxaline or dpq = dipyrido[3,2-*d*:2',3'-*f*]quinoxaline) was investigated by CD and UV/Vis spectroscopy and by gel electrophoresis. Complexes $[RhCl(tap)(Tpm)]^{2+}$ and particularly $[RhCl(dpq)(Tpm)]^{2+}$ (IC_{50} = 0.43 and 0.37 μM, respectively) are potent cytotoxic agents toward the human cancer cell lines MCF-7 (breast cancer) and HT-29 (colon cancer) [74].

Cobalt

The water-stable redox-active cobalt(II) complexes [$CoCl_2(H_2O)(Tpm^{Py})$], [$Co(Tpm^{OH})_2$]·[$Co(Tpm^{OH})(H_2O)_3$]2$(Cl)_6\cdot 6H_2O$ and [$CoCl_2(H_2O)(Tpm^{OMs})$] [23] showed moderate cytotoxicity *in vitro* towards HCT116 colorectal carcinoma and HepG2 hepatocellular carcinoma human cancer cell lines.

Copper

The cytotoxic activity of mixed-ligand copper(I) complexes comprising scorpionate and phosphine ligands [CuL(Tpm)][BF_4] (L = PTA or PCN = tris(cyanoethyl)phosphine) has been evaluated by the MTT test against a panel of several human tumor cell lines including examples of breast (MCF-7), colon (HCT-15 and LoVo), lung (A549), cervix (A431) and ovarian (2008 and its cisplatin resistant variant, C13*) carcinoma, melanoma (A375) and promyelocytic leukemia (HL60). These copper complexes generally show an *in vitro* antitumor activity comparable to that of cisplatin [75].

Silver

[Ag(Tpms)] or [Ag(Tpms)L] (L = PPh_3, PCy_3 or PTA) (PCy_3 = tricyclohexylphosphane) exhibit a pronounced antiproliferative activity against human malignant melanoma (A375), which is often higher than that of $AgNO_3$, which has been used as a control, following the same order of activity inhibition on DNA: [Ag(Tpms)(PCy_3)] > [Ag(Tpms)(PPh_3)] > [Ag(Tpms)] > [Ag(Tpms)(PTA)] > $AgNO_3$. [76]

Zinc

The synthesis and characterization of the mononuclear zinc hydroxo complex [Zn(OH)$Tp^{tBu,iPr}$] obtained by using a sterically hindered tris(pyrazol-1-yl)borate ligand, its reaction products with uracil and its halo derivatives have been reported. The halogenated uracil complex Zn[$Tp^{tBu,iPr}$](5-fluorouracilate) exhibits significant antitumor activity against Dalton's Lymphoma tumour system [77].

4.2. Antimicrobial Activity

Ruthenium

A series of half-sandwich ruthenium(II) complexes containing Tp and the water-soluble phosphane PTA, [RuX(Tp)$(PPh_3)_{2-n}$$(PTA)_n$] (n = 2, X = Cl, n = 1, X = Cl, I, NCS or H), [Ru(Tp)(PPh_3)(PTA)L][PF_6] (L = NCMe or PTA), [RuCl(Tp)$(m\text{-}PTA)_2$][CF_3SO_3]$_2$ and [RuX(Tp)(PPh_3)(m-PTA)][CF_3SO_3] (X = Cl or H) have been synthesized and their DNA binding properties (by using a mobility shift assay) and antimicrobial activity have been evaluated. Among the compounds used those with the methylated PTA phosphane ligand (m-PTA) were the most active ones against prokaryotic microorganisms, in spite of the low water solubility found in these derivatives. On the other hand, compounds [RuH(Tp)$(PPh_3)_2$(PTA)] and [RuH(Tp)(PPh_3)(m-PTA)][CF_3SO_3] showed a high interaction capability with DNA, probably due to their ability for establishing hydrogen bonding. [78]

Silver

The silver(I) complexes containing a highly fluorinated hydrotris(pyrazol-1-yl)borate [Ag{$Tp^{(CF3)2}$}(THF) and [Ag{$Tp^{(CF3)2}$}($OSMe_2$)] (THF =

tetrahydrofuran; $OSMe_2$ = dimethyl sulfoxide) exhibit good antimicrobial properties against the Gram-positive pathogenic bacterium *Staphylococcus aureus*, showing a greater activity than that of silver nitrate or of a currently used silver antimicrobial, silver(I) sulfadiazine. Furthermore, it appear that the fluorinated substituents on the tris(pyrazol-1-yl)borate are important for this high level of growth inhibition [79, 80].

The silver(I) complexes of formulae [Ag(Tpms)] and [Ag(Tpms)L], which exhibit antiproliferative activity (see above), were screened for their *in vitro* antimicrobial activities against the standard strains *Enterococcus faecalis* (ATCC 29922), *Staphylococcus aureus* (ATCC 25923), *Staphylococcus pneumonia* (ATCC 49619), *Staphylococcus pyrogenes* (SF37), *Staphylococcus sanguinis* (SK36), *Staphylococcus mutans* (UA159), *Escherichia coli* (ATCC 25922), and the fungi *Candida albicans* (ATCC 24443). They display effective antimicrobial activity against the series of bacteria and fungi and some of them are potential candidates for antiseptic or disinfectant drugs [80].

4.3. Models of Metalloenzimes

Complexes of diverse scorpionate ligands were used to mimic the activity of enzymes containing various metals, usually providing an environment similar to that of three coordinated histidine nitrogens. For that purpose, several homoscorpionates with a variety of 3-substituents have been used. [2]

Vanadium

The oxidovanadium(IV) (3,5-dimethylpyrazolyl)borate complexes $[VO(Tp^{3,5-Me})(3,5\text{-}Me_2pz)(HOOC\text{-}CH_2CH_2COO)]$, $[VO(Tp^{Me2})(C_5H_4N\text{-}COO)]$ [81] and a series of complexes containing phenolate ligands of general structure $[VO(Tp^{Me2})(OAr)_2]$ [82] were synthesized, characterized and discussed as possible models for the active site in vanadate-dependent haloperoxidases.

Manganese

Mimicking the activity of manganese superoxide dismutase and of various binuclear manganese enzymes active in redox functions was approached with the monomeric Tp benzoate Mn(II) complex $[Mn(Tp^{iPr2})(OBz)]$ (OBz = benzoate) and related binuclear complexes [83, 84].

Molybdenum and Tungsten

Molybdenum and tungsten enzymes catalyze a wide range of two-electron oxidation and reduction reactions that generally involve oxygen atom transfer [85].

Various molybdenum Tp complexes were employed to model the mononuclear molybdenum cofactors of molybdopterin enzymes, such as xanthine oxidase-dehydrogenase, sulfite oxidase, nitrate reductase, and dimethyl sulfoxide reductase [86-90]. Similar complexes were also employed in mimicking sulfur-containing tungsten enzymes, such as aldehyde oxidoreductase, formate dehydrogenase, and formylmethanofuran dehydrogenase [91, 92].

Recently, three classes of known metal sulfide clusters that resemble the NFe_7MoS_9 core of FeMo-co, the active site of nitrogenase, have been assessed theoretically. The three model systems possess the $X^cFe_4S_4$ face which is the key active site of that cofactor. The most promising functional models are based on the clusters $[(Tp)Mo(\mu_3\text{-}S)_3Fe_2(Fe\text{-}L)S^c(\mu\text{-}S)_2(Fe\text{-}L)Fe_2(\mu_3\text{-}S)_3Mo(Tp)]$.[93]

Dioxido-Mo(VI) complexes of the type $[Mo(L)O_2(SPh)]$ (L = Tp^{Me2} or Tp^{iPr}) and related derivatives model systems mimic the complete reaction sequence of sulfite oxidase, in which oxygen from water is ultimately incorporated into the product [94].

The Tp^{Me2}-molybdenum (VI) complex $[MoO_2Cl(Tp^{Me2})]$ (with triphenylphosphine as the reducing cofactor) was shown to be an efficient catalyst of nitroglycerin degradation (nitrates, including nitroglycerin, are clinically important vasodilators) believed to be biotransformed *in vivo* to NO, producing significant amounts of NO. $Mo(Tp^{Me2})/Ph_3P$ is an excellent enzyme model system, showing the feasibility of nitrate biotransformation mediated by a molybdoenzyme [95].

Iron

In the area of iron-containing enzymes, several complexes of general structure $[(FeTp)_2(\mu\text{-}O)(\mu\text{-}OOCR)_2]$ (R = H, Me or Ph) were prepared [96] and the geometry of $[Fe_2O(OOCR)_2]$ core was found to resemble closely that of the oxo-bridged diiron enzyme hemerythrin.

The spectral and magnetic properties of $\{(TpFe)_2(\mu\text{-}OH)(\mu\text{-}OAc)_2\}^+$ differed sufficiently from the oxo-bridged complex to suggest that oxo, rather than hydroxo, bridges are present in the methemerythrins. [97] The structure of $[Fe(Tp^{iPr2})(OBz)(MeCN)]$ (OBz = benzoate) was regarded as a synthetic model for the dioxygen binding site of nonheme iron proteins, based on its electronic spectral changes under argon and/or oxygen.[98]

Nickel

The nickel complex [Ni(Tp^{Me2})(OEt-cysteinato)] has been studied as a model to the nickel component of the active site in several hydrogenase enzymes, which participate in the bio-generation of hydrogen and methane, as well as in nitrogen fixation [48].

Copper

A variety of copper tris(pyrazol-1-yl)borate complexes has been synthesized in order to model properties of several copper-containing metalloproteins, the [Cu(Tp^{Ph})(pterin)] complex being useful in revealing the binding mode at the copper site in phenylalanine hydroxylase [99].

The structurally characterized peroxo-bridged dicopper complex $[CuTp^{Me2}]_2(\mu\text{-}O_2)$ was used as a synthetic model of oxyhaemocyanin (a reversibly binding oxygen carrier) and oxytyrosinase (a monooxygenase oxidizing phenols) [100].

Zinc

The Zn(II) complex [Zn($Tp^{tBu,Me}$)(OH)] was demonstrated to be a functional model of carbonic anhydrase, by reacting rapidly and reversibly with carbon dioxide, forming the bicarbonate complex [Zn($Tp^{tBu,Me}$)(OCO_2H)] and slowly converted to the dinuclear $[ZnTp^{tBu,Me}]_2(CO_3)$ in which the carbonate ion bridged the two zinc centres in a symmetric, unidentate mode.[101, 102] Similar complexes were also employed in mimicking liver alcohol dehydrogenase (LADH) [49].

Functional modelling of cobalamine-independent methionine synthase with pyrazol-1-ylborate-zinc-thiolate complexes was aimed by employing a series of new pyrazol-1-ylborate-zinc-thiolate complexes, $Tp^{Ph,Me}$Zn-SR and Tp^{Me2}Zn-SR (R = Et or CH_2Ph), including two homocysteine derivatives. Their reactions with methyl iodide under mild conditions resulted in the formation of the methyl thioethers MeSR, including two methionine derivatives, and $Tp^{R',Me}$Zn-I in all cases [103].

The first example of a dinuclear zinc complex bridged solely by a phosphate ligand, resembling the alkaline phosphatase enzyme was $[ZnTp^{iPr2}]_2[OP(O)(OC_6H_4NO_2)O]$, obtained by treatment of Tp^{iPr2}ZnOH with mono(*p*-nitrophenyl)phosphate [104].

Various octahedral divalent metal [M(Tpms^{iPr})$_2$] (M = Zn, Cu, Ni, Co, Fe) complexes are good structural models for facial *N,N,O*-coordination in 2-His-1-carboxylate enzymes [62].

REFERENCES

[1] Trofimenko S., Scorpionates: The Coordination Chemistry of Polypyrazolylborates Ligands; Imperial College Press: London (1999).

[2] Pettinari C., Scorpionates II: Chelating Borate Ligands, Imperial College Press, World Scientific Pub. (2008).

[3] Pettinari C., Santini C., Polypyrazolylborate and Scorpionate Ligands, in: McCleverty J.A., Meyer T.J. (Eds.), *Comprehensive Coordination Chemistry II: From Biology to Nanotechnology*, vol. 1, Elsevier Pergamon (APS), Amsterdam, (2003) 159-210.

[4] Pettinari C., Pettinari R., *Coord. Chem. Rev.* 249 (2005) 525-543.

[5] Bigmore H.R., Lawrence C.S., Mountford P., Tredget C.S., *Dalton Trans.* (2005) 635-651.

[6] Trofimenko S., *J. Am. Chem. Soc.* 89 (1967) 3170-3177.

[7] Trofimenko S., *Chem. Rev.* 93 (1993) 943-980.

[8] Trofimenko S., *Adv. Chem. Series* (1976) 289-299.

[9] Trofimenko S., *Progr. Inorg. Chem.* 34 (1986) 115-210.

[10] Brunker T.J., Cowley A.R., O'Hare D., *Organomet.* 21 (2002) 3123-3138.

[11] Macchioni A., Bellachioma G., Cardaci G., Gramlich V., Ruegger H., Terenzi S., Venanzi L.M., *Organomet.* 16 (1997) 2139-2145.

[12] Trofimenko S., *J. Am. Chem. Soc.* 92 (1970) 5118-5126.

[13] Alegria E.C.B.A., Martins L.M.D.R.S., Haukka M., Pombeiro A.J.L., *Dalton Trans.* (2006) 4954-4961.

[14] Silva T.F.S., Luzyanin K.V., Kirilova M.V., Silva M.F.C.G, Martins L.M.D.R.S., Pombeiro A.J.L., *Adv., Synth. Cat.* 352 (2010) 171-187.

[15] Martins L.M.D.R.S., Alegria E.C.B.A., Smoleński P., Kuznetsov M.L., Pombeiro A.J.L., *Inorg. Chem.* 52 (2013) 4534-4546.

[16] Almeida M.P., Martins L.M.D.R.S., Carabineiro S.A.C., Lauterbach T., Rominger F., Hashmi A.S.K., Pombeiro A.J.L., Figueiredo J.L., *Catal. Sci. Technol.* 3 (2013) 3056-3069.

[17] Ellis D.D., Jeffery J.C., Jelliss P.A., Kautz J.A., Stone F.G.A., *Inorg. Chem.* 40 (2001) 2041-2050.

[18] Wanke R., Silva M.F.C.G., Lancianesi S., Silva T.F.S., Martins L.M.D.R.S., Pettinari C., Pombeiro A.J.L., *Inorg. Chem.*, 49 (2010) 7941-7952.

[19] Reger D.L., Grattan T.C., Brown K.J., Little C.A., Lamba J.J.S., Rheingold A.L., Sommer R.D., *J. Organomet. Chem.* 607 (2000) 120-128.

[20] Pombeiro A.J.L., Martins L.M.D.R.S., *Alegria E.C.B.A.*, PT103681: award date: 4/12/2009.

[21] Alegria E.C.B.A., Guedes da Silva M.F.C., Kuznetov M.L., Cunha S.M.P.R.M., Martins L.M.D.R.S., Pombeiro A.J.L., *Dalton Trans.* 41 (2012) 13876-13890.

[22] Rocha B.G.M., Wanke R., Guedes da Silva M.F.C., Luzyanin K.V., Martins L.M.D.R.S., Smolénski P., Pombeiro A.J.L., *J. Organomet. Chem.* 714 (2012) 47-52.

[23] Silva T.F.S., Martins L.M.D.R.S., Guedes da Silva M.F., Fernandes A.R., Silva A., Borralho P.M., Santos S., Rodrigues C.M.P., Pombeiro A.J.L., *Dalton Trans.* 41 (2012) 12888-12897.

[24] Marchetti F., Pettinari C., Pettinari R., Cerquetella A., Martins L.M.D.R.S., Guedes da Silva M.F., Silva T.F.S., Pombeiro A.J.L., *Organomet.* 30 (2011) 6180-6188.

[25] Silva T.F.S., Silva M.F.C.G., Mishra G.S., Martins L.M.D.R.S., Pombeiro A.J.L., *J. Organomet. Chem.*, 696 (2011) 1310-1318.

[26] Dinoi C., Silva M.F.C.G., Alegria E.C.B.A., Smoleński P., Martins L.M.D.R.S., Poli R., Pombeiro A.J.L., *Eur. J. Inorg. Chem.* 16 (2010) 2415-2424.

[27] Silva T.F.S., Mishra G.S., Silva M.F.G., Wanke R., Martins L.M.D.R.S., Pombeiro A.J.L., *Dalton Trans.* 42 (2009) 9207-9215.

[28] Wanke R., Smoleński P., Guedes da Silva M.F.C., Martins L.M.D.R.S., Pombeiro A.J.L., *Inorg. Chem.* 47 (2008) 10158-10168.

[29] Silva T.F.S., Alegria E.C.B.A., Martins L.M.D.R.S., Pombeiro A.J.L., *Adv., Synth. Cat.* 350 (2008) 706-716.

[30] Aubagnac J.-L., Claramunt R.M., Elguero J., Gilles I., Sanz D., Trofimenko S., Virgili A., *Bull. Chem. Soc. Belg.* 104 (1995) 473-482.

[31] Trofimenko S., *J. Am. Chem. Soc.* 89 (1967) 6288-6294.

[32] Reger D.L., Tarquini M.E., *Inorg. Chem.* 21 (1982) 840-842.

[33] Reger D. L., Semeniuc R.F., Little C.A., Smith M.D., *Inorg. Chem.* 45 (2006) 7758-7769.

[34] Sanchez-Mendez A., Ortiz A.M., de Jesus E., Flores J.C., Gomez-Sal P., *Dalton Trans.* (2007) 5658-5669.

[35] Bigmore H.R., Dubberley S.R., Kranenburg M., Lawrence S.C., Sealey A.J., Selby J.D., Zuideveld M.A., Cowley A.R., Mountford P., *Chem. Commun.* (2006) 436-438.

[36] Kläui W., Berghahn M., Rheinwald G., Lang H.R., Angew. *Chem., Int.* Ed. 39 (2000) 2464-2466.

[37] Esteruelas M.A., Oro L.A., Claramunt R.M., Lopez C., Lavandera J.L., Elguero J., *Organomet. Chem.* 366 (1989) 245-255.

[38] Byers P.K., Canty A.J., Honeyman R.T., Claremont R.M., Lopez C., Lavandera J.L., Elguero J., *Gazz. Chim. Ital.* 122 (1992) 341-344.

[39] Reger D.L., Watson R.P., Smith M.D., *Inorg. Chem.* 45 (2006) 10077-10087.

[40] Reger D.L., Semeniuc R.F., Smith M.D., *Dalton Trans.* (2008) 2253-2260.

[41] Reger D.L., Grattan T.C., *Synthesis-Stuttgart* (2003) 350-356.

[42] Byers P.K., Carr N., Stone F.G.A., *J. Chem. Soc., Dalton* (1990) 3701-3708.

[43] Trofimenko S., Calabrese J.C., Domaille P.J., Thompson J.S., *Inorg. Chem.* 28 (1989) 1091-1101.

[44] Edwards P.G., Harrison A., Newman P.D., Zhang W.J., *Inorg. Chim. Acta* 359 (2006) 3549-3556.

[45] Amoroso A.J., Jeffery J.C., Jones P.L., McCleverty J.A., Rees L., Rheingold A.L., Sun Y.M., Takats J., Trofimenko S., Ward M.D, Yap G.P.A., *J. Chem. Soc., Chem.* (1995) 1881-1887.

[46] Mehn M.P., Fujisawa K., Hegg E.L., Que L., *J. Am. Chem. Soc.* 125 (2003) 7828-7842.

[47] Wang L.Y., Chambron J.C., *Org. Lett.* 6 (2004) 747-750.

[48] Kim K., Lippard S. J., *J. Am. Chem. Soc.* 118 (1996) 4914-4915.

[49] Bergquist C., Parkin G., *Inorg. Chem.* 38 (1999) 422-423.

[50] Mirica L.M., Ottenwaelder X., Stack T.D.P., *Chem. Rev.* 104 (2004) 1013-1046.

[51] Olson M.D., Rettig S.J., Storr A., Trotter J., Trofimenko S., *Acta Crystallogr., Sect.* C 47 (1991) 1543-1544.

[52] Calabrese J.C., Domaille P.J., Thompson J.S., Trofimenko S., *Inorg.Chem.* 29 (1990) 4429-4437.

[53] Nabika M., Kiuchi S., Miyatake T., Okamoto K.I., Fujisawa K., *J. Mol. Catal.* 269 (2007) 163-168.

[54] Kaim W., Titze C., Schurr T., Sieger M., Lawson M., Jordanov J., Rojas D., Garcia A.M., Manzur J., Z. *Anorg. Allg. Chem.* 631 (2005) 2568-2574.

[55] Fujisawa K., Ono T., Aoki H., Ishikawa Y., Miyashita Y., Okamoto K., Nakazawa H., Higashimura H., *Inorg. Chem. Commun.* 7 (2004) 330-332.

[56] Krummenacher I., Ruegger H., Breher F., *Dalton Trans.* (2006) 1073-1081.

[57] Benisvy L., Wanke R., Kuznetsov M.L., Guedes da Silva M.F.C., Pombeiro A.J.L., *Tetrahedron* 65 (2009) 9218-9223.

[58] Bigmore H.R., Meyer J., Krummenacher I., Rugger H., Clot E., Mountford P., Breher F., *Chem. Eur. J.* 14 (2008) 5918-5934.

[59] Kuzu I., Krummenacher I., Meyer J., Armbruster F., Breher F., *Dalton Trans.* (2008) 5836-5865.

[60] Kläui W., Berghahn M., Frank W., Reiss G.J., Schonherr T., Rheinwald G., Lang H., *Eur. J. Inorg. Chem.* (2003) 2059-2070.

[61] Tisato F., Refosco F., Bandoli G., Pilloni G., Corain B., *Inorg. Chem.* 40 (2001) 1394-1396.

[62] Papish E.T., Taylor M.T., Jernigan F. E., Rodig M.J., Shawhan R. R., Yap G.P.A., Jove F. A., *Inorg. Chem.* 45 (2006) 2242-2250.

[63] Chenskaya T.B., Berghahn M., Kunz P.C., Frank W., Kläui W., *J. Mol. Struct.* 829 (2007) 135-148.

[64] Smolenski P., Dinoi C., da Silva M.F.G., Pombeiro A.J.L., *J. Organomet. Chem.* 693 (2008) 2338-2344.

[65] García-Orozco I., Quijada R., Vera K., Valderrama M., *J. Mol. Catal. A: Chem*, 260 (2006) 70-76.

[66] Reger D.L., Wright T.D., Semeniuc R.F., Grattan T.C., Smith M.D., *Inorg. Chem.* 40 (2001) 6212-6219.

[67] Niesel J., Pinto A., N'Dongo H.W.P., Merz K., Ott I., Gust R., Schatzschneider U., *Chem. Commun.* (2008) 1798-1800.

[68] Reger D.L., Gardinier J.R., Grattan T.C., Smith M.R., Smith M.D., New *J. Chem.* 27 (2003) 1670-1677.

[69] Reger D.L., Foley E.A., Semeniuc R.F., Smith M.D., *Inorg. Chem.* 46 (2007) 11345-11355.

[70] Singh A.K., Aggarwal V., Singh U.P., Mehtab S., Talanta 77 (2008) 718-726.

[71] Xing Y.H., Sun Z., Zou W., Song J., Aoki K., Ge M.F., *J. Struct. Chem.* 47 (2006) 913-922.

[72] Garcia-Fernandez A., Diez J., Manteca A.C., Sanchez J., García-Navas R., Sierra B.G., Mollinedo, Gamasa M.P., Lastra E., *Dalton Trans.* 39 (2010) 10186-10196.

[73] Foxon S.P., Metcalfe C., Adams H., Webb M., Thomas J.A., *Inorg. Chem.* 46 (2007) 409-416.

[74] Bieda R., Ott I., Gust R., Sheldrick W., *Eur. J. Inorg. Chem.* (2009) 3821-3831.

[75] Porchia M., Dolmella A., Gandin V., Marzano C., Pellei M., Peruzzo V., Refosco F., Santini C., Tisato F., *Eur. J. Med. Chem.* 59 (2013) 218-226.

[76] Pettinari C., Marchetti F., Lupidi G., Quassinti L., Bramucci M., Petrelli D., Vitali L.A., Guedes da Silva M.F.C., Martins L.M.D.R.S., Smoleński P., Pombeiro A.J.L., *Inorg. Chem.*, 50 (2011) 11173-11183.

[77] Tyagi S., Sharma C.L., Singha S.M., Singh U.P., *Indian J. Chem.*, 43A (2004) 83-87.

[78] Gárcia-Fernández A., Díez J., Manteca A., Sánchez J., Gamasa M.P., Lastra E., *Polyhedron* 27 (2008) 1214-1228.

[79] van Waasbergen L.G., Fajdetic I., Fianchini M., Rasika Dias H.V., *J. Inorg. Biochem.* 101 (2007) 1180-1183.

[80] Rasika Dias H.V., Batdorf K.H., Fianchini M., Diyabalanage H.V.K., Carnahan S., Mulcahy R., Rabiee A., Nelson K., van Waasbergen L.G., *J. Inorg. Biochem.* 100 (2006) 158-160.

[81] Cao Y.Z., Li Z.P., Xing Y.H., Wei D.M., Pu Z.F., Ge M.F., Shi Z., *Chem. Journ. Chin. Univ.-Chin.* 32 (2011) 1025-1032.

[82] Holmes S., Carrano C.J., *Inorg. Chem.* 30 (1991) 1231-1235.

[83] Kitajima N., Osawa M., Tamura N., Moro-oka Y., Hirano T., Hirobe M., Nagano T., *Inorg. Chem.* 32 (1993) 1879-1880.

[84] Kitajima N., Osawa M., Tanaka M., Moro-oka Y., *J. Am.Chem. Soc.* 113 (1991) 8952-8953.

[85] Metal Ions in Biological Systems In: Molybdenum and Tungsten. Their Roles in Biological Processes. Sigel A, Sigel H, (eds.) Vol. 39. Dekker; New York (2002) 759.

[86] Xiao Z., Young C.G., Enemark J.H., Wedd A.G., *J. Am.Chem. Soc.* 114 (1992) 9194-9195.

[87] Xiao Z., Bruck M.A., Doyle C., Enemark J.H., Grittini C., Gable R.W., Wedd A.G., Young C.G., *Inorg. Chem.* 34 (1995) 5950-5962.

[88] Xiao Z., Bruck M.A., Enemark J.H., Young C.G., Wedd A.G., *Inorg. Chem.* 35 (1996) 7508-7515.

[89] Xiao Z., Gable R.W., Wedd A.G., Young C.G., *J. Am.Chem. Soc.* 118 (1996) 2912-2921.

[90] Eagle A.A., Laughlin L.J., Young C.G., Tiekink E.R.T, *J. Am. Chem. Soc.* 114 (1992) 9195-9197.

[91] Eagle A.A., Harben S.M., Tiekink E.R.T., Young C.G., *J. Am. Chem. Soc.* 116 (1994) 9749-9750.

[92] Thomas S., Tiekink E.R.T., Young C.G., *Organometallics* 15 (1996) 2428-2430.

[93] Dance I., *Dalton Trans.*, 39 (2010) 2972-2983.

[94] Rajapakshe A., Snyder R.A., Astashkin A.V., Bernardson P., Evans D.J., Young C.G., Evans D.H., Enemark J.H., *Inorg. Chim. Acta.* 362(12) (2009) 4603-4608.

[95] Murray J., Macartney D., Thatcher G.R.J., *Organic Lettters*, 3 (2001) 3635-3638.

[96] Armstrong W.H., Spool A., Papaephthymiou G.C., Frankel R.B., Lippard S.J., *J. Am. Chem. Soc.* 106 (1984) 3653-3667.

[97] Armstrong W.H., Lippard S.J., *J. Am. Chem. Soc.* 106 (1984) 4632-4633.

[98] Kitajima N., Fukui H., Moro-oka Y., *J. Am. Chem. Soc.* 112 (1990) 6402-6403.

[99] Perkinson J., Brodie S., Yoon K., Mosny K., Caroll P.J., Morgan T.V., Burgmayer S.J.N., *Inorg. Chem.* 30 (1991) 719-727.

[100] Bérces A., *Inorg. Chem.* 36 (1997) 4831-4837.

[101] Tashian R.E., *Bioessays* 10 (1990) 186-192.

[102] Alsfasser R., Ruf M., Trofimenko S., Vahrenkamp H., *Chem. Ber.* 126 (1993) 703-710.

[103] Brand U., Rombach M., Seebacher J., Vahrenkamp H., *Inorg. Chem.* 40 (2001) 6151-6157.

[104] Hikichi S., Tanaka M., Moro-oka Y., Kitajima N., *J. Chem. Soc., Chem. Commun.* (1992) 814-815.

In: Ligands
Editors: Paula Gawryszewska et al.
ISBN: 978-1-63117-143-7

Chapter 5

COORDINATION ENGINEERING WITH 1,3,5-TRIAZA-7-PHOSPHAADAMANTANE (PTA) TOWARDS BIOLOGICALLY ACTIVE SILVER-ORGANIC FRAMEWORKS

Piotr Smoleński*
Faculty of Chemistry, University of Wrocław, Wroclaw, Poland

ABSTRACT

The series of silver(I) coordination polymers $[Ag(N\text{-}N)(\mu\text{-}PTA)]_n(X)_n$ and discrete monomers $[Ag(N\text{-}N)(PTA)_2](X)$ and [Ag(Tpms)(PTA)] {N-N = polypyridine: 2,2'-bipyridine, 4,4'-di-*tert*-butyl-2,2'-bipyridine, 1, 10-phenanthroline, 2,9-dimethyl-1,10-phenanthroline and 1,10-phenanthroline-5,6-dione, Tpms - tris(pyrazol-1-yl)methanesulfo-nate; X = NO_3, PF_6} were generated at ambient conditions with 1,3,5-triaza-7-phosphaadamantane (PTA) and the corresponding polypyridines or Tpms.

Derivatives of PTA: PTA=O and PTA=S, were also used as a new versatile N,O,S-building block towards engineering functional metal-organic frameworks (MOFs). Depending on the type of the starting silver (I) salt, different distinct silver-organic 3-D networks driven by multiply bridging PTA=O and PTA=S spacers were self-assembled. All compounds were obtained as air and light stable solids, and characterized

* Corresponding author: Piotr Smoleński. Faculty of Chemistry, University of Wrocław, ul. F. Joliot-Curie 14, 50-383, Wroclaw, Poland. E-mail: piotr.smolenski@chem.uni.wroc.pl.

by IR, ^{1}H and $^{31}P\{^{1}H\}$ NMR spectroscopy, ESI^{+}-MS and elemental analyses. The crystal structures were determined by single crystal X-ray diffraction analysis. Apart from representing unusual examples of metal-PTA derivatives they also feature solubility in water. Selected compounds were additionally tested for their biological properties, and were found to exhibit significant antibacterial and antifungal activities, screened *in vitro* against the standard strains of *Staphylococcus aureus*, *Staphylococcus pyogenes*, *Staphylococcus pneumoniae*, *Staphylococcus sanguinis*, *Staphylococcus mutans*, *Enterococcus faecalis*, *Pseudomonas aeruginosa*, *Escherichia coli* and *Candida albicans*. Furthermore, some compounds show a pronounced antiproliferative activity against human malignant melanoma, and the effects on the inhibition of tumor cells *in vitro* are in agreement with DNA-binding studies.

Keywords: 1,3,5-triaza-7-phosphaadamantane, silver complexes, coordination polymers, MOFs, antimicrobial activity, antiproliferative activity, antitumor activity

1. INTRODUCTION

The coordination chemistry of aqua-soluble aminophosphine - 1,3,5-triaza-7-phosphaadamantane (PTA, Scheme 1) and its derived species has received increased interest in recent years because of its interesting steric and electronic properties. This is mainly due to the good solubility of transition metal PTA complexes in water, thus making possible their efficient application in aqueous phase catalysis, as water-soluble antitumor agents and photoluminescent materials [1-7]. In particular, it has been shown that Rh and Ru complexes with PTA and its derivatives (Scheme 1) are active and selective catalysts for the hydrogenation and hydroformylation of alkenes [1].

The physical and chemical properties of PTA are somewhat related to hexamethylenetetramine (HMT).

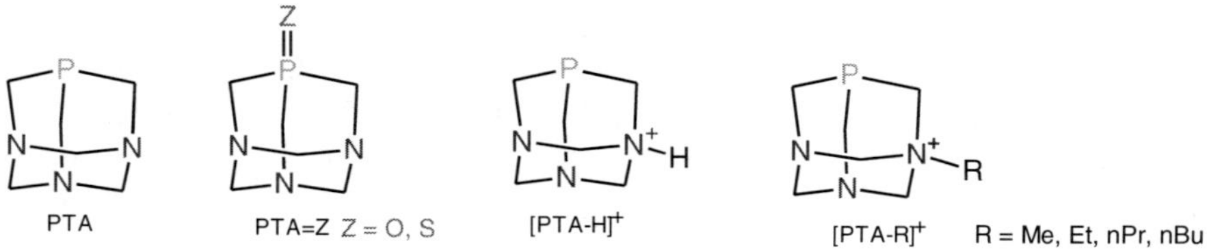

Scheme 1. Molecular structures of PTA and its derivatives.

PTA is air-stable, and soluble in water, methanol, ethanol and chloroform. This is a small (cone angle of PTA = 103°), basic phosphine, similar in electronic and steric properties to PMe_3, but in contrast to the latter, is air stable in the solid state and in solutions [1]. In molecule of 1,3,5-triaza-7-phosphaadamantane there are 4 donor atoms: one phosphorus and three nitrogen, and therefore the molecule is able to coordinate with hard and soft metals, according to HSAB theory (Scheme 2).

Additionally, one unique feature of PTA is the ability to undergo regioselective protonation at the nitrogen center at lower pH of aqua solutions, as well as N-alkylation with alkyl halides (Scheme 1). The resulting salts [PTA-R]I (R = Me, Et, *n*Pr, *n*Bu) are air and water stable, but as expected due to their stronger ionic character, [PTA-Me]I is less soluble in organic solvents than PTA [8, 9]. Regioselective protonation or N-alkylation results also in change of σ-donor and π-acceptor properties of metal complexes with ionic PTA derivatives [10, 11].

Using classic P-coordination modes of PTA ligands to metal center (relatively soft) through the phosphorus atom, a large series of new transition metal complexes have been synthesized and investigated in last decade, also finding some attractive applications [1]. However, since the last review on the coordination chemistry of PTA in 2009, some new complexes with PTA and derivatives have been synthesized and examined, towards new potential catalytic, biological and luminescent applications: Rhodium catalysts of reversible dehydrogenation of organic compounds [12] and isomerization of allylic alcohols [13] were prepared and investigated, while new ruthenium compounds were tested as active catalysts for the selective hydration of nitriles to amides [14]. New iron complexes with PTA were characterized, and their ability to catalytically produce dihydrogen from acetic acid was examined electrochemically [15]. From the other side, the cytotoxic properties of the alkynyl complexes of gold against human cancer cells were analyzed, and moreover, some of the cytotoxic compounds also show luminescence at room temperature [16].

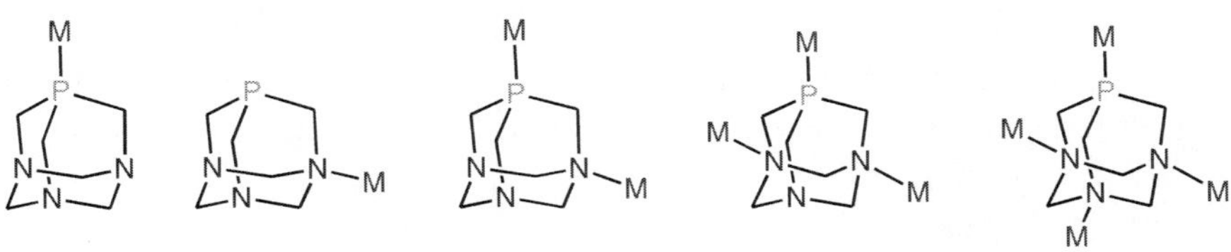

Scheme 2. Possible coordination modes of PTA.

Other, ruthenium - arene complexes with perfluorinated alkyl chains and PTA derivatives were tested as more cytotoxic media than the parent, RAPTA complex [17]. In fact, since the discovery of the antitumor activity of cisplatin (*cis*-$[PtCl_2(NH_3)_2]$) for the treatment of several human tumours, many complexes of platinum and other metals have been tested for their potential antitumor properties in recent decades.

A series of compounds showing promising prospects were the phosphine complexes with metals of the 11 Group. Copper(I) complexes reported by us first of the type $[Cu(PTA)_4](NO_3)$ and $[Cu(PTA\text{-}H)_4](NO_3)_5$ are very promising for potential biological applications, because of their stability in water in wide pH range.

Therefore analogous complexes with $[Cu(PTA)_4](NO_3)$ and $[Cu(PTA\text{-}H)_4](NO_3)_5$ using a ^{64}Cu isotope were positively tested as potential radio-pharmaceuticals, especially because of rapid cell-association with EMT-6-cells, which was observed by PET imaging [18]. Complexes of the type $[Cu(PTA)_4]X$ (X – counterion) were successfully compared with cisplatin against a panel of human tumour cell lines (mostly comparable with those elicited by cisplatin) [19].

Some of the compounds appeared to be more powerful than cisplatin towards human cervix and non-small lung carcinoma cells and confirm the high potential of water-soluble copper complexes with PTA and its derivatives as a new class of non-platinum anticancer drugs.

Among the metal complexes used for the treatment of several diseases [20], some silver complexes are currently useful as antibacterial agents in therapeutic procedures [21, 22]. In view of their antimicrobial properties and low toxicity, silver complexes have been used in a variety of applications such as water purification, eye-drops and anti-infective coatings in medical devices [23-30].

Unfortunately, many silver complexes are sparingly soluble in common solvents, especially water, which limits their uses in oil treatments and creams, as for the most used topical antibacterial agent silversulfadiazine [31]. Silver complexes are also interesting because of their antimicrobial activity and other properties, which can be tuned by changing the number and type of ligands.

For example, the silver(I) imidazolate complex has excellent antibacterial and antifungal properties [32], in contrast to PPh_3 adduct, which does not show antimicrobial activity [33].

At least, silver(I) sulfadiazine, a polymeric compound releasing silver ions slowly, is one of the most common silver based antimicrobial agents for treating infections in severe burn sites [29, 30, 34-37].

2. Silver(I) Coordination Polymers – Methods of Construction

The design and construction of coordination polymers have been recently intensively investigated, as demonstrated by the very fast growth in the number of articles in this field [38]. It is known that crystal engineering of coordination polymers depends not only on the choice of organic building blocks and metal center but also appropriate conditions of syntheses. An organic linker can be aimed to have particular functional groups with donating atoms and coordination sites having appropriate steric geometries [38]. Moreover, the ligands used are responsible for the unique structure and properties of particular products.

In fact, many coordination polymers involve problematic and expensive multistage syntheses: high temperature and pressure, are applicable only to particular combinations of metals, require organic, building blocks often with high cost e.g., crown ether, calixarene, cavitand, porphyrine, cryptand, dendrimer, cyclodextrin may require inert atmosphere, air-sensitive chemicals and toxic reagents [39].

The commonly used technique for the syntheses of coordination polymers is fine-tuning of the reaction and the crystallization conditions, such as solvothermal method [40]. These are typically carried out in autoclaves under pressure, and at relatively high temperatures. Many of these processes involve C-C bond formation [41], the transformation of carboxylic acids, the formation of heterocycles from small molecules [42], as well as the redox conversion of metal species [43, 44], and also constitute an important family of metal oxides, which have been investigated extensively for use as building blocks in the synthesis of hybrid materials.

Silver(I) coordination polymers constitute a separate class of these compounds and are interesting not only because of their curious structural variety, but also due to their rich physicochemical properties, which could find applications in: luminescent materials [45], conductive materials - silver(I) coordination polymers typically exhibit electrical conductivity [46], therapeutic agents – the antimicrobial activities of Ag^{+}-X (X=P, N, O, and S) bonding complexes were investigated in aqueous media, showing that Ag^{+} complexes containing easily replaceable Ag-N/P/O bonds play important roles in exhibiting an effective and wide spectrum of antimicrobial activities [28, 45, 47], anion or guest exchange and gas sorption [49] and supramolecular chirality, etc.

The physical and chemical properties of coordination polymers might be tuned and improved by appropriate design of bridging ligands and/or using various noncoordinating backbones. Multidentate bridging ligands with different coordination geometries, containing two or more nitrogen, and/or sulfur, oxygen and phosphorus donor groups, and with a distinct ability to bind metal ions, are widely used [50]. A variety of coordination architectures have been produced using neutral or ionic ligands. In fact, a majority of coordination polymers have been obtained by using the coordinative ability of pyridyl or carboxylate functional groups with transition-metal ions. For example, 4,4'-dipyridyl is a simple neutral ligand that has been used as a bidentate building block for coordination polymers [51]. On the other hand, polycarboxylic acids display completely different connectivity than that of the dipyridyl ligands, due to the useful coordination ability of carboxylate and to the sensitivity of such groups to acidity [52]. By changing the nature of ligands it is possible to obtain porous structures with the free volume up to 90% of the unit cell [53]. Therefore, these compounds are believed to be able to find applications as high scale sorbents.

The backbones and substructures of organic ligands ensure an important influence on the structure and properties of the metal complexes, and ligand choice might be a key point for constructing functional coordination architectures. Thus, a slight modification in the ligand backbone by adding a noncoordinating functional group could also play an important role in structural assemblies of coordination frameworks [54-57].

3. Coordination Engineering with PTA. P,N-Coordination Modes of PTA

Coordination engineering with PTA is the design and synthesis of transition metal complexes bearing PTA for optimization of their physical and chemical properties towards useful applications. Using the unique properties of PTA it is possible to change and control the hydrophilic and electronic properties. PTA seems to be a versatile ligand for coordination chemistry, because of rich coordination modes, compared to other alkyl and aryl phosphines or aminophosphines. The advantage of PTA coordination modes can be extended using different metals with different affinities to P- and N-donating atoms of its cage-like molecule, and can lead to interesting homo- and heterometallic metal-organic frameworks with advanced properties.

Additionally, some physical and chemical properties of coordination polymers may be tuned and improved by appropriate design of bridging ligands/counterions and/or by using various noncoordinating backbones. The above mentioned features of the coordination chemistry of PTA form the basis for qualification of coordination engineering with PTA towards potential applications or new metal-organic frameworks.

4. Silver(I) Tris(Pyrazolyl) Methanesulfonate (Tpms) and Polypyridine (N-N) Complexes with PTA

The tris(1-pyrazolyl)methanesulfonate species $^{-}O_3SC(pz)_3$ (Tpms) was designed by analogy with tris(1-pyrazolyl)borate (Tp), both being able to act as monoanionic C_{3v}-symmetrical nitrogen-donor ligands [58-60]. However, unlike the tris(1-pyrazolyl)borate derivatives, insoluble in water and unstable towards hydrolysis, Tpms has a methanesulfonate unit which imparts an increased solubility in polar solvents and a very good stability in aqueous media over a wide range of pH [59]. The tris(1-pyrazolyl)methanesulfonate coordinated as a N_3 tripodal ligand can also be compared to the isoelectronic pentahapto cyclopentadienyl (Cp) ligand [60].

However, as shown recently [61], the Tpms ligand can act not only as a tripodal but also as a bipodal ligand. In the former case, it can exhibit either a N,N,N or a N,N,O coordination, whereas as a bipodal ligand it can coordinate through either N,N or N,O [61, 62] (Scheme 3). Tpms and analogous species having different substituents in the 3-position of the pyrazole rings exhibit a marked coordination versatility, acting either as a tripodal or a bipodal ligand (*i.e.*, with N_3-, N_2O-, N_2- or NO- coordination modes) with the possibility of involving the sulfonate moiety in the coordination, hence being suitable ligands for the synthesis of structural models for N,N,O-binding in metalloenzymes [63].

The aromatic *N,N*-ligands such as 2,2'-bipyridine, 1,10-phenanthroline and their various derivatives (Scheme 4) have been the subject of intense research due to their versatile coordination chemistry and recognized function in diverse biological systems [64-66]. In contrast to Cu-polypyridine derivatives [67], the biological properties of the analogous silver(I) complexes have only been investigated in a few studies [68].

A series of Ag-polypyridine complexes stabilized by polar tertiary phosphines have also been reported [64, 69], but their poor solubility in polar solvents and light sensitivity eventually prevented the application of these compounds as bioactive materials.

The silver complex with the Tpms ligand was obtained from the reaction of $AgBF_4$ with Tpms, in the presence of the water-soluble 1,3,5-triaza-7-phosphaadamantane, PTA) [63a].

A treatment of $AgNO_3$ or $AgPF_6$ with poly-pyridile, N-N (Scheme 4) and PTA ligands leads to obtaining discrete complexes or coordination polymers, depending on the ratio N-N/PTA, wherein μ-PTA acts as an N,P-linker [63b]. Their antimicrobial activity was also reported.

We have also investigated their antiproliferative activity against human malignant melanoma cells.

4.1. Synthesis and Spectroscopic Determination of the Silver Complexes with Tpms 1-3

The complex [Ag(Tpms)] (1) has been synthesized by reaction of the tris (pyrazoly-1-yl)methanesulfonate lithium salt, Li(Tpms), with $AgBF_4$ in methanol, at room temperature. 1 is a greyish-rose solid, stable in air and in solution. 1. The IR spectrum of 1 displays the expected bands of the pyrazolyl rings at 1522 cm^{-1}, together with several bands likely due to the SO_3 group vibration modes at 1268, 1251, 1199, 1061, 1056 and 1046 cm^{-1} [70, 63a].

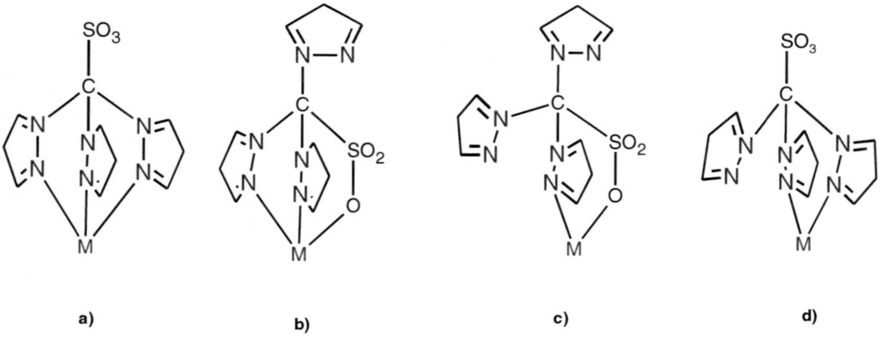

Scheme 3. Possible coordination modes for the Tpms ligand: a) tripodal N,N,N; b) tripodal N,N,O; c) bipodal N,O; d) bipodal N,N.

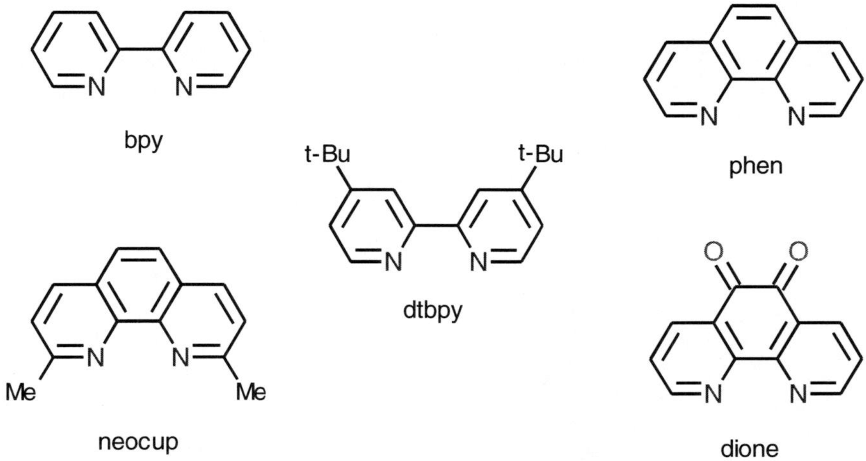

Scheme 4. Structural Formulas of Polypyridines.

The positive ESI-MS spectrum carried out in acetonitrile shows isotopic clusters, due to $[Ag(MeCN)]^+$, $[Ag(MeCN)_2]^+$, $[(Ag)_2(Tpms)(MeCN)]^+$ and $[(Ag)_2(Tpms)(MeCN)_2]^+$, while in the negative ESI-MS spectrum the peaks are assigned to $[Ag(Tpms)_2]^-$ and $[(Ag)_2(Tpms)_3]^-$. By treatment of an MeOH solution of $AgBF_4$ with PTA at room temperature, $[Ag(PTA)(MeOH)][BF_4]$ (2) is obtained. It is a brown solid, soluble in water, acetonitrile, acetone and DMSO. The IR spectrum of 2 confirms the presence of the coordinated PTA ligand [63c] and the $[BF_4]^-$ counterion [71].

The ^{1}H NMR spectrum of 2 in D_2O shows two types of methylene protons, typical of the PTA ligand. One of them, assigned to the P–CH_2–N moiety, occurs as a singlet at δ 4.3 whereas the other one, corresponding to the N–CH_2–N group, displays an AB spin system centered at δ 4.7 (J_{AB} = 13 Hz), attributed to the N–CH_{ax}–N and N–CH_{eq}–N protons [63c]. The $^{31}P\{^1H\}$ NMR spectrum exhibits a singlet at δ -78, the ^{107}Ag-^{31}P and ^{109}Ag-^{31}P spin-spin coupling not being observed due to the fast ligand exchange [63, 72]. The ESI-MS spectra of 2 carried out in water and acetonitrile show peaks due to silver species always containing the PTA ligand.

By reaction of a methanol solution of $AgBF_4$ with Li(Tpms) and PTA at room temperature, the compound [Ag(PTA)(Tpms)] (3) has been synthesized (Scheme 5).

The ^{1}H NMR spectrum of 3 in DMSO-d_6 similarly to 2, shows two types of methylene protons for the coordinated PTA.

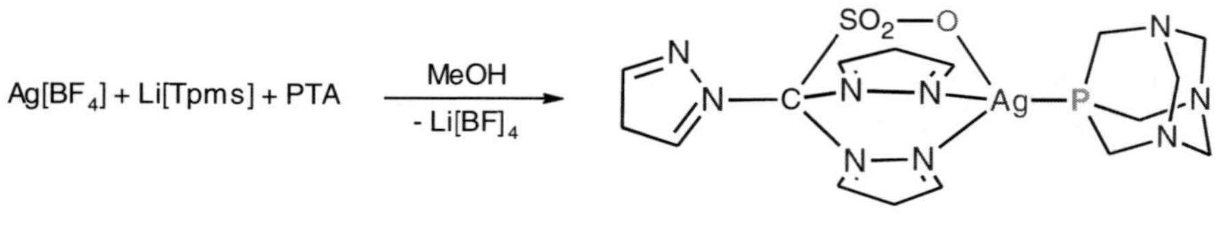

Scheme 5. Preparation of 3.

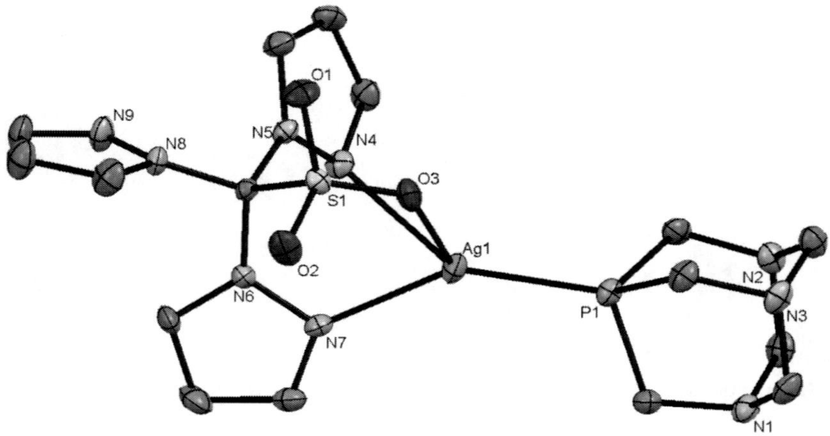

Figure 1. Ortep diagram of [Ag(Tpms)(PTA)] (3). Hydrogen atoms were omitted for clarity.

One of them, occurs as a singlet at 4.23 ppm while the second one displays an AB spin system centered at 4.50 ppm (J_{AB} = 15 Hz) [63c]. Similarly to the other silver Tpms derivatives, the ^{1}H NMR spectrum of 3 reveals the presence of three equivalent pyrazolyl rings (three singlets at δ 8.1, 7.6 and 6.5 corresponding to the pyrazolyl protons in positions 5, 3 and 4). The $^{31}P\{^{1}H\}$ NMR spectrum, similarly to 2, shows a broad singlet at δ -84, and no ^{107}Ag-^{31}P or ^{109}Ag-^{31}P spin-spin coupling was observed, likely due to the fast ligand exchange [72].

This effect was previously reported for other PTA-Ag(I) complexes [63]. The positive ESI-MS spectrum in the acetonitrile of 3 shows peaks for ($[Ag(CH_3CN)_2]^+$), ($[(Ag)_2(Tpms)(CH_3CN)_2]^+$), ($[(Ag)_2(Tpms)(CH_3CN)(PTA)]^+$) and ($[(Ag)_2(Tpms)(PTA)_2]^+$), in accordance with probable aggregation in solution and formation of dinuclear species. The X-ray quality crystals of 3 were obtained from a methanol solution. An ortep diagram is shown in Figure 1.

4.2. Synthesis and Spectral Characterization of the Silver Complexes with Polypyridines 4-14

Using the PTA ligand as a versatile *N,P*-building block, and extending these studies to polypyridine-type *N,N*-ligands, namely 2,2'-bipyridine (bpy), 4,4'-di-*tert*-butyl-2,2'-bipyridine (dtbpy), 1,10-phenanthroline (phen), 2,9-dimethyl-1,10-phenanthroline (neocup) and 1,10-phenanthroline-5,6-dione (dione) (Scheme 4), we reported the facile self-assembly generation, characterization and biological activity of the series of new coordination polymers $[Ag(N\text{-}N)(\mu\text{-}PTA)]_n(X)_n$ (4, 5, 7−11, 13, 14) {N-N = bpy (4−6), dtbpy (7), neocup (8, 9), phen (10−12), dione (13, 14); X = NO_3 (4, 7, 8, 10, 12, 13), PF_6 (5, 7, 9, 11, 14)} and monomers $[Ag(N\text{-}N)(PTA)_2](X)$ (6, 12) [63b]. Selected compounds (4, 6, 8, 10, 12 and 13) were also evaluated for their antibacterial, antifungal and antiproliferative activity, as well as DNA binding properties.

Treatment of AgX (X = NO_3 or PF_6) with a two-fold molar excess of bpy or phen, in MeOH solution at ~25 °C, followed by the addition of an equimolar amount of PTA (i.e., Ag/N-N/PTA molar ratio of 1:2:1), leads to $[Ag(bpy)(\mu\text{-}PTA)]_n(X)_n$ {X = NO_3 (4), PF_6 (5)} and $[Ag(phen)(\mu\text{-}PTA)]_n(X)_n$ {X = NO_3 (10), PF_6 (11)} coordination polymers, wherein μ-PTA acts as an *N,P*-linker (Scheme 6). The reactions of $AgNO_3$ using bpy/phen and PTA in equimolar amounts (Ag/N-N/PTA molar ratio of 1:1:1 or 1:2:2) afford discrete mononuclear $[Ag(N\text{-}N)(PTA)_2](NO_3)$ {N-N = bpy (6), phen (12)} complexes, in which two PTA moieties act as *P*-bound monodentate ligands. In contrast, the same reactions employing the sterically hindered dtbpy, neocup and dione as *N,N*-ligands in equivalent amounts relatively to AgX and PTA (i.e. Ag/N-N/PTA molar ratio of 1:1:1), also give rise to the formation of resembling coordination polymers with the general formula $[Ag(N\text{-}N)(\mu\text{-}PTA)]_n(X)_n$ {N-N = dtbpy (7), neocup (8, 9), dione (13, 14); X = NO_3 (8, 13), PF_6 (7, 9, 14)}. The products 4−14 have been isolated as air and light stable, white (4−12) or orange (13, 14) microcrystalline solids in ca. 45−80% yields based on $AgNO_3$ or $AgPF_6$, and characterized by IR, ^1H and $^{31}P\{^1H\}$ NMR spectroscopy, ESI^+-MS, elemental analyses and single-crystal X-ray diffraction (for 4 as a representative example). The proposed formulations of four-coordinate silver(I) compounds 4−14 either as coordination polymers (5, 7−11, 13, 14) bearing one μ-PTA linker or discrete complexes (6, 12) bearing two terminal PTA ligands are evident from the 1:1 and 1:2 Ag/PTA ratios observed in these compounds, respectively, being also supported by the above-mentioned methods and literature background [63].

A noteworthy feature of 4–14 concerns their solubility in water, with the $S_{25°C}$ values ranging from 4 to 18 mg·mL^{-1}. In addition, these compounds are soluble in DMSO and other polar solvents.

The solution ^{1}H NMR spectra of 4–14 exhibit two types of methylene protons in the PTA region. The first type, P–CH$_2$–N, occurs as a doublet at δ 4.3–4.4 ppm with the respective ^{2}J(PH) values of 2–3 Hz, which are rather low in comparison with that of 10 Hz in the uncoordinated PTA [1]. The second type, N–CH$_2$–N, displays an AB spin system centred at δ 4.3–4.6 ppm that is assigned to the N–CH$_{ax}$–N and the N–CH$_{eq}$–N protons, as reported in other compounds [1, 63b]. In the aromatic region, the ^{1}H NMR spectra of 4–14 also show the resonances due to the phenyl rings of bpy, dtbpy, phen, neocup and dione coordinated to silver(I). Furthermore, the broad singlets ascribed to the Me and *t*-Bu protons of neocup (8, 9) and dtbpy (7) are observed at δ 2.9–2.9 and 1.4 ppm in their spectra, respectively.

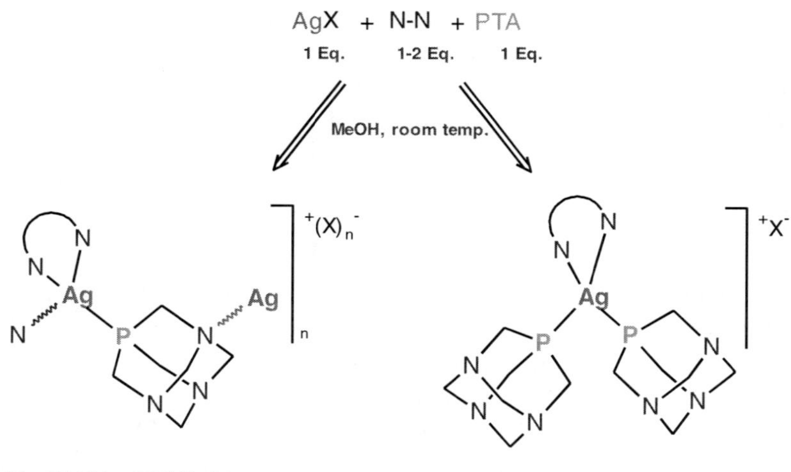

[Ag(N-N)(μ-PTA)]$_n$(X)$_n$
(4,5,7-11,13,14)

[Ag(N-N)(PTA)$_2$](X)
(6,12)

N-N	bpy	bpy	bpy	dtbpy	neocup	neocup	phen	phen	phen	dione	dione
X$^-$	NO_3	PF_6	NO_3	PF_6	NO_3	PF_6	NO_3	PF_6	NO_3	NO_3	NO_3
Comp.	4	5	6	7	8	9	10	11	12	13	14

Scheme 6. Simplified representation of syntheses with the structural formulas of compounds 4–14.

Integration of the ^{1}H NMR spectra confirmed the 1:1 molar ratio of coordinated N-N/PTA ligands for polymeric structures of the type $[Ag(N\text{-}N)(\mu\text{-PTA})]_n(X)_n$ (4, 5, 7−11, 13, 14), and the 1:2 molar ratio in mononuclear complexes $[Ag(N\text{-}N)(PTA)_2](X)$ (6, 12). The $^{31}P\{^{1}H\}$ NMR spectra of 4−14 are typical for silver-PTA derivatives, showing the corresponding broad singlets in the δ range from −77 to −85 ppm, which are shifted downfield in comparison with the uncoordinated PTA. In addition, the spectra of 5, 7, 9, 11 and 14 reveal characteristic septets at δ −144 ppm due to PF_6^- anions. The ^{107}Ag-^{31}P and ^{109}Ag-^{31}P spin-spin coupling was not observed in all cases due to the fast ligand exchange [72]. In general, the NMR data support the preservation of the coordination environments around Ag^I atoms upon dissolution of 4−14 in the deuterated solvents.

The IR spectra of 4−14 exhibit absorptions due to typical vibrations of polypyridine and PTA ligands.

Hence, several ν(CH) bands are detected in the 2960−2855 cm^{-1} range, whereas a number of bands in the 1500−500 cm^{-1} range are associated to ν(C−X) (X = C, N, P) vibrations [1, 73] In 13−14, the additional strong bands are observed at ~1690 cm^{-1} due to ν(C=O) of the dione ligand [74]. which is coordinated through only nitrogen atoms. The IR spectra of compounds containing NO_3^- or PF_6^- anion also show characteristic strong and broad bands in the 1385−1365 and 844−837cm^{-1} regions, respectively.

The formulations of 4−14 are further confirmed by the ESI^+-MS spectra of their solutions in H_2O or MeCN, showing $[Ag(N\text{-}N)(PTA)(solvent)]^+$ and/or $[Ag(N\text{-}N)(solvent)]^+$ peaks, with the expected isotopic patterns. In 8 and 9, the loss of the neocup ligand also occurs as attested by the detection of the $[Ag(PTA)(CH_3CN)]^+$ fragments. Besides, peaks due to the $[Ag(bpy)(PTA)_2]^+$ and $[Ag(phen)(PTA)_2]^+$ ionic fragments are seen in the spectra of mononuclear complexes 6 and 12, respectively [63].

As a representative of the series of coordination polymers 4, 5, 7−11, 13 and 14, the structure of 4·MeOH is composed of the cationic $[Ag(bpy)(\mu\text{-PTA})]^+$ block, NO_3^- counter anion and crystallization MeOH molecule per formula unit (Figure 2a).

The four-coordinate Ag1 atoms exhibit a significantly distorted tetrahedral $\{AgN_3P\}$ geometry filled by nitrogen atoms of the bidentate bpy ligand, and the phosphorus and nitrogen atoms from two adjacent μ-PTA linkers. The cagelike μ-PTA linkers adopt alternating P- or N-coordination, thus assembling the $[Ag(bpy)]^+$ units into infinite 1D linear chains with a pitch of 6.915(3) Å [Ag···Ag], being equal to the *a* unit cell dimension (Figure 2b-c).

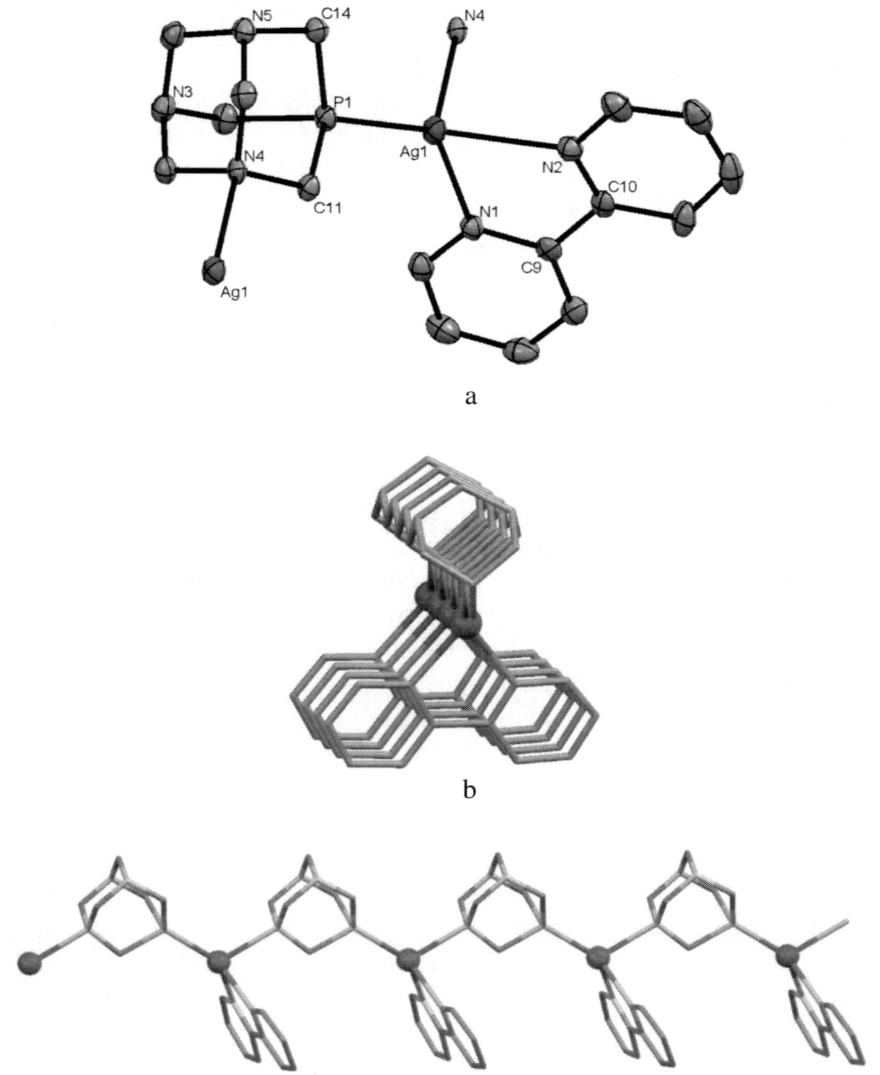

Figure 2. Structural fragments of 4 showing (a) ellipsoid plot with partial labeling scheme, (b) side and (c) front views of a linear 1D metal-organic chain. H atoms, NO_3^- anions are omitted for clarity.

4.3. Antibacterial Activity

The antimicrobial activities of the compounds 1-3 against bacteria and fungi were examined qualitatively by agar diffusion tests (Table 1). While the lithium salts of Tpms show no activity, the silver derivatives possess a wide spectrum of effective antibacterial and antifungal activities. In general, compounds bearing the PTA ligand (2 and 3) show a significantly higher antibacterial activity than silver nitrate ($P < 0.01$ by Fisher's least significant difference procedure).

Among Gram-positive bacteria, only *Enterococcus faecalis* did significantly respond to any of the compounds [63a]. This result is not surprising in view of the well known general tolerance toward many antibacterial agents that characterize this species. For the two Gram-negative species tested, our results demonstrate that compound 2 is less active against *Eschericha coli* but with a diameter of the inhibition zone still wider than that of the comparison test compound. The active compounds 2 and 3 significantly inhibit the growth of the *Pseudomonas aeruginosa* strain. The recorded diameters were quite wide (17 mm), even if a general higher susceptibility toward silver is clear from the diameter of the silver nitrate (14 mm against a mean value of 12). It is possible that *Pseudomonas* is basically more susceptible to the action of the silver ions. Lastly, compound 3 shows a very good activity against *Candida albicans*, with a mean diameter size of 20 mm.

The disk diffusion method was applied to screen the activity of compounds 4, 6, 8, 10, 12 and 13 towards various bacterial species and a single yeast species. Among the bacterial strains, six Gram-positives and two Gram-negatives were considered [63b]. The tested compounds were active against all species. The antimicrobial activity analysis was extended to measure the minimum inhibitory concentration and minimum bactericidal concentration of each derivative against a subset of species. The Gram-positive *S. aureus* and the Gram-negative *E. coli* were selected for their importance and diffusion as main human pathogens, while the Gram-positive *E. faecalis* and the Gram-negative *P. aeruginosa* were selected for their natural or acquired ability to become antibiotic resistant. Table 2 shows the results of these experiments.

Compounds 8 and 13 were the most active with MICs of 4−8 μM, which are two or more dilutions lower than those of the reference salt $AgNO_3$ (MICs = 16−64 μM). The unique exception was represented by *P. aeruginosa*, for which similar activities were recorded for the obtained silver compounds and $AgNO_3$.

Table 1. Disc diffusion test of compounds 1-3 against a panel of reference bacterial strains. Diameters of zone inhibition are expressed in mm [63a]

Compound	1	2	3	$AgNO_3$
S. aureus ATCC 25923	11.0	14.0	15.0	11.0
S. pyogenes SF370	12.0	15.0	16.0	13.0
S. pneumonae ATCC 49619	11.0	15.0	16.0	13.0
S. sanguins SK36	12.0	14.0	16.0	11.0
S. mutans UA159	11.0	14.0	16.0	12.0
E. faecalis ATCC 29212	10.0	9.0	10.0	11.0
P. aeruginosa ATCC 27853	15.0	17.0	17.0	14.0
E. coli ATCC 25922	11.0	130	15.0	11.0
C. albicans ATCC 24433	12.0	17.0	20.0	14.0

Interestingly, in the cases of *E. faecalis* and *P. aeruginosa*, there was a difference between the activity levels determined by the disk diffusion and the broth microdilution methods.

The first method showed that all the polypyridine Ag-PTA compounds with the exception of 12 were more active than $AgNO_3$. In the case of the microdilution method, only two of them, i.e. 8 and 13, were more active than the $AgNO_3$ reference. This result could reflect a different ability of compounds to diffuse on a surface-like system (disk-agar plate) in respect to being active in solution in an aqueous system (microdilution broth).

4.4. Antineoplastic Activity

Biomedical applications of Ag^+ complexes are related mostly to their antibacterial action, which appears to involve interaction with DNA.

A similar interaction of the metal with DNA may also be the cause of the antitumor action of some of those complexes.

The cytotoxic activities of the Ag^+ compounds 1–13 as 50% inhibitory concentration (IC_{50}) values against human malignant melanoma cells (A375) are shown in Table 5 and compared with $AgNO_3$. The cytotoxic activities of the complexes follow the sequence 8>13>1>3> $AgNO_3$>10>12>>2>6>4, in line with the binding activity of these complexes on DNA. The neocup and phen ligands are found to have IC_{50} values slightly higher than or similar to the corresponding compounds (Table 3).

Table 2. Antimicrobial activity MIC [μM] of selected polypyridine Ag-PTA compounds against a representative panel of pathogens [63b]

Compound	*S. aureus*	*E. faecalis*	*E. coli*	*P. aeruginosa*	*C. albicans*
4	64	64	32	32	32–64
6	32	64	32	16	32–64
8	8	8	8	16–32	8
10	32	64	32	32	4–8
12	32–64	64	>128	32	8–16
13	8	8–16	4–8	16	4–8
$AgNO_3$	16–64	16–64	16–32	8–16	4

Table 3. Cytotoxic activity (IC_{50}) of compounds 1-13 on A375 human malignant melanoma cell line (confidence interval = 95%, the IC_{50} value is the concentration of compound that affords a 50% reduction in cell growth, after 72 h of incubation) [63a-b]

Compound	IC_{50} [μM]
1	1.62 (0.87 – 2.65)
2	12.35 (10.32 – 15.24)
3	1.90 (0.75 – 3.26)
4	24.30 (18.67 – 31.62)
6	18.47 (14.50 – 23.52)
8	0.37 (0.28 – 0.48)
10	4.93 (4.35 – 5.59)
12	5.43 (4.97 – 5.87)
13	0.39 (0.31 – 0.50)
$AgNO_3$	2.54 (1.91 – 2.88)

The cytotoxicity of the free neocup, phen and dione ligands has been reported in the literature [75]. It has been suggested that polypyridyl ligands can impart tumor cell death by forming a cytotoxic metal complex in situ or by sequestering metal ions vital to cell survival [75]

5. 3-D Coordination Polymers of Silver with PTA=O and PTA=S and Their Antimicrobial Activity

Following our research projects on the design of new metal-organic networks and on the coordination chemistry of PTA and derived ligands, we have recently reported a new series of Ag(I) coordination polymers obtained via a unique synthetic entry allowing the controlled N,P-coordination mode of PTA [63] and the first metal complex bearing 1,3,5-triaza-7-phospha-adamantane-7-oxide (PTA=O, Scheme 1) and 1,3,5-triaza-7-phospha-adamantane-7-sulfide (PTA=S, Scheme 1) with coordination through sulfur atom [63g, 76]. As an extension of these studies, the present work also aims at opening up the application of PTA=Z (Z = O, S) as a new and still unexplored N,O- and N,S-building block for the engineering of metal-organic frameworks.

5.1 Synthesis and Characterization of the Silver Coordination Polymers with PTA=O and PTA=S 6-19

The reaction of silver(I) nitrate with PTA=O as the main ligand and spacer, in equimolar amounts at 20 °C in $MeOH/CH_2Cl_2$ medium, followed by alkalization with NH_4OH, leads to the self-assembly formation of the polymer 16, in which PTA=O acts as a μ_3-spacer (Scheme 7) [77].

However, a similar synthetic procedure with silver(I) sulfate as the metal source and in $MeCN/H_2O$ medium results in the generation of the distinct and more complex polymer 17, wherein the coordination network is driven by neutral μ_5-PTA=O spacers and μ_2-SO_4 linkers. These MOFs have been isolated as air- and light-stable colorless microcrystalline solids in moderate yields. Their molecular structures have been established by single-crystal X-ray crystallography, and supported by elemental analysis, and ESI-MS(±), IR and NMR spectroscopies.

The IR spectra of 16 and 17 exhibit related features with vibrations in the 2970–2870 and 1440–400 cm^{-1} ranges due to the PTA=O cores, showing in particular the characteristic intense ν(P=O) bands [1, 76] at 1165 (16) and 1144 (17) cm^{-1}. In addition, $\nu(NO_3)$ or $\nu(SO_4)$ bands are detected at 1384 and 1138 cm^{-1}, respectively [73].

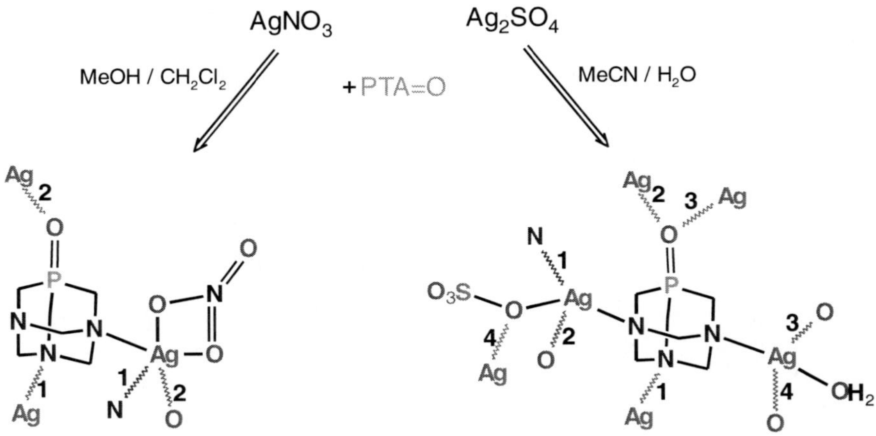

16

17

Scheme 7. Preparation of $[Ag(NO_3)(N_2,O\text{-PTA=O})]_n$ (16) and $[Ag_2(\mu_2\text{-}SO_4)(\mu_5\text{-}N_3,O\text{-PTA=O})(H_2O)]_n$ (17). Numbers 1, 2, 3 and 4 correspond to extensions of polymeric motifs.

The ESI-MS(+) and ESI-MS(−) spectra of aqueous solutions of both compounds show the $[Ag(PTA{=}O)_2]^+$, $[Ag(PTA{=}O)(H_2O)_2]^+$, $[Ag(PTA{=}O)]^+$, $[PTA{=}O + H]^+$ and characteristic $[Ag(NO_3)_2]^-$ or $[Ag(SO_4)]^-$ fragments with expected isotopic distribution patterns. The ^1H and $^{31}P\{H\}$ NMR spectra also expose the expected resonances at typical chemical shifts for the PTA=O moieties [63g, 77].

Reaction of $AgNO_3$ or Ag_2SO_4 with PTA=S, in $MeCN/H_2O$ medium at ~25 °C, gives 3D silver-organic frameworks, $[Ag(\mu_3\text{-PTA=S})]_n(NO_3)_n{\cdot}nH_2O$ (18) or $[Ag_4(\mu_4\text{-PTA=S})(\mu_5\text{-PTA=S})(\mu_2\text{-}SO_4)_2(H_2O)_2]_n{\cdot}2nH_2O$ (19) (Scheme 8) as colourless or pale brown air-stable crystalline solids, respectively [78]. Their molecular structures have been established by single crystal X-ray diffraction and supported by IR and NMR spectroscopy, ESI-MS(±) and elemental analysis. Compounds 18 and 19 reveal unprecedented N_2S-coordination modes of PTA=S and represented the first MOFs derived from this cagelike building block. The IR spectra of 18 and 19 exhibit a set of bands typical for the PTA=S cages, at 715−700 and 634−616 cm^{-1} due to ν(P=S) vibrations [76a-b, 78].

In addition, $\nu(NO_3)$ or $\nu(SO_4)$ bands are observed at 1378 cm^{-1} and 1111−1057 cm^{-1}, respectively [76a-b, 78] The ^{1}H NMR spectra of 18 and 19 reveal a set of the expected resonances for the methylene protons of the PTA=S ligands, while the $^{31}P\{^1H\}$ NMR spectra exhibit the corresponding singlets at δ −14.2 (18) or −13.7 (19).

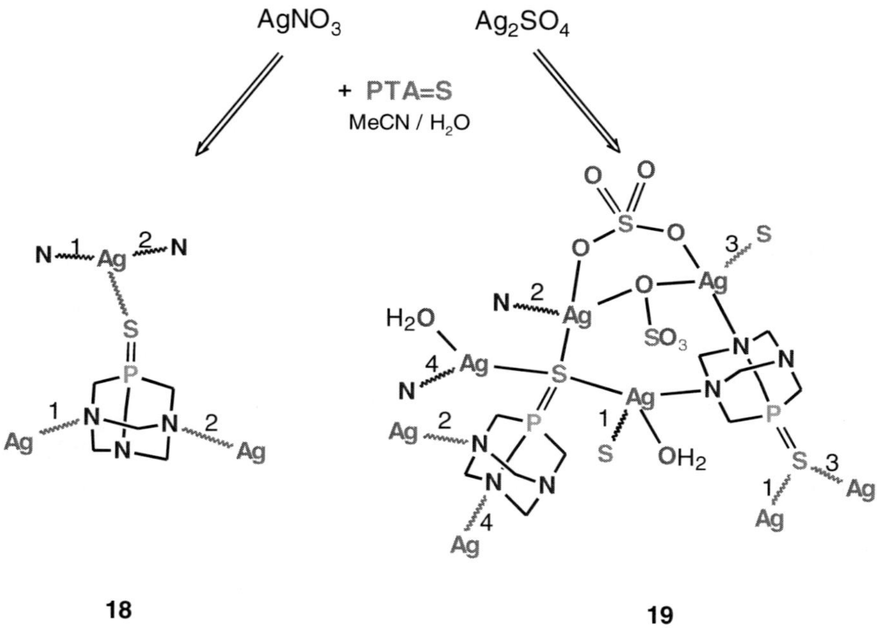

Scheme 8. Schematic representation of the self-assembly synthesis and structural formulae of 18 and 19 (crystallization H_2O molecules and nitrate anion are omitted for clarity). Numbers 1, 2, 3 and 4 correspond to extensions of polymeric motifs.

The ESI-MS(+) spectra of both compounds show intense $[Ag(PTA{=}S)_2]^+$ peaks with expected isotopic distribution pattern, while the characteristic fragments in the MS(−) mode concern the $[Ag(NO_3)_2]^-$ species in 18 and the $[Ag_4(PTA{=}S)_2(SO_4)_2(H_2O)_3 - H]^-$ and $[Ag(PTA{=}S)(SO_4)]^-$ moieties in 19.

The structure of 16 (Figure 3) is composed of the Ag1 atom, one μ_3-PTA=O moiety and one nitrate ligand.

The coordination environment of each four-coordinate Ag1 atom is occupied by three alternately N- and O-bound μ_3-PTA=O ligands and one weakly bound terminal NO_3^- ligand (Figure 3a), forming a significantly distorted $\{AgO_2N_2\}$ tetrahedral secondary building unit.

These units are bridged by μ_3-PTA=O spacers generating three kinds of the Ag1-PTA=O-Ag1-PTA=O helical chain motifs running along the *a*, *b* and *c* axis if one considers the connections of the adjacent Ag1 atoms via the N13/ O1, N11/ N13 and N11/O1 donor pairs of PTA=O, respectively.

Interestingly, if seen along the *a* axis (Figure 3b), this network resembles a "basket weave" topology.

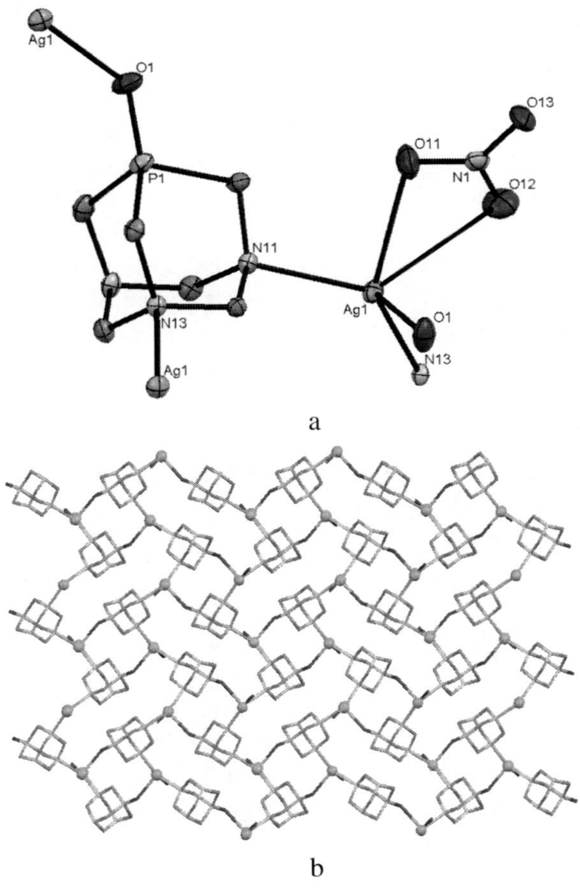

a

b

Figure 3. Structural fragments of $[Ag(NO_3)(N_2,O\text{-PTA=O})]_n$ (16) showing: (a) ellipsoid plot with partial labeling scheme, (b) helical 3D framework (hydrogen atoms and some counterions are omitted).

The crystal structure of 17 (Figure 4) features a more complicated (in comparision to 16) 3D non-interpenetrating network owing to the presence of the two symmetry non-equivalent Ag1 and Ag2 atoms, one μ_5-PTA=O and one μ_2-SO_4 moiety, and one terminal H_2O ligand (at the Ag2) per formula unit.

The Ag1 atom adopts a rather distorted $\{AgO_2N_2\}$ tetrahedral environment filled by the N11, N13 and O1 atoms from three adjacent μ_5-PTA=O spacers and the O11 atom of the bridging sulfate ligand (Figure 4a). The Ag2 atom also forms a significantly distorted $\{AgO_3N\}$ tetrahedral with the N12 and O1 donors coming from μ_5-PTA=O, and the O11 and O10 atoms from μ_2-SO_4 and H_2O ligands, respectively. In the construction of 17, a main role is thus played by the μ_5-PTA=O spacers that unprecedentedly act as pentadentate ligands. It is interesting to note that owing to the presence of the phosphine oxide moiety, the 3D network of 17 significantly differs from the 2D layers created in the reaction of silver(I) sulfate with hexamethylenetetramine as a spacer. The central part of $[Ag(\mu_3\text{-PTA=S})]_n(NO_3)_n \cdot nH_2O$ (18) is composed of the Ag1 atom and the μ_3-PTA=S ligand (Figure 5a). The three-coordinate Ag1 atoms display a distorted $\{N_2S\}$ trigonal pyramidal site, filled by the two nitrogen and S1 atoms from three adjacent μ_3-PTA=S spacers [78]. These adopt an alternating N,N- and S-coordination and form three different Ag1−PTA=S−Ag1−PTA=S helical chain motifs considering the binding of the adjacent Ag1 atoms through the S1/N12, N11/N12 and S1/N11 donor pairs of μ_3-PTA=S (Figure 5b). The resulting helical chains are multiply assembled via a third donor atom of μ_3-PTA=S, creating an infinite helical 3D silver-organic framework (Figure 5b-c).

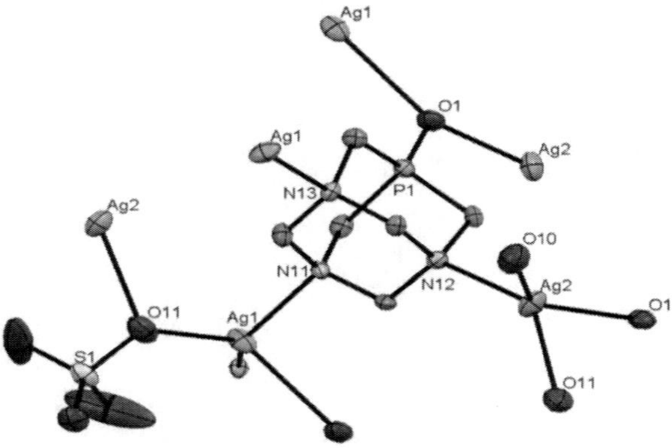

a

Figure 4. (Continued).

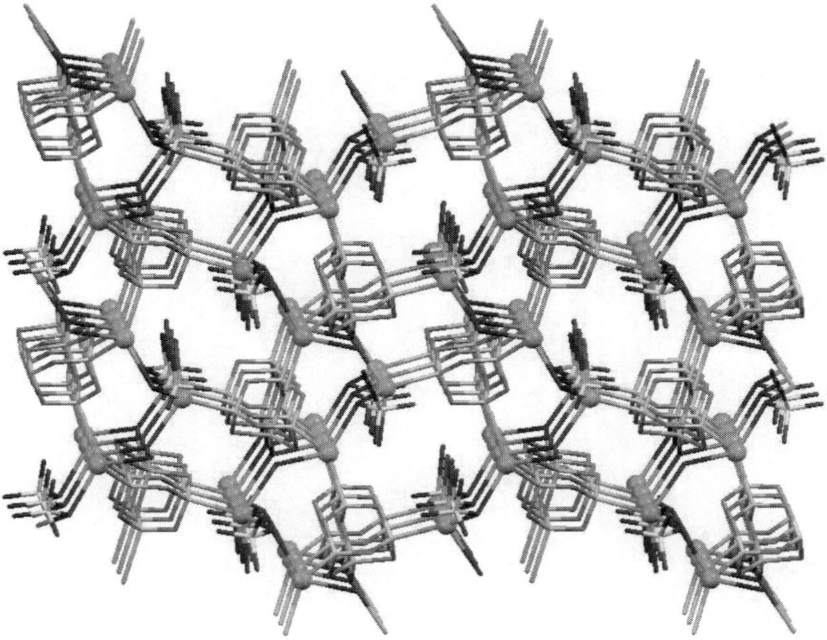

b

Figure 4. Structural fragments of 17 showing: (a) ellipsoid plot with partial labelling scheme, (b) complex 3D framework (hydrogen atoms are omitted).

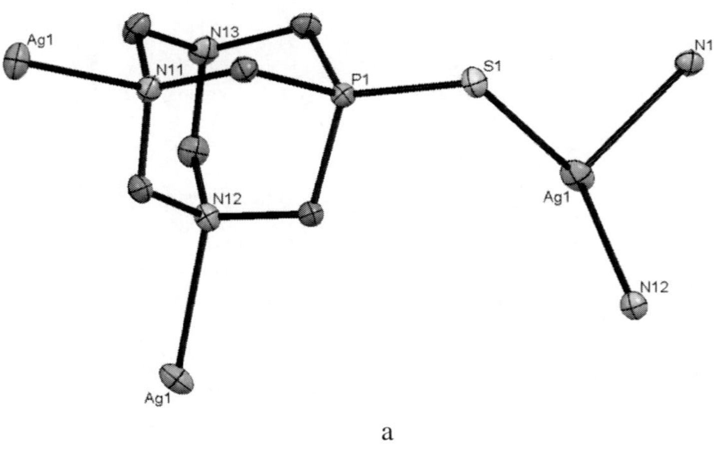

a

Figure 5. (Continued).

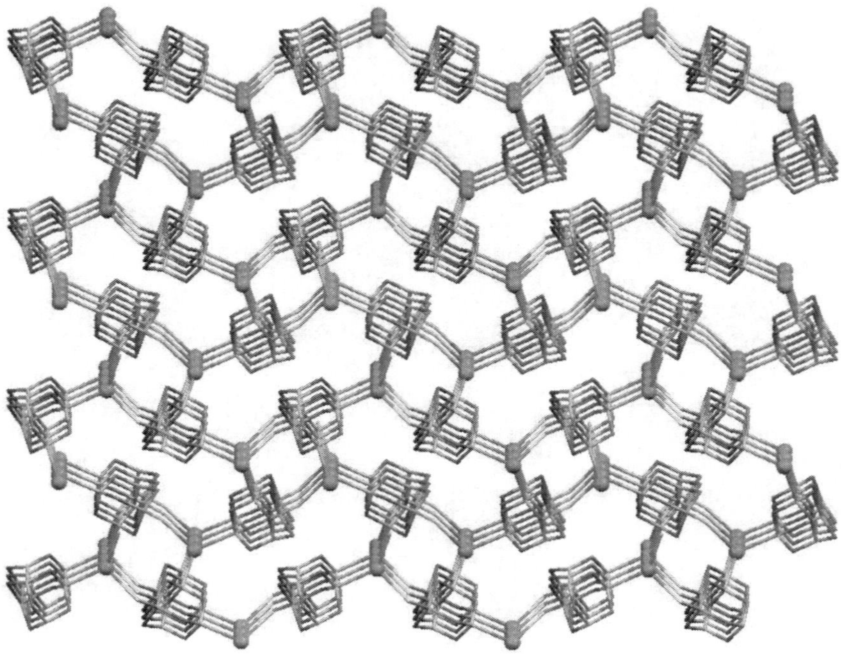

b

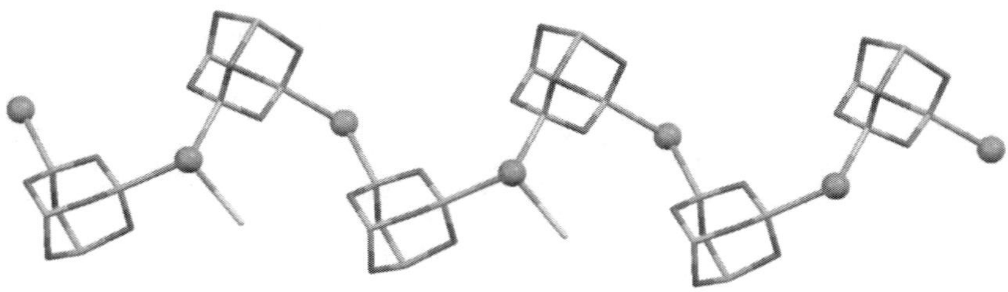

c

Figure 5. Structural fragments of 18 showing: (a) ellipsoid plot with partial atom labeling scheme, 3D metal-organic framework (b) composed of repeating helical 1D chain motifs (c). Further details: H atoms are omitted for clarity.

Structure 19 in contrast to the 18 shows a significantly more complex 3D silver-organic framework due to the presence of four symmetry non-equivalent silver atoms, two different μ_4- and μ_5-PTA=S spacers, two distinct μ_2-η^1- and μ_2-η^2-SO_4 ligands, as well as two terminal aqua ligands and two crystallization water molecules per asymmetric unit (Figure 6a).

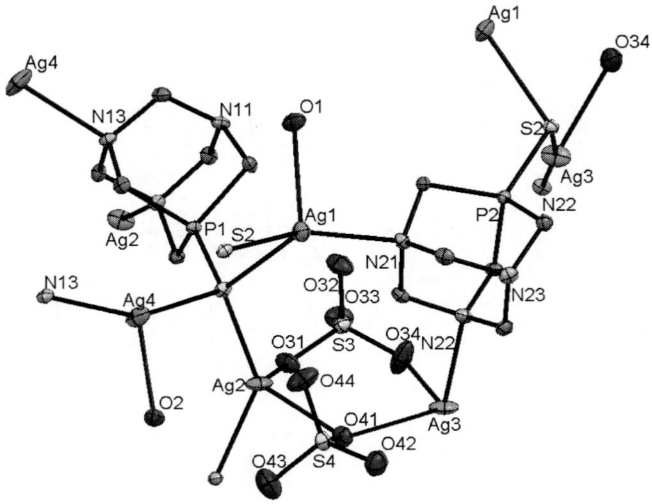

a

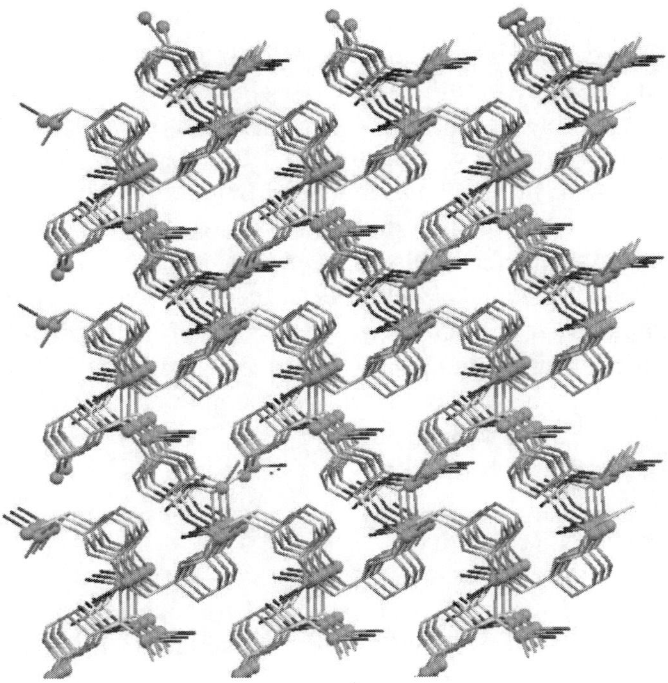

b

Figure 6. (Continued).

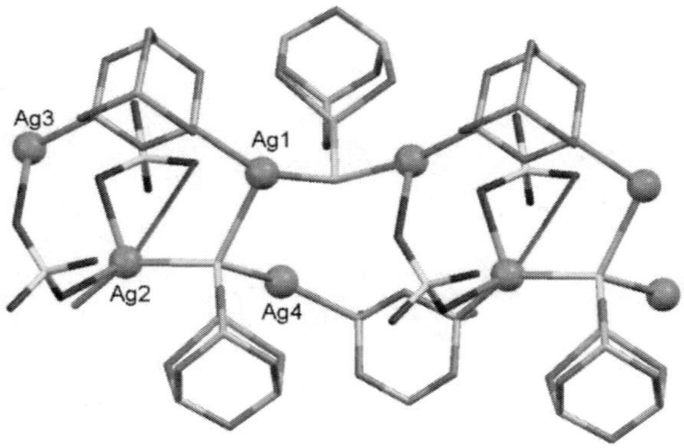

c

Figure 6. Structural fragments of 19 showing: (a) ellipsoid plot with partial atom labeling scheme, intricate 3D metal-organic framework (b) composed of repeating tubular 1D chain motifs (c). Further details: all H atoms and H_2O molecules are omitted for clarity.

The four-coordinate Ag1 atoms show $\{NOS_2\}$ environment, with a geometry intermediate between the trigonal pyramid and seesaw, filled by the N21, S1 and S2 atoms from two μ_4-PTA=S and one μ_5-PTA=S spacers, and the O1 atom of H_2O ligand. The Ag2 atoms are structurally similar but the Ag3 atoms have a seesaw geometry, both adopting $\{NO_2S\}$ environments occupied by a pair of the N and S atoms from the μ_5-PTA=S or μ_4-PTA=S spacers, and the O atoms from the μ_2-η^1-SO_4 and μ_2-η^2-SO_4 linkers. The tri-coordinate Ag4 atoms form a distorted {NOS} trigonal planar location taken by the N13 and S1 atoms of two adjacent μ_5-PTA=S moieties and the H_2O ligand. An unique feature of 19 is the presence of both μ_4- and μ_5-PTA=S spacers which, along with sulfate linkers, assemble different silver nodes into an intricate 3D framework composed of repeating tubular 1D chain motifs running along the *a* axis (Figure 6b) [78].

5.2. Antibacterial Activity

The obtained compounds were found to exhibit antibacterial and antifungal activity screened in vitro against *Escherichia coli* and *Pseudomonas aeruginosa* (Gram-negative bacteria), *Staphylococcus aureus* (a Gram-positive bacterium), as well as *Candida albicans* (a fungus). For these tests, the serial dilutions method with Antibiotic Broth [79] was employed, and the obtained results are given in Table 4 as the minimum inhibitory concentration (MIC, μg mL^{-1}) of compound that fully inhibits microbial growth. All compounds were most active against Gram-negative bacteria and the MIC values obtained for 16-18 (4−7 μg mL^{-1}) were substantially lower than those of 19.

Table 4. Antimicrobial activity MIC [µg mL^{-1}] of 3-D coordination polymers of silver with PTA=O and PTA=S (16-19) against a representative panel of pathogens [77, 78]

Strains	MIC [µg mL^{-1}]					
	16	17	18	19	PTA=O	PTA=S
Escherichia coli	6	6	4	20	>2200	>600
Pseudomonas aeruginosa	7	6	5	20	>1760	>600
Staphylococcus aureus	20	20	20	40	>1100	>600
Candida albicans	30	20	30	>60	>1100	>600

A similar trend was observed for *S. aureus* with the MIC values of 20 (for 16-18 and 40 μg mL^{-1} for 19, respectively. Concentrations of the compounds required to inhibit the growth of *C. albicans* were noticeably higher. For the compound 19, even the highest concentration examined (60 μg mL^{-1}) was not effective. The observed antibacterial and antifungal activity of 16-19 is presumably associated with the presence of bioactive silver nodes, since the efficiency of free PTA=O and PTA=S was very low (MIC > 1100 and 600 μg mL^{-1}, respectively). The antimicrobial activity exhibited by 18 is higher than that of the related silver MOFs derived from PTA=O [77] and is also noticeably stronger than the activity of some other Ag(I) coordination compounds [28, 80].

CONCLUSION

The presented report indicates silver, a metal of the 11 Group as a good candidate for the generation of coordination polymers, due to the affinity of Ag atoms to both hard and soft bases, as well as rich physicochemical properties. However, for practical application, it seems appropriate to use the ligands which can stabilize the coordination structures of silver compounds towards not only air stability and water-solubility, but also towards light-resistance. Herein it was demonstrated that PTA and PTA=Z (Z = O, S) ligands appear to be good tools for the syntheses of stable and useful discrete silver(I) complexes, as well as one-, two- and three-dimensional coordination polymers with silver.

Additionally, it was shown that the introduction of inorganic ligands/ counterions represents a strategy to tune and control the assembly of coordination polymers and their physical/chemical properties. Many compounds described herein exhibit good photoluminescence and, moreover, significant antibacterial and antifungal activities. Their advantage also consists in good solubility and stability in water solutions as well as resistance to light.

Derivatives 1 and 2, likely polynuclear in the solid state, dissociate in solution into mononuclear and dinuclear species. These silver(I) complexes display antibacterial activity which is compared with that of $AgNO_3$ used as the control. In particular, 2 and 3 also show a high activity toward *C. albicans*, which is higher than that of $AgNO_3$. Considering the structure of the two complexes we can observe that the presence of the {Ag(PTA)} fragment in solution seems to be important for the selective activity on *C. albicans*. These findings demonstrate a possible relationship between structure/activity and selectivity for silver(I) complexes, which can be further explored toward a more selective antibacterial activity.

The series of PTA-driven 1D coordination polymers $[Ag(N\text{-}N)(\mu\text{-}PTA)]_n(X)_n$ (4, 5, 7−11, 13, 14), as well as discrete silver(I) complexes $[Ag(N\text{-}N)(PTA)_2](X)$ (6, 12), have been generated by a facile self-assembly method from simple silver(I) salts, PTA and the corresponding N-N ligands (bpy, dtbpy, phen, neocup and dione). To our knowledge, 4−14 not only represent unique examples of polypyridine silver-PTA derivatives, but they also extend the still limited number of compounds wherein an unconventional *N,P*-coordination mode of PTA has been observed. All the products 1−14 represent rare examples of air and light stable Ag phosphine derivatives that also show solubility in water, a feature of particular significance for the development of new materials for biomedical applications.

In fact, some of our tris(pyrazolyl)methanesulfonate and polypyridine Ag-PTA coordination polymers and complexes exhibit significant antibacterial, antifungal and antiproliferative activities screened in vitro against different bacteria, fungi and human malignant melanoma cells, respectively. The interaction mode between DNA and a series of different silver compounds has been investigated by spectrophotometric methods, and the obtained results suggest that all derivatives having a strong binding affinity to DNA. The in vitro cytotoxic activity studies are consistent with the results of the DNA binding investigation. The antimicrobial activity observed for the most active compounds 8 and 13 was at least two times higher than that of $AgNO_3$ used as a control.

We have also shown that derivative of PTA: PTA=O and PTA=S can be used as a new versatile N,O,S-building block towards engineering the functional metal-organic frameworks (MOFs). Depending on the type of the starting silver(I) salt different distinct silver-organic 3D frameworks 16-19 driven by multiply bridging PTA=O and PTA=S spacers have been self-assembled. These compounds not only feature novel coordination modes of PTA=S cages and different net topologies, but also exhibit notable anti-bacterial and antifungal activity, thus widening the still limited family of silver bioMOFs. Besides, the present study extends an underexplored application of PTA and its derivatives as cage-like building blocks in crystal engineering and for the synthesis of new biologically active materials.

ACKNOWLEDGMENTS

This work was supported by NCN/KBN (Grant No. 2012/07/B/ST5/00885 and N204 280438). Special thanks are due to Aleksandra Zatajska (Faculty of Chemistry, University of Wrocław) for assistance in preparation of drawings.

REFERENCES

[1] (a) Phillips, A. D., Gonsalvi, L., Romerosa, A., Vizza, F., Peruzzini, M., *Coord. Chem. Rev*. 248 (2004) 955; (b) Bravo, J., Bolaño, J., Gonsalvi, S., Peruzzini, M., *Coord. Chem. Rev*. 254 (2009) 555.

[2] Joó, F., *Aqueous Organometallic Catalysis*, Kluwer Academic Publishers, Dordrecht, 2001.

[3] (a) Renfrew, A. K., Phillips, A. D., Tapavicza, E., Scopelliti, R., Rothlisberger, U., Dyson, P. J., *Organometallics* 28 (2009) 5061; (b) Miranda, S., Vergara, E., Mohr, F., de Vos, D., Cerrada, E., Mendía, A., Laguna, M., *Inorg. Chem*. 47 (2008) 5641; (c) Scolaro, C., Geldbach, T. J., Rochat, S., Dorcier, A., Gossens, C., Bergamo, A., Cocchietto, M., Tavernelli, I., Sava, G., Rothlisberger, U., Dyson, P. J., *Organometallics* 25 (2006) 756.

[4] Scolaro, C., Bergamo, A., Brescacin, L., Delfino, R., Cocchietto, M., Laurenczy, G., Geldbach, T. J., Sava, G., Dyson, P. J., *J. Med. Chem.* 12 (2005) 4161.
[5] Serli, B., Zangrando, E., Gianferrara, T., Scolaro, C., Dyson, P. J., Bergamo, A., Alessio, E., *Eur. J. Inorg. Chem.* (2005) 3423.
[6] Mohr, F., Sanz, S., Tiekink, E. R. T., Laguna, M., *Organometallics* 25 (2006) 3084.
[7] Mohr, F., Cerrada, E., Laguna, M., *Organometallics* 25 (2006) 644.
[8] (a) Pruchnik, F. P., Smoleński, P., Gałdecka, E., Gałdecki, Z., *New J. Chem.* (1998) 1395; (b) Pruchnik, F. P., Smoleński, P., *Appl. Organomet. Chem.*, 13 (1999) 829; (c) Pruchnik, F. P., Smoleński, P., Gałdecka, E., Gałdecki, Z., *Inorg. Chim. Acta*, 293 (1999) 110.
[9] Smoleński, P., Pruchnik, F. P., Ciunik, Z., Lis, T., *Inorg. Chem.* 42 (2003) 3318.
[10] Wanke, R., Smoleński, P., Guedes da Silva, M. F. C., Martins, L. M. D. R. S., Pombeiro, A. J. L., *Inorg. Chem.*, 47 (2008) 10158.
[11] Smoleński, P., Pombeiro, A. J. L., *Dalton Trans.* (2008) 87.
[12] Smoleński, P., *J. Organomet. Chem.* 696 (2011) 3867.
[13] Ahlsten, N., Lundberg, H., Martín-Matute, B., *Green Chem.*, 12 (2010) 1628.
[14] Cadierno, V., Díez, J., Francos, J., Gimeno, J., *Chem. Eur. J.* 16 (2010) 9808.
[15] Vannucci, A. K., Wang, S., Nichol, G. S., Lichtenberger, D. L., Evans, D. H., Glass, R. S., *Dalton Trans.*, 39 (2010) 3050.
[16] Vergara, E., Cerrada, E., Casini, A., Zava, O., Laguna, M., Dyson, P. J., *Organometallics*, 29 (2010) 2596.
[17] Renfrewa, A. K., Juillerat-Jeanneret, L., Dyson, P. J., *J. Organometal. Chem.*, 696 (2011) 772.
[18] Alidori, S., Lobbia, G. G., Papini, G., Pellei, M., Porchia, M., Refosco, F., Tisato, F., Lewis, J. S., Santini, C., *J. Biol. Inorg. Chem.* 13 (2008) 307.
[19] (a) Porchia, M., Benetollo, F., Refosco, F., Tisato, F., Marzano, C., Gandin, V., *J. Inorg. Biochem.* 103 (2009) 1644; (b) Santini, C., Pellei, M., Papini, G., Morresi, B., Galassi, R., Ricci, S., Tisato, F., Porchia, M., Rigobello, M. P., Gandind, V., Marzano, C., *J. Inorg. Biochem.* 105 (2011) 232.
[20] Orvig, C., Abrams, M. J., *Chem. Rev.* 99 (1999) 2201.
[21] Marambio-Jones, C., Hoek, E. M. V., *J. Nanopart. Res.* 12 (2010) 1531.

[22] (a) Klasen, H. J., *Burns* 26 (2000) 117; (b) Klasen, H. J., *Burns* 26 (2000) 131.
[23] Farrell, N., In: *Comprehensive Coordination Chemistry*, 2nd ed.; McCleverty, J. A., Meyer, T. J., Eds.; Elsevier Pergamon: Oxford, UK, 9 (2004) 809.
[24] Clement, J. L., Jarret, P. S., *Met. Based Drugs* 1 (1994) 467.
[25] Tambe, S. M., Sampath, L., Modak, S. M., *J. Antimicrob. Chemother.* 47 (2001) 589.
[26] Jakupec, M. A., Unfried, P., Keppler, B. K., *Rev. Physiol. Biochem. Pharmacol.* 153, (2005) 101.
[27] Nomiya, K., Takahashi, S., Noguchi, R., Nemoto, S., Takayama, T., Oda, M., *Inorg. Chem.* 39 (2000) 3301.
[28] Kasuga, N. C., Sugie, A., Nomiya, K., *Dalton Trans.* (2004) 3732.
[29] Chung, J. Y., Herbert, M. E., *West. J. Med.* 175 (2001) 205.
[30] Fuller, F. W., Parrish, M., Nance, F. C., *J. Burn Care Rehabil.* 15 (1994) 213.
[31] Fox, C. L. Jr., *Surg. Gynecol. Obstet.* 157 (1983) 82.
[32] Nomiya, K., Tsuda, K., Sudoh, T., Oda, M., *J. Inorg. Biochem.* 68 (1997) 39.
[33] (a) Nomiya, K., Tsuda, K., Kasuga, N. C., *J. Chem. Soc., Dalton Trans.* (1998) 1653; (b) Nomiya, K., Noguchi, R., Oda, T., *Inorg. Chim. Acta* 298 (2000) 24; (c) Nomiya, K., Tsuda, K., Tanabe, Y., Nagano, H., *J. Inorg. Biochem.* 69 (1998) 9.
[34] Bult, A., Plug, C. M., *Anal. Profiles Drug Substances* 13 (1984) 553.
[35] Nangia, A. K., Hung, C. T., Lim, J. K. C., *Drugs Today* (1987) 21.
[36] Deshpande, L. M., Chopade, B. A., *Biometals* 7 (1994) 49.
[37] Kawai, K., Suzuki, S., Tabata, Y., Taira, T., Ikada, Y., Nishimura, Y., *J. Biomed. Mater. Res.* 57 (2001) 346.
[38] (a) *Encyclopedia of Supramolecular Chemistry*, Eds.: Atwood, J. L., Steed, J. W., Taylor and Francis (2004); (b) Batten, S. R., Turner, D. R., Neville, S. M., *Coordination Polymers: Design, Analysis and Application*, Royal Society of Chemistry (2009); (c) *Design and Construction of Coordination Polymers*, Eds.: Hong, M.-C., Chen, L., Willey (2009); (d) *Metal-Coordination Polymers*, Eds.: Abd-El-Aziz, A. S., Carraher, Jr., C. E., Pittman, C. U., Jr., Zeldin, M., Willey (2005).
[39] (a) Zheng, S. L., Nygren, C. L., Messerschmidt, M., Coppens, P., *Chem. Commun.* (2006) 3711; (b) Jansen, M., *Angew. Chem., Int. Ed.* 26 (1987) 1098; (c) Kitagawa, S., Kitaura, R., Noro, S., *Angew. Chem., Int. Ed.* 43 (2004) 2334; (d) Moulton, B., Zaworotko, M. J., *Chem. Rev.* 101 (2001)

1629; (e) Rao, C. N. R., Natarajan, S., Vaidhyanathan, R., *Angew. Chem., Int. Ed.* 43 (2004) 1466.
[40] Hagrman, P. J., Hagrman, D., Zubieta, J., *Angew. Chem., Int. Ed.* 38 (1999) 2638.
[41] Feller, R. K., Forster, P. M., Wudl, F., Cheetham, A. K., *Inorg. Chem.* 46 (2007) 8717.
[42] Yue, Y.-F., Wang, B.-W., Gao, E.-Q., Fang, C.-J., He, C., Yan, C.-H., *Chem. Commun.* (2007) 2034.
[43] Zhai, Q. G., Lu, C. Z., Zhang, Q. Z., Wu, X. Y., Xu, X. J., Chen, S. M., Chen, L. J., *Inorg. Chim. Acta* 359 (2006) 3875.
[44] Chen, L. J., He, X., Xia, C. K., Zhang, Q. Z., Chen, J. T., Yang, W. B., Lu, C. Z., *Cryst. Growth Des.* 6 (2006) 2076.
[45] (a) Khlobystov, A. N., Blake, A. J., Champness, N. R., Lemenovskii, D. A., Majouga, A. G., Zyk, N. V., Schröder, M., *Coord. Chem. Rev.* 222 (2001) 155; (b) Seward, C., Jia, W.-L., Wang, R.-Y., Enright, G. D., Wang, S., *Angew. Chem., Int. Ed.* 43 (2004) 2933.
[46] (a) Sagatys, D. S., Dahlgren, C., Smith, G., Bott, R. C., White, J. M. J., *Chem. Soc., Dalton Trans.* (2000) 3404; (b) Sun, D. F., Cao, R., Weng, J. B., Hong, M. C., Liang, Y. C., J. *Chem. Soc., Dalton Trans.* (2002) 291; (c) Lin, P., Henderson, R. A., Harrington, R. W., Clegg, W., Wu, C.-D., Wu, X.-T., *Inorg. Chem.* 43 (2004) 181; (d) Rao, C. N. R., Ranganathan, A., Pedireddi, V. R., Raju, A. R., *Chem. Commun.* (2000) 39.
[47] Mäkinen, S. K., Melcer, N. J., Parvez, M., Shimizu, G. K. H., *Chem. Eur. J.* 7 (2001) 5176.
[49] May, L. J., Shimizu, G. K. H., *Chem. Mater.* 17 (2005) 217.
[50] (a) Prior, T. J., Rosseinsky, M. J., *Chem. Commun.* (2001) 495; (b) Du, M., Bu, X. H., Guo, Y. M., Liu, H., Batten, S. R., Ribas, J., Mak, T. C. W., *Inorg. Chem.* 41 (2002) 4904; (c) Zhang, X. M., Fang, R. Q., Wu, H. S., *J. Am. Chem. Soc.* 127 (2005) 7670.
[51] (a) MacGillivray, L. R., Groeneman, R. H., Atwood, J. L., *J. Am. Chem. Soc.* 120 (1998) 2676; (b) Fujita, M., Kwon, Y. J., Sasaki, O., Yamaguchi, K., Ogura, K., *J. Am. Chem. Soc.* 117 (1995) 7287.
[52] (a) Groeneman, R. H., MacGillivray, L. R., Atwood, J. L., *Chem. Commun.* (1998) 2735; (b) Guilera, G., Steed, J. W., *Chem. Commun.* (1999) 1563; (c) Groeneman, R. H., MacGillivray, L. R., Atwood, J. L., *Inorg. Chem.* 38 (1999) 208; (d) Li, H. L., Davis, C. E., Groy, T. L., Kelley, D. G., Yaghi, O. M., *J. Am. Chem. Soc.* 120 (1998) 2186; (e)

Vodak, D. T., Braun, M. E., Kim, J., Eddaoudi, M., Yaghi, O. M., *Chem. Commun*. (2001) 2534.
[53] Furukawa, H., Miller, M. A., Yaghi, O. M., *J. Mater. Chem*. 17 (2007) 3197.
[54] Awaleh, M. O., Badia, A., Brisse, F., *Cryst. Growth Des*. 6 (2006) 2674.
[55] Li, J. R., Bu, X. H., Jiao, J., Du, W. P., Xu, X. H., Zhang, R. H., *Dalton Trans*. (2005) 464.
[56] Li, J. R., Zhang, R. H., Bu, X. H., *Cryst. Growth Des*. 3 (2003) 829.
[57] Zou, R.-Q., Liu, C.-S., Shi, X.-S., Bu, X.-H., *J. Cryst. Eng. Comm*. 7 (2005) 722.
[58] Kläui, W., Schramm, D., Peters, W., Rheinwald, G., Lang, H., *Eur. J. Inorg. Chem*. (2001) 1415.
[59] Kläui, W., Berghahn, M., Rheinwald, G., Lang, H., *Angew. Chem*. 112 (2000) 2590; *Angew. Chem., Int*. Ed. 39 (2000) 2464.
[60] Trofimenko, S., *Chem. Rev*. 93 (1993) 943.
[61] (a) Herrick, R. S., Brunker, T. J., Maus, C., Crandall, K., Cetin, A., Ziegler, C. J., *Chemm. Commun*. (2006) 4330; (b) Alegria, E. C. B., Martins, L. M. D. R. S., Haukka, M., Pombeiro, A. J. L., *Dalton Trans*. (2006) 4954; (c) Santini, C., Pellei, M., Lobbia, G. G., Cingolani, A., Spagna, R., Camalli, M., *Inorg. Chem. Comm*. 5 (2002) 430.
[62] Kläui, W., Berghahn, M., Frank, W., Reiβ, G. J., Schönherr, T., Rheinwald, G., Lang, H., *Eur. J. Inorg. Chem*. (2003) 2059.
[63] (a) Pettinari, C., Marchetti, F., Lupidi, G., Quassinti, L., Bramucci, M., Petrelli, D., Vitali, L. A., Guedes da Silva, M. F. C., Martins, L. M. D. R. S., Smoleński, P., Pombeiro, A. J. L., *Inorg. Chem*., 50 (2011) 11173; (b) Smoleński, P., Jaros, S. W., Pettinari, C., Lupidi, G., Quassinti, L., Bramucci, M., Vitali, L. A., Petrelli, D., Kochel, A., Kirillov, A. M., *Dalton Trans*., 42 (2013) 6572; (c) Dinoi, C., Guedes da Silva, M. F. C., Alegria, E. C. B. A., Smoleński, P., Martins, L. M. D. R. S., Poli, R., Pombeiro, A. J. L., *Eur. J. Inorg. Chem*. (2010) 2415; (d) (f) Kirillov, A. M., Smoleński, P., Guedes da Silva, M. F. C., Pombeiro, A. J. L., *Acta Cryst*., (2008) E64, o496-o497; (e) Smoleński, P., Kirillov, A. M., Guedes da Silva, M. F. C., Pombeiro, A. J. L., *Acta Cryst*. (2008) E64, o556; (f) Kirillov, A. M., Smoleński, P., Guedes da Silva, M. F. C., Pombeiro, A. J. L., *Acta Cryst*. (2008) E64, m603-m604; (g) Jaremko, L., Kirillov, A. M., Smoleński, P., Lis, T., Pombeiro, A. J. L., *Inorg. Chem*., 47 (2008) 2922; (h) Smoleński, P., Dinoi, C., Guedes da Silva, M. F. C., Pombeiro, A. J. L., *J. Organomet. Chem*., 693 (2008) 2338; (i) Lis, A., Guedes da Silva, M. F. C., Kirillov, A. M., Smoleński, P.,

Pombeiro, A. J. L., *Cryst. Growth Des.*, 10 (2010) 5244; (j) Lis, A., Guedes da Silva, M. F. C., Kirillov, A. M., Smoleński, P., Pombeiro, A. J. L., *Cryst. Growth Des.*, 10 (2010) 5245; (k) Kirillov, A. M., Wieczorek, S., Guedes da Silva, M. F. C., Sokolnicki, J., Smoleński, P., Pombeiro, A. J. L., *Cryst. Eng. Comm.*, 13 (2011) 6329.

[64] See the Cambridge Structural Database (CSD, version 5.32, Aug. 2011): F. H. Allen, *Acta Crystallogr.* B58 (2002) 380.

[65] (a) Kaes, C., Katz, A., Hosseini, M. W., *Chem. Rev.* 100 (2000) 3553. (b) Ye, B. H., Tong, M. L., Chen, X. M., *Coord. Chem. Rev.* 249 (2005) 545. (c) Accorsi, G., Listorti, A., Yoosaf, K., Armaroli, N., *Chem. Soc. Rev.* 38 (2009) 1690.

[66] Gielen M., Tiekink, E. R. T., *Metallotherapeutic Drugs and Metal-based Diagnostic Agents: the Use of Metals in Medicine*, John Wiley and Sons (2005).

[67] For selected examples, see: (a) Kumar, R. S., Arunachalam, S., *Eur. J. Med. Chem.* 44 (2009) 1878; (b) Kumar, R. S., Arunachalam, S., Periasamy, V. S., Preethy, C. P., Riyasdeen, A., Akbarsha, M. A., *J. Inorg. Biochem.* 103 (2009) 117; (c) Tarushi, A., Psomas, G., Raptopoulou, C. P., Kessissoglou, D. P., *J. Inorg. Biochem.* 103 (2009) 898; (e) Śliwińska, U., Pruchnik, F. P., Pelińska, I., Ułaszewski, S., Wilczok, A., Zajdel, A., *J. Inorg. Biochem.* 102 (2008) 1947.

[68] (a) Rowan, R., Moran, C., McCann, M., Kavanagh, K., *Biometals* 22 (2009) 461; (b) Yilmaz, F., Yilmaz, V. T., Karakaya, H., Bueyuekguengoer, O., *Z. Naturforsch. B.* 63 (2008) 134; (c) Mostafa, S. I., Badria, F. A., *Metal-Based Drugs* (2008) 723634.

[69] (a) Hung-Low, F., Klausmeyer, K. K., Gary, J. B., *Inorg. Chim. Acta* 362 (2009) 426; (b) Balakrishna, M. S., Venkateswaran, R., Mobin, S. M., *Inorg. Chim. Acta* 362 (2009) 271; (c) Wang, X.-J., Gui, L.-C., Ni, Q.-L., Liao, Y.-F., Jiang, X.-F., Tang, L.-H., Zhang, Z., Wu, Q., *Cryst. Eng. Comm.* 10 (2008) 1003; (d) Effendy, Marchetti, F., Pettinari, C., Skelton, B. W., White, A. H., *Inorg. Chim. Acta* 360 (2007) 1424; (e) Di Nicola, C., Effendy, Marchetti, F., Pettinari, C., Skelton, B. W., White, A. H., *Inorg. Chim. Acta* 360 (2007) 1433. (f) Effendy, J. V., Hanna, C., Pettinari, R., Pettinari, C., Santini, Skelton, B. W., White, A. H., *Inorg. Chem. Commun.* 10 (2007) 571.

[70] Chenskaya, T. B., Berghahn, M., Kunz, P. C., Frank, W., Kläui, W., *J. Mol. Struct.* 829 (2007) 135.

[71] Rosenthal, M. R. J., *Chem. Educ.* 50 (1973) 331.

[72] Siegert, U., Hahn, H., Lang, H., *Inorg. Chim. Acta* 363 (2010) 944.

[73] Nakamoto, K., *Infrared and Raman Spectra of Inorganic and Coordination Compounds*, 5th ed.; Wiley: New York (1997).

[74] Hadadzadeh, H., Olmstead, M. M., Rezvani, A. R., Safari, N., Saravani, H., *Inorg. Chim. Acta* 359 (2006) 2154.

[75] (a) Wein, A. N., Stockhause, A. T., Hardcastle, K. I., Saadein, M. R., Peng, S. B., Wang, D., Shin, D. M., Chen, Z. G., Eichler, J. F., *J. Inorg. Biochem.* 105 (2011) 663; (b) Deegan, C., Coyle, B., McCann, M., Devereux, M., Egan, D. A., *Chem. Biol. Interact.* 164 (2006) 115; (c) Messori, L., Abbate, F., Marcon, G., Orioli, P., Fontani, M., Mini, E., Mazzei, T., Carotti, S., O'Connell, T., Zanello, P., *J. Med. Chem.* 43 (2000) 3541; (d) Sigman, D. S., Landgraf, R., Perrin, D. M., Pearson, L., *Met. Ions Biol. Syst.* 33 (1996) 485; (e) Torti, S. V., Torti, F. M., Whitman, S. P., Brechbiel, M. W., Park, G., Planalp, R. P., *Blood* 92 (1998) 1384.

[76] (a) Smoleński, P., Kochel, A., *Polyhedron*, 29 (2010) 1561; (b) Frost, B. J., Lee, W.-C., Pal, K., Kim, T. H., VanDerveer, D., Rabinovich, D., *Polyhedron*, 29 (2010) 2373.

[77] Kirillov, A. M., Wieczorek, S., Lis, A., Guedes da Silva, M. F. C., Florek, M., Król, J., Staroniewicz, Z., Smoleński, P., Pombeiro, A. J. L., *Cryst. Growth Des.*, 11 (2011) 2711.

[78] Jaros, S. W., Smoleński, P., Guedes da Silva, M. F. C., Florek, M., Król, J., Staroniewicz, Z., Pombeiro, A. J. L., Kirillov, A. M., *Cryst. Eng. Comm.*, 15 (2013) 8060.

[79] Grove, D. C., Randall, W. A., *Assay Methods of Antibiotic. A Laboratory Manual*; Medical Encyclopedia, New York (1955).

[80] (a) Berchel, M., Gall, T. L., Denis, C., Hir, S. L., Quentel, F., Elleouet, C., Montier, T., Rueff, J.-M., Salaun, J.-Y., Haelters, J.-P., Hix, G. B., Lehn, P., Jaffres, P.-A., *New J. Chem.*, 35 (2011) 1000; (b) Liu, Y., Xu, X., Xia, Q., Yuan, G., He, Q., Cui, Y., *Chem. Commun.*, 46 (2010) 2608; (c) Onodera, K., Kasuga, N. C., Takashima, T., Hara, A., Amano, A., Murakami, H., Nomiya, K., *Dalton Trans.* (2007) 3646.

In: Ligands
Editors: Paula Gawryszewska et al.
ISBN: 978-1-63117-143-7

Chapter 6

ARYLHYDRAZONES OF METHYLENE ACTIVE NITRILES AS PROMISING LIGANDS AND STARTING MATERIALS FOR ORGANIC SYNTHESIS

Kamran T. Mahmudov[1,2,*]***, Maximilian N. Kopylovich***[1,†]
and Armando J. L. Pombeiro[1,‡]
[1] Centro de Química Estrutural, Complexo I, Instituto Superior Técnico, Universidade de Lisboa, Lisbon, Portugal
[2] Department of Chemistry, Baku State University, Xalilov Baku, Azerbaijan

ABSTRACT

This chapter summarizes existing information on the synthesis of arylhydrazones of methylene active nitriles (AMANs) and their use as starting materials for the generation of new organic and coordination compounds. Reactions of AMANs with different nucleophiles lead to triazoles, aminopyrazoles, arylazoaminopyrazoles, dicyanopyridazines and other compounds of pharmaceutical significance, depending on the reaction conditions and reagents used. A specific hydrogen bond assisted

* Email: kamran_chem@yahoo.com, kamran_chem@mail.ru.
† Email: kopylovich@yahoo.com.
‡ Email: pombeiro@ist.utl.pt.

regioselective activation of a cyano group in a AMAN affords a variety of amidines, carboxamides and iminoesters. Synthesis and structural details of several mono-, tetra- and polynuclear metal complexes of Na^{I}, Ag^{I}, Ni^{II}, Cu^{II} and $Mn^{II,III}$ with AMAN ligands are also described. The H-bond and coordination induced *E/Z*-isomerisation of AMANs can serve for the creation of new molecular switches. Antiferromagnetic interactions are observed in $Mn^{II,III}$-AMAN complexes, while Cu^{II}-AMAN complexes are shown to be effective catalyst precursors for the selective oxidation of alcohols to the corresponding carbonyl compounds, and for the diastereoselective nitroaldol (Henry) reaction. Thus, AMANs and their complexes and derivatives can be applied in different fields of modern chemistry, e.g., as pharmaceuticals, molecular magnets, switches and catalyst precursors.

Keywords: Arylhydrazones of methylene active nitriles, template condensation, coordination compounds, *E*/*Z* isomerization, H-bond assisted reactions

1. INTRODUCTION

Aryldiazonium salts have been extensively used for years [1,2], and are of interest for industry, organic and medicinal chemistries [1−3]. They also can be used to functionalize single wall nanotubes [4] or in the synthesis of binding polymers, biomacromolecules and nanoparticles [5]. Furthermore, the reaction of diazonium salts with aromatic molecules ($ArNH_2$, ArOH, etc.) and methylene active compounds yields the respective azo derivatives and arylhydrazones of methylene active compounds, which are of tremendous importance as dyes [1,2] and ligands in coordination chemistry [6]. Apart from its synthetic role, the azo-coupling is an important reaction, e.g., in analytical chemistry for the determination of some anions [1−3].

On the other hand, methylene active nitriles are versatile compounds of rich reactivity [7,8]. They are used extensively as reactants or reaction intermediates, since the methylene group and either one or both cyano groups can take part in different types of reactions to give a variety of addition products and heterocyclic compounds. This versatile reactivity makes them important chemicals in medical, industrial, and coordination chemistries.

Coupling of aryldiazonium salts with methylene active nitriles (for example, malononitrile, benzoylacetonitriles, 2-cyano thioacetamide, ethyl cyanoacetate, etc., see below) leads to multifunctionalized arylhydrazones of methylene active nitriles (AMANs) [6–32]. AMANs, along with their synthetic potential, are convenient model compounds to study the enol-azo ⇌ hydrazone tautomeric equilibrium and *E/Z*-hydrazone isomerisation (Scheme 1) [6,15].

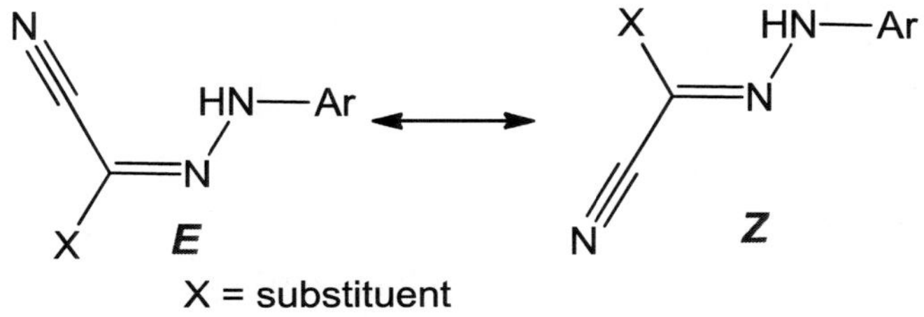

Scheme 1. *E/Z*-isomerisation in AMANs.

These and other related properties can be used to design new compounds with required biological, catalytic or spectrochemical properties. Thus, some AMANs were shown to possess antimicrobial activity, e.g., against *Staphylococcus epidermidis* or *Pseudomonas aeruginosa* [11]. Antitumor evaluation of another AMAN compound against human tumor cells lines, namely breast adenocarcinoma, non-small cell lung cancer and CNS cancer, showed that it exhibits higher inhibitory effects than the reference doxorubicin [32]. Moreover, other publications demonstrate the extensive synthetic potential of AMANs for the preparation of diverse compounds of biological or catalytic activity, or with promising magnetic properties.

This chapter is meant to consider systematically the published data on the synthesis, structure, and reactions of AMANs and to analyze the possible routes for their use in organic and inorganic synthesis.

2. Synthesis of AMANs

Until now, several methods for the synthesis of AMANs are reported [6–32]. One of the most common and widely used methods is the Japp-Klingemann reaction, in which methylene active nitriles react with diazonium salts in basic medium (Scheme 2) [6–18,20–32]. This method has been utilized for the synthesis of a variety of multifunctionalized AMANs (Table).

$$ArNH_2 + NaNO_2 + 2HCl \xrightarrow{0\text{-}5\,^{o}C} [ArNN]^{+}Cl^{-} + NaCl + 2H_2O \quad (1)$$

$$(CN)(X)CH_2 + base \longrightarrow (CN)(X)CH^{-}(Hbase)^{+} \quad (2)$$

$$[ArNN]^{+}Cl^{-} + (CN)(X)CH^{-}(Hbase)^{+} \xrightarrow{0\text{-}5\,^{o}C} ArNHN{=}C(CN)(X) + (Hbase)Cl \quad (3)$$

$$X = \text{substituent;} \quad base = CH_3COO^{-}, OH^{-}$$

Scheme 2. Japp-Klingemann synthesis of AMANs.

Table. Examples of known AMANs

Compound	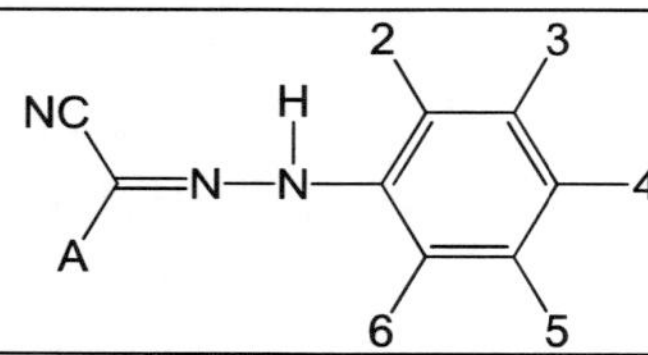					
	A	2	3	4	5	6	Ref.
1	CN	H	H	H	H	H	[12,13]
2	CN	COOH	H	H	H	H	[14,15]
3	CN	SO_3H	H	H	H	H	[15,16]
4	CN	H	H	CH_3	H	H	[13,17]
5	CN	H	H	NO_2	H	H	[18]
6	CN	$NHSO_2C_4F_9$	H	H	H	H	[19]
7	CN	H	H	F	H	H	[13]
8	CN	H	H	Cl	H	H	[13]
9	CN	H	H	Br	H	H	[13]
10	CN	H	H	I	H	H	[13]
11	CN	H	H	CH_2NH_2	H	H	[13]
12	CN	H	H	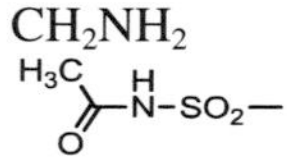	H	H	[13]
13	CN	H	H	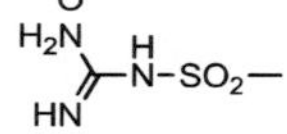	H	H	[13]
14	CN	H	H	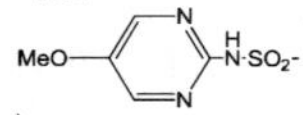	H	H	[13]
15	CN	H	H	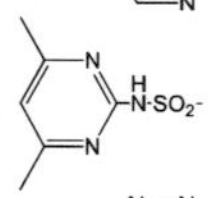	H	H	[13]
16	CN	H	H	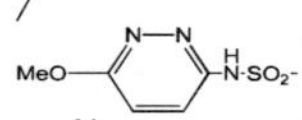	H	H	[13]
17	CN	H	H	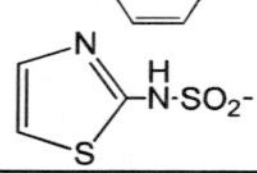	H	H	[13]

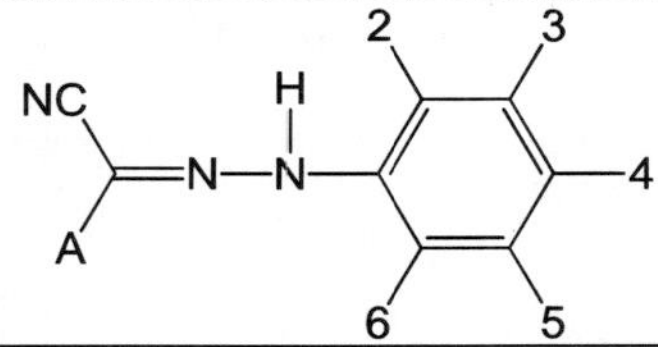

Compound	A	2	3	4	5	6	Ref.
18	CN	H	H	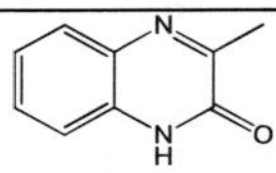	H	H	[13]
19	CN	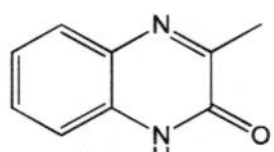	H	H	H	H	[13]
20	CN	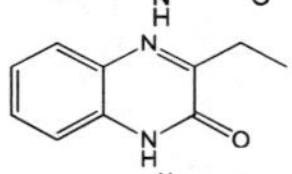	H	H	H	H	[13]
21	CN	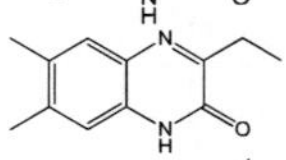	H	H	H	H	[13]
22	CN	Ar = 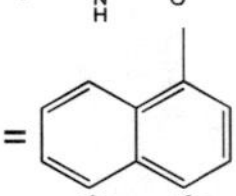					[13]
23	CN	Ar = 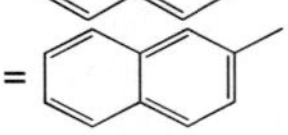					[13]
24	CN	Ar = 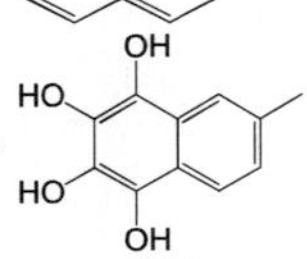					[13]
25	CN	Ar = 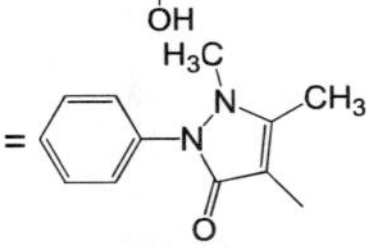					[13]
26	CN	Ar = 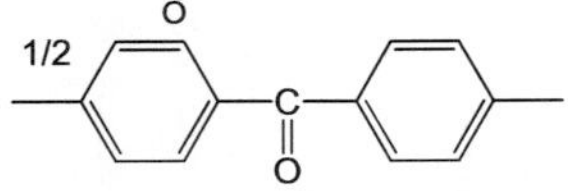					[13]
27	CN	Ar = 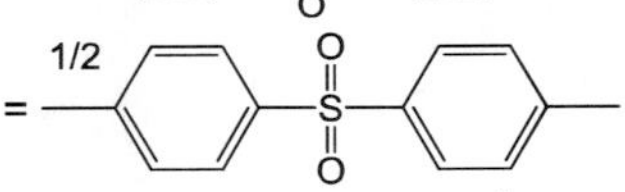					[13]
28	CN	Ar = 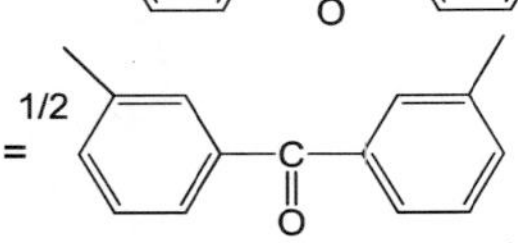					[13]
29	CN	Ar = 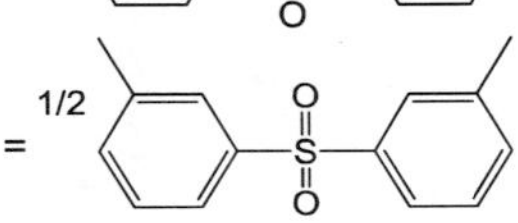					[13]
30	C(=O)CH_3	CH_3	H	H	H	H	[20]
31	C(=O)CH_3	Br	H	H	H	H	[20]
32	C(=O)CH_3	NO_2	H	H	H	H	[21]
33	C(=O)CH_3	OCH_3	H	H	H	H	[21]

Table. (Continued)

Compound	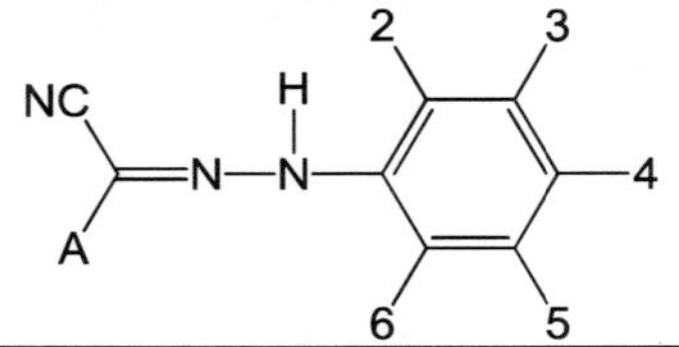						
	A	2	3	4	5	6	Ref.
34	$C(=O)CH_3$	$COCH_3$	H	H	H	H	[21]
35	$C(=O)CH_3$	H	H	CN	H	H	[21]
36	$C(=O)CH_3$	H	H	CH_3	H	H	[21]
37	$C(=O)CH_3$	H	H	$COCH_3$	H	H	[21]
38	$C(=O)CH_3$	H	H	H	H	H	[22]
39	$C(=O)C_6H_5$	H	H	H	H	H	[22]
40	$C(=O)C_6H_5NO_2(-p)$	H	H	H	H	H	[22]
41	$C(=O)C_2H_5$	H	H	H	H	H	[23]
42	$C(=O)C_2H_5$	Cl	H	H	H	H	[24]
43	$C(=O)C_2H_5$	H	H	Cl	H	H	[24]
44	$C(=O)C_2H_5$	H	H	CH_3	H	H	[24]
45	CHO	H	H	H	H	H	[24]
46	C_6H_5NN	H	H	H	H	H	[24]
47	$C_6H_5NH_2(-o)$	H	H	H	H	H	[25]
48	$C(=NH)CH_3$	H	H	H	H	H	[26]
49	$C(=NH)CH_3$	Cl	H	H	H	H	[26]
50	$C(=NH)CH_3$	CH_3	H	H	H	H	[26]
51	$C(=NH)CH_3$	OCH_3	H	H	H	H	[26]
52	$C(=NH)CH_3$	H	Cl	H	H	H	[26]
53	$C(=NH)CH_3$	H	CH_3	H	H	H	[26]
54	$C(=NH)CH_3$	H	OCH_3	H	H	H	[26]
55	$C(=NH)CH_3$	H	H	Cl	H	H	[26]
56	$C(=NH)CH_3$	H	H	CH_3	H	H	[26]
57	$C(=NH)CH_3$	H	H	OCH_3	H	H	[26]
58	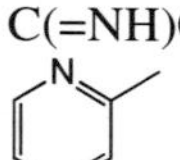	H	H	Cl	H	H	[27]
59	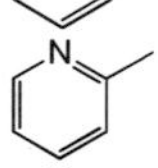	H	H	NO_2	H	H	[27]
60	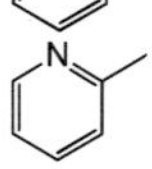	H	H	OCH_3	H	H	[27]
61	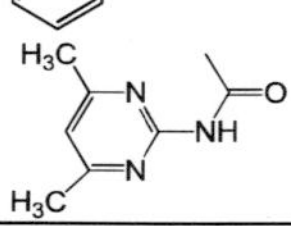	H	H	H	H	H	[28]

Compound	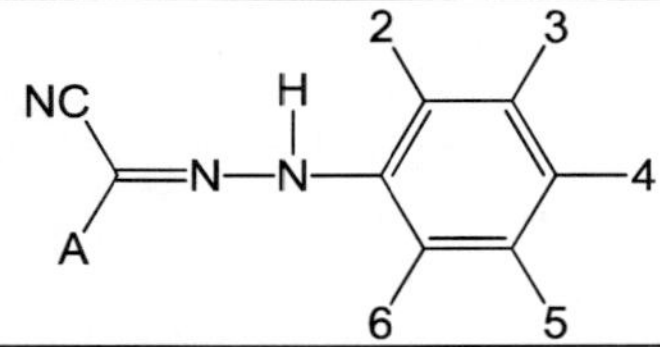						
	A	2	3	4	5	6	Ref.
62	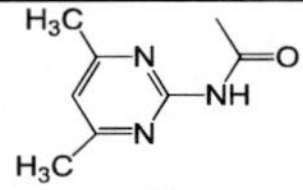	H	H	Cl	H	H	[28]
63	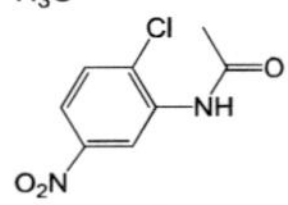	H	H	H	H	H	[28]
64	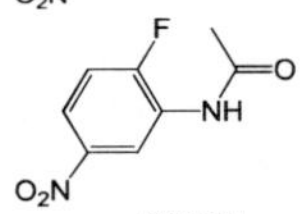	H	H	H	H	H	[28]
65	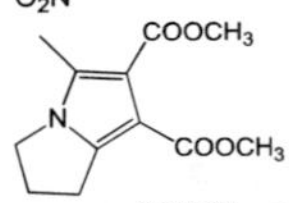	H	H	H	H	H	[29]
66	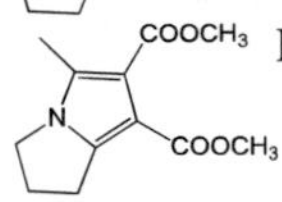	H	H	NO_2	H	H	[29]
67	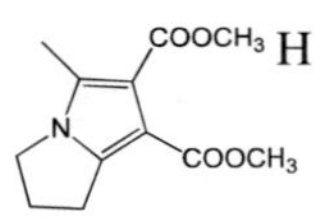	H	H	Cl	H	H	[29]
68	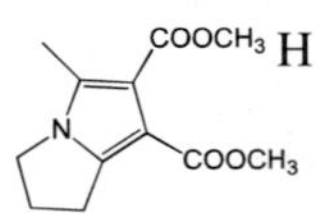	H	H	CF_3	H	H	[29]
69	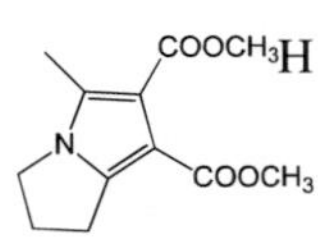	H	H	$COOC_2H_5$	H	H	[29]
70	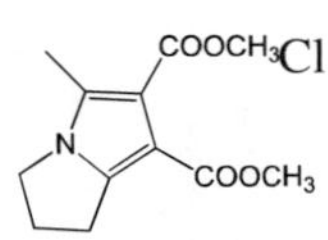	Cl	H	Cl	H	H	[29]
71	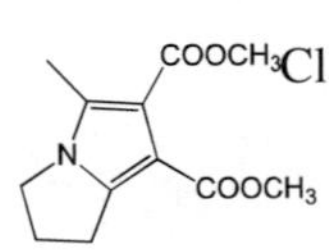	Cl	H	CH_3	H	H	[29]

Table. (Continued)

Compound	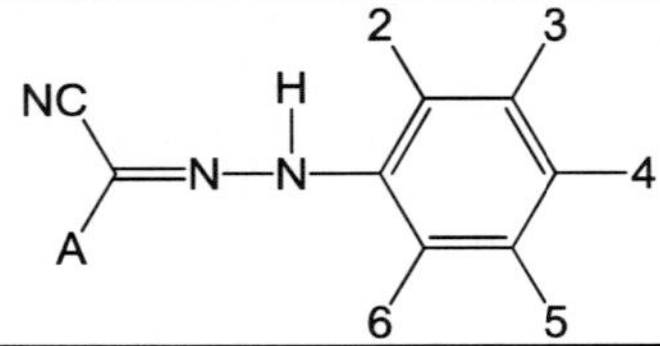						
	A	2	3	4	5	6	Ref.
72	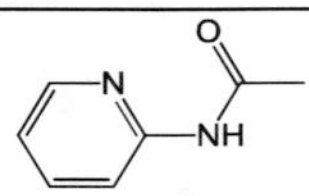	Ar = 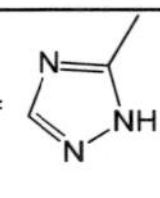					[30]
73	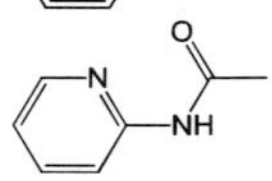	Ar = 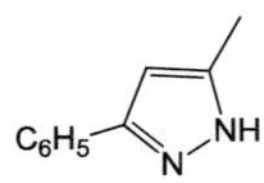					[30]
74	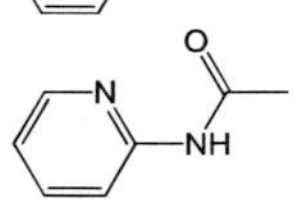	Ar = 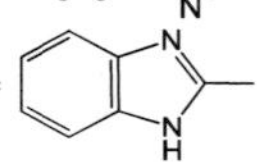					[30]
75	$C(=O)OC_2H_5$	Ar = 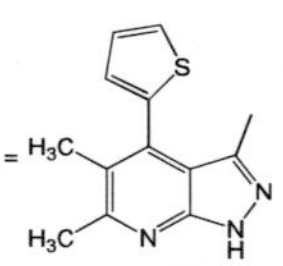					[31]
76	$C(=O)OC_2H_5$	Ar = 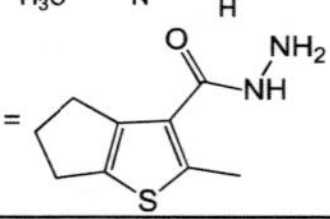					[32]

Another, less common, method of the synthesis involves the treatment of *n*-nonafluorobutylsulfonylbenzotriazole with methylene active nitriles (Scheme 3) [19,33]. Toluene or THF are used as solvents for the reaction, which were found to be more convenient to handle especially in the case of solid substrates. The synthesis of the starting material, i.e., *n*-nonafluorobutylsulfonylbenzotriazole is not trivial, and possibly this has hampered the wide use of the preparative procedure.

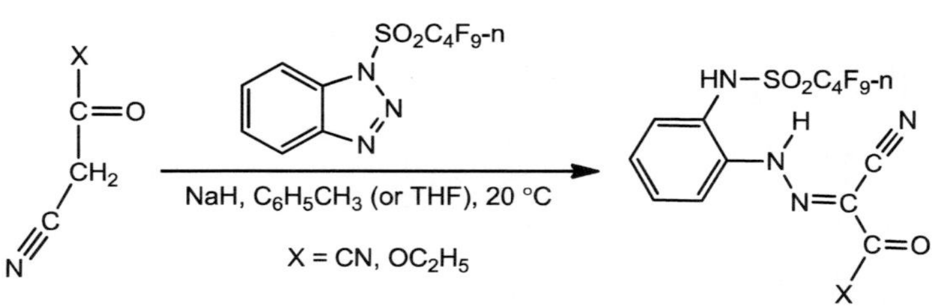

Scheme 3. Synthesis of AMANs from *n*-nonafluorobutylsulfonylbenzotriazole.

The cycloaddition of 3-alkylsulfanyl-2-arylazo-3-(pyrrolidin-1-yl)-acrylonitriles with dimethyl acetylenedicarboxylate was carried out in benzene at reflux, to yield AMANs with the 2,3-dihydro-1H-pyrrolizine moiety (Scheme 4) [29]. It was noted that the complete conversion is reached after 0.5–2.0 h and accompanied by the formation of a mixture of products that were difficult to separate.

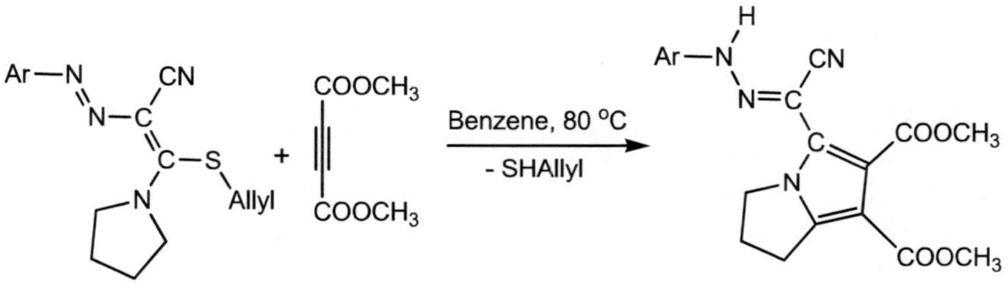

Scheme 4. Cycloaddition of 3-alkylsulfanyl-2-arylazo-3-(pyrrolidin-1-yl)-acrylonitriles with dimethyl acetylenedicarboxylate.

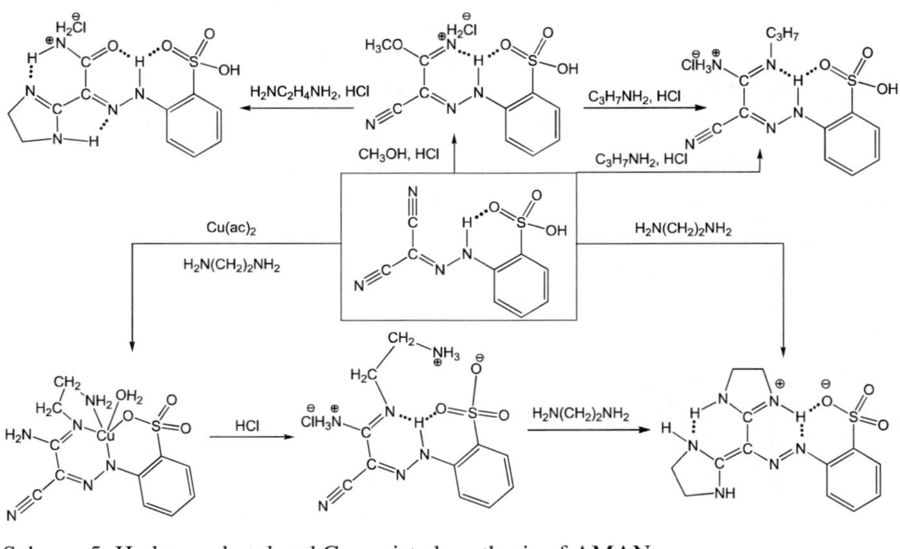

Scheme 5. Hydrogen bond and Cu-assisted synthesis of AMANs.

For the easy metal-ion or hydrogen-bond mediated syntheses of a variety of substituted AMANs, arylhydrazones of malononitrile have been used as starting materials. Metal ions (Mn^{II}, Cu^{II}) or intramolecular hydrogen bonding assisted nucleophilic attack of different nucleophiles on one or both cyano groups leading to the corresponding unsymmetrical AMANs (Scheme 5)

[14−16]. Interestingly, the metal-mediated nucleophilic addition to one of the nitrile moieties leaves the other cyano-group unreacted, thus creating possibilities for further attack with another nucleophile [15,16]. Addition to both cyano groups has been reported in the acid catalyzed reactions with nucleophiles such as ethylenediamine, hydrazine and hydroxylamine [14,34].

3. ORGANIC TRANSFORMATIONS OF AMANS

As it was mentioned above, the functionalization of AMANs makes them versatile starting materials for the synthesis of various organic compounds. Owing to the different reactivities of their parts, AMANs combine synthetic possibilities of nitriles, hydrazones and methylene active fragment, along with other less reactive moieties such as aryls or carbonyls (see Scheme 1 and Table).

The nitrile group has a particular chemical versatility, and numerous known additions to the C≡N moiety offer attractive routes for the creation of many new organic compounds [9,35,36]. In pure organic chemistry, the nitrile group is generally little reactive, but the coordination to a transition metal can significantly enhance its reactivity [9,37,38]. Thus, metal-mediated and metal-catalyzed nucleophilic additions of ammonia or amines to the CN group of RCN species are employed for the generation of various amidines, which, in turn, serve as useful building blocks for the syntheses of many biologically important compounds [9,13].

The possibility of tunable regioselective activation of a dinitrile towards nucleophilic attack in **3** is reported [16]. In **3**, intramolecular hydrogen bond systems assist specific nucleophilic attacks to the cyano moieties leading to a variety of amidines, carboxamides and iminoesters depending on the nucleophiles and conditions used (Scheme 5).

The reactivity of the hydrazone carbon atom of AMANs can also be used for the synthesis of many substances, namely of pharmaceutical significance [34]. For instance, chemical transformations such as Mannich coupling reactions give rise to an immense variety of heterocycles. Further, the hydrazone nitrogen atom can be attacked by acylating and alkylating agents. It seems that hard nucleophiles attack preferentially a nitrogen atom, while the soft ones preferentially react with a carbon atom. Thus, **38** or **39** react with hydroxylamine hydrochloride to yield an amidooxime which is further cyclized into the corresponding oxadiazole upon refluxing in acetic anhydride (Scheme 6) [22]. When refluxed in DMF in the presence of piperidine, the corresponding 1,2,3-triazoleamine was formed. The latter was acylated to *N*-(5-acetyl-2-phenyl-2*H*-imidazol-4-yl)acetamide by addition of acetic anhydride, while treatment of 1,2,3-triazoleamine with malononitrile in refluxing DMF yields 1,2,3-triazolo [4,5-*b*]pyridine (Scheme 6) [22].

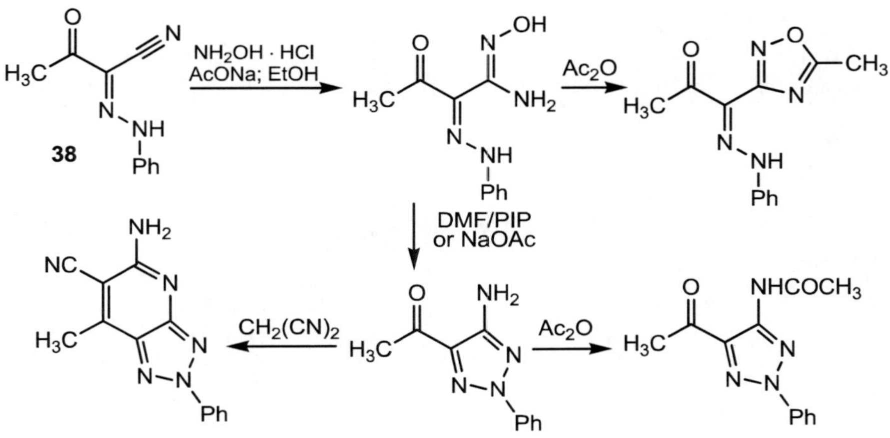

Scheme 6. Organic transformations of **38**.

The reactions with hydrazines have been extensively studied due to the great potential of the formed arylazo heteroaromatics in dye industry; many of arylazopyrazolones and aminoarylazopyrazoles are successfully commercialized. It is also well established that AMANs react with hydrazine derivatives to yield arylazoaminopyrazoles (Scheme 7) [39–41].

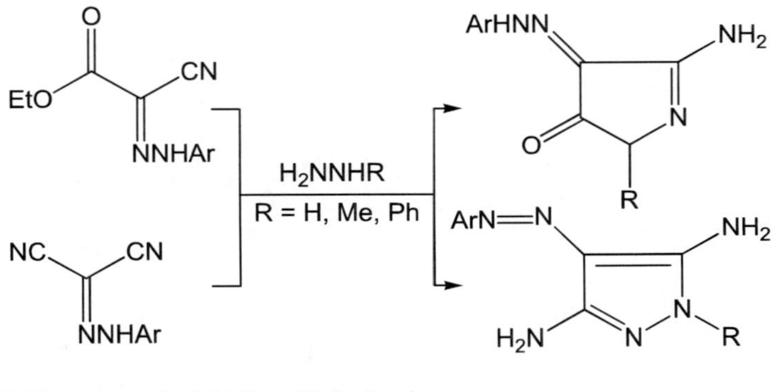

Scheme 7. Reactions of AMANs with hydrazines.

The methylene group in 2-aminoprop-1-ene-1,1,3-tricarbonitrile couples with aromatic diazonium salts to yield substituted AMANs that readily cyclize into dicyanopyridazines (Scheme 8a) [34,42]. The condensation of AMANs with ethyl cyanoacetoacetate also leads to dicyanopyridazines (Scheme 8b) [34]. This type of reaction has been extensively applied to the synthesis of various functionally substituted dicyanopyridazines. The reaction can be also conducted under microwave irradiation to afford highly pure dicyanopyridazines in a much shorter time [43].

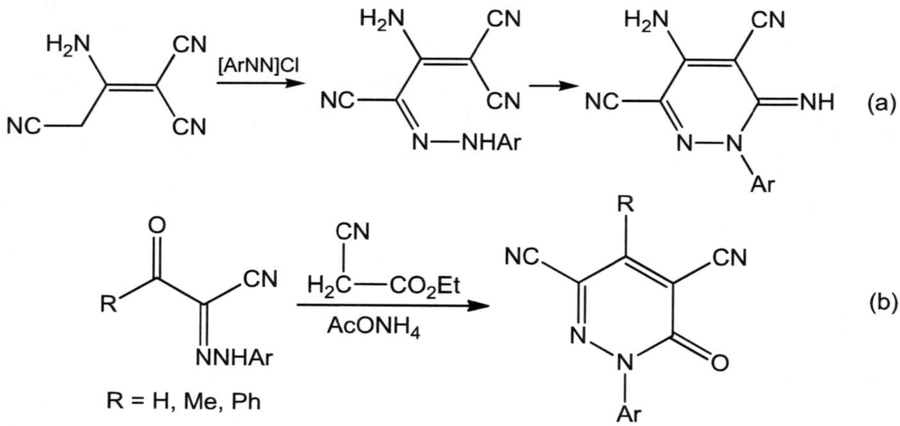

Scheme 8. Synthesis of dicyanopyridazines.

Treatment of AMANs with alkylhalides affords N-alkyl derivatives (Scheme 9a) [44,45]. The reaction of AMANs with chloroacetone, chloroacetonitrile or ethyl chloroacetate leads to aminopyrazoles (Scheme 9b) [46].

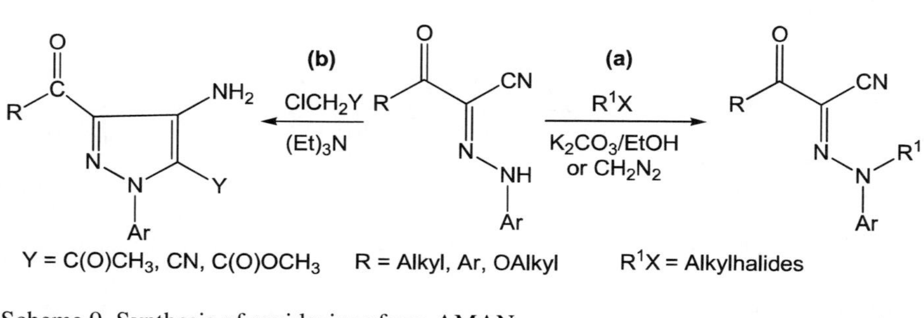

Scheme 9. Synthesis of pyridazines from AMANs.

The reaction of AMANs with a Grignard reagent leads to an unstable intermediate, which is readily dehydrated by reflux in acetic acid to yield 3,3-diphenyl-2-(phenyldiazenyl) acrylonitrile (Scheme 10) [47].

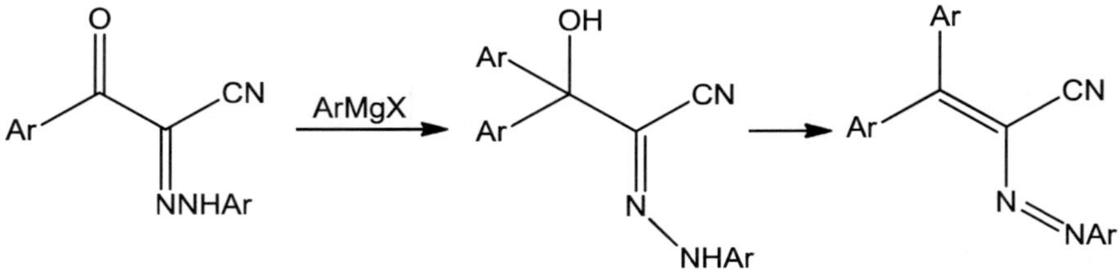

Scheme 10. Reaction of AMANs with a Grignard reagent.

4. Coordination Compounds Derived from AMANs

Arylhydrazones are among the well-recognized ligands which are able to form a wide variety of coordination compounds with different metals [6]. Nitriles can also coordinate to many metal ions, be transformed on the matrix of the metal or serve as starting materials for a number of alicyclic and heterocyclic compounds [9,22,34,37,38,43]. At the same time, only few coordination compounds with AMAN as ligands or ligand precursors have been prepared and fully structurally characterized [12,14,15,48]. In the presence of metal ions, additional electron-donating atoms can be incorporated by nucleophilic or electrophilic addition to the C≡N bond of the forming new ligands [14,15]. Upon coordination, the transformations of the nitrile group(s) can generate additional ligating sites, thus enabling the preparation of new coordination compounds of interesting confined and overall structures [14,15].

Thus, in the silver(I) complexes **77–80** (Scheme 11), a variety of coordination modes of the AMAN ligands is observed [12]. In **78** and **79**, the phenylhydrazonide anion is coordinated via the hydrazone nitrogen atom. **77** is a coordination polymer, where the phenylhydrazone coordinates as a tridentate ligand through both nitrile nitrogen atoms and the hydrazone nitrogen atom. In **80** a free, non-coordinated phenylhydrazonide anion is disclosed.

The relatively simple Ni^{II}-AMAN complex **81** (Scheme 11) was obtained upon reaction of the corresponding AMAN with nickel acetate in methanol solution [48]. The coordination environment of the nickel ion is of a distorted tetragonal geometry.

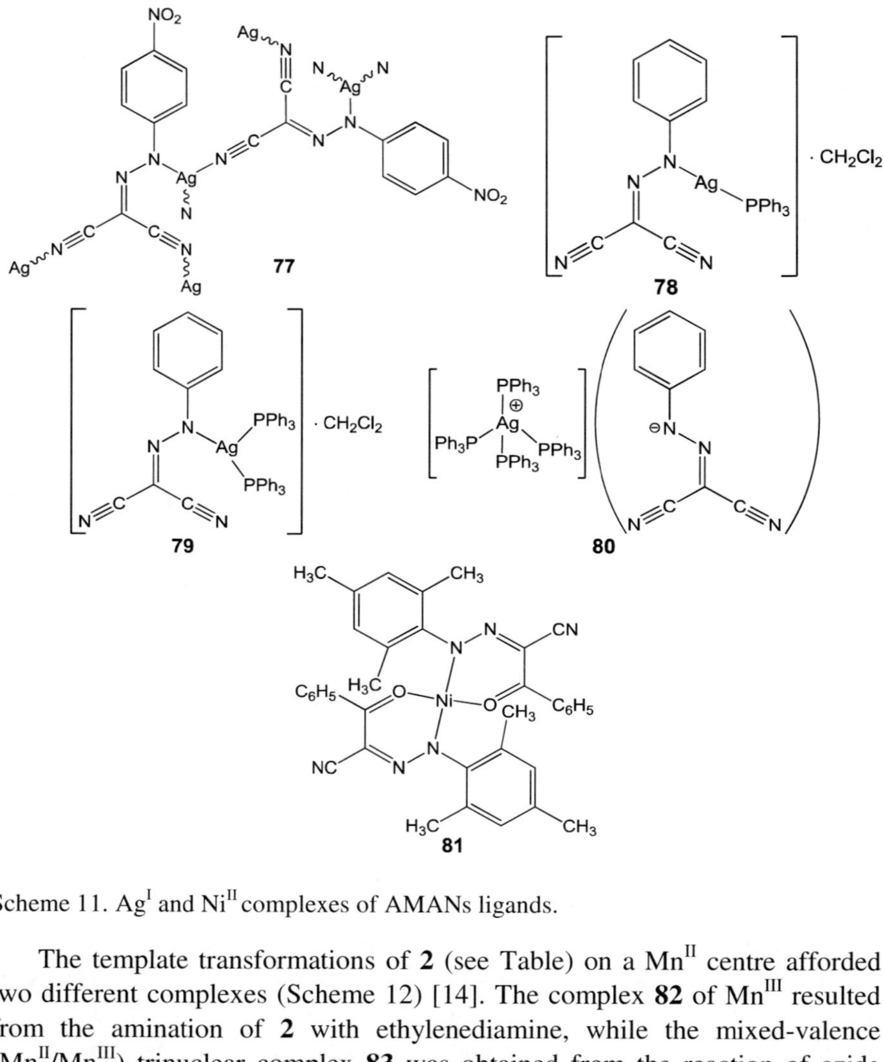

Scheme 11. Ag^{I} and Ni^{II} complexes of AMANs ligands.

The template transformations of **2** (see Table) on a Mn^{II} centre afforded two different complexes (Scheme 12) [14]. The complex **82** of Mn^{III} resulted from the amination of **2** with ethylenediamine, while the mixed-valence (Mn^{II}/Mn^{III}) trinuclear complex **83** was obtained from the reaction of azide with **2**. In this case, a [2+3] cycloaddition between N_3^- and a nitrile of **2** occurs with formation of a tetrazole *cis* to the amino group. As result, a new anionic AMAN ligand is created under mild conditions on the Mn^{II} centre. This tetrazole synthesis contrasts to the known tetrazole formations which mostly involve rather harsh hydrothermal conditions [49,50].

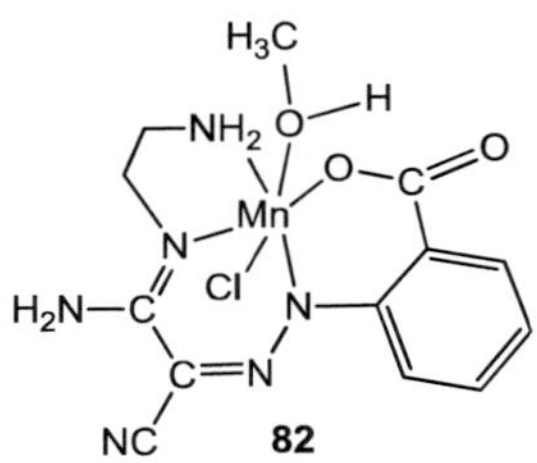

82

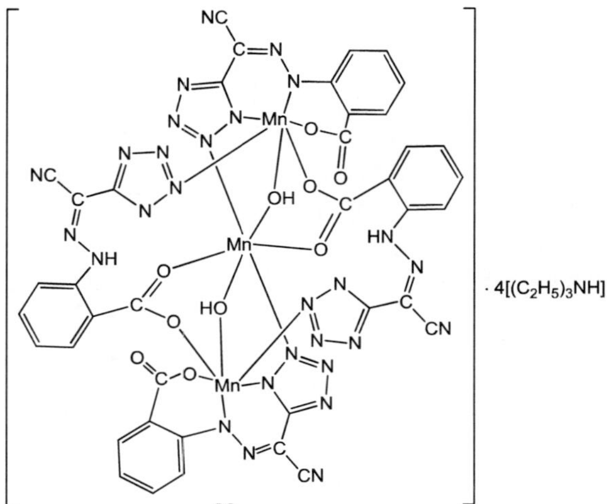

83

Scheme 12. Mn^{II}-AMAN complexes.

83 is a stable, trinuclear manganese complex; the octahedral coordination environment of the central Mn is filled with an oxygen atom from the $AMAN^{2-}$ carboxylate, a hydroxide, a nitrogen atom from the tetrazole of another $AMAN^{3-}$ and their symmetry equivalents. The preparations of **82** and **83** illustrate how *in situ* template transformations of AMANs can be used for the synthesis of new ligands with certain chelating and/or bridging sites. As a result of these *in situ* syntheses, a wide variety of coordination compounds of different nuclearity and with different supramolecular architectures can be easily prepared.

Conventional organic synthetic routes to the ligands produced upon such template transformations are hard to design. In contrast, upon both syntheses of **82** and **83**, the products of the nucleophilic attack on only one CN moiety are stabilized upon coordination, whereas the other cyano group is left unreacted. Several attractive complexes of different nuclearity were thus synthesized, but the difficulty of separating the formed ligands and the problem of their stabilization in the free state arise. Nevertheless, such a synthetic strategy can be useful in the preparation of new AMAN coordination compounds and their organic derivatives. Extended networks can also be produced, and, e.g., the presence of intramolecular and intermolecular antiferromagnetic interactions with $J_1/k_B = -9.5(1)$ K, $g = 1.95(2)$, and $zJ'/k_B = -0.37(5)$ K was found in **83** [14].

To facilitate the complex formation and ligand liberation a supporting group (e.g., sulfonyl or carboxyl) can be introduced into AMANs in the proximity of a cyano moiety that undergoes nucleophilic attack (Scheme 13) [15]. These supporting groups facilitate the formation of the complexes, and various structures and topologies can be thus obtained depending on the AMANs and nucleophiles used (Scheme 13). Furthermore, the new moieties can participate in intra- and intermolecular hydrogen bonding, thus facilitating the ligand liberation and stabilization (Scheme 5).

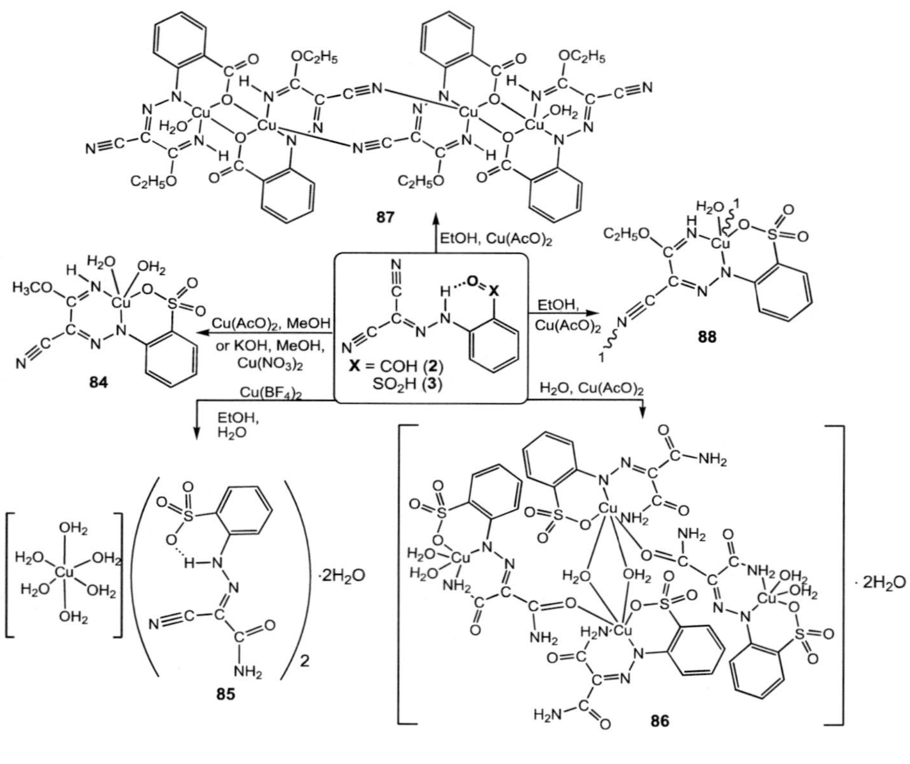

Scheme 13. Reactions of **2** and **3** with Cu^{II}.

Based on these principles, successful syntheses lead to the mononuclear **84**, **85**, tetranuclear **86**, **87** and polymeric **88** complexes (Scheme 13) of different nuclearities and types of structures which depend on AMANs, the nucleophile and conditions (solvents, temperature and time) used in the reactions [15]. The synthetic procedure involves a combination of Cu^{II} acetate, AMAN and a protic nucleophile such as an amine, water or an alcohol

(Scheme 13). The conversions of a cyano to amino/imino group occurs in these cases, while the second cyano group usually remains unreacted possibly due to stabilization of the product of one nucleophilic attack upon coordination to Cu^{II}.

The complexes **84–88** can be used as catalyst precursors for the aerobic or peroxidative oxidations of primary and secondary alcohols into the corresponding carbonyl compounds. Thus, primary benzyl alcohols were effectively oxidized to the respective benzaldehydes using O_2/TEMPO (TEMPO = 2,2,6,6-tetramethylpiperidine-1-oxyl) and *tert*-butylhydroperoxide (TBHP), whereas secondary alcohols were oxidized to ketones under microwave irradiation with TBHP as oxidant [15]. The same copper complexes were shown to be effective catalysts for the diastereoselective conversion of the carbonyl products to β-nitroalkanols, with predominance of the *threo* diastereoisomer [15].

5. *E/Z* Isomerisation in AMANs and Their Complexes

Besides their synthetic importance, AMANs are also convenient objects to create and study molecular switches due to their versatile *E/Z*-hydrazone isomerisations [15,51–53]. Formally, the intramolecular resonance assisted hydrogen bonding (RAHB) in carbonyl-containing AMANs should stabilize a *Z*-hydrazone configuration (Scheme 14). However, all the structurally characterized AMANs exist in the solid state in *E*-hydrazone form [20,21,34]. It was recently found that **89** exists in DMSO solution as a mixture of the two isomeric E- and Z-hydrazone forms (Scheme 14), the proton NMR sifts of the N–H···O protons, characteristic for the two- and three-centered RAHB systems, being observed at 12.71 and 14.03 ppm, respectively [54]. It was concluded that the observed low field NH signal in the 1H-NMR is due to an extensive delocalization of the nitrogen atom lone pair with a resulting increase of electronic delocalization at the hydrazone carbon atom. Thus, AMANs prefer to exist in anti (*E*) rather than in syn (*Z*) conformations; the stereoelectronic factors seem to be more important in governing conformations in these systems than the hydrogen bonding interactions.

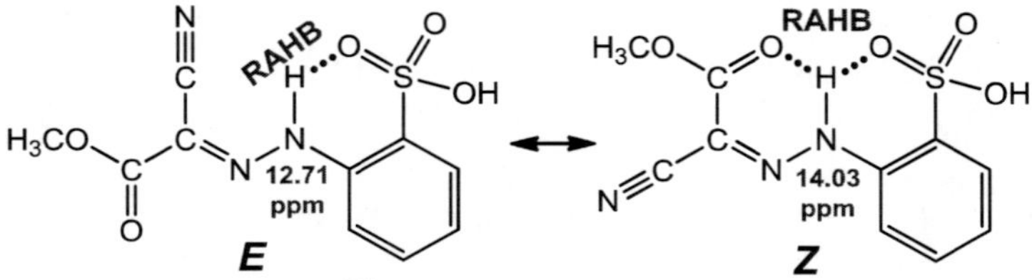

Scheme 14. *E/Z*-hydrazone isomerization in AMAN.

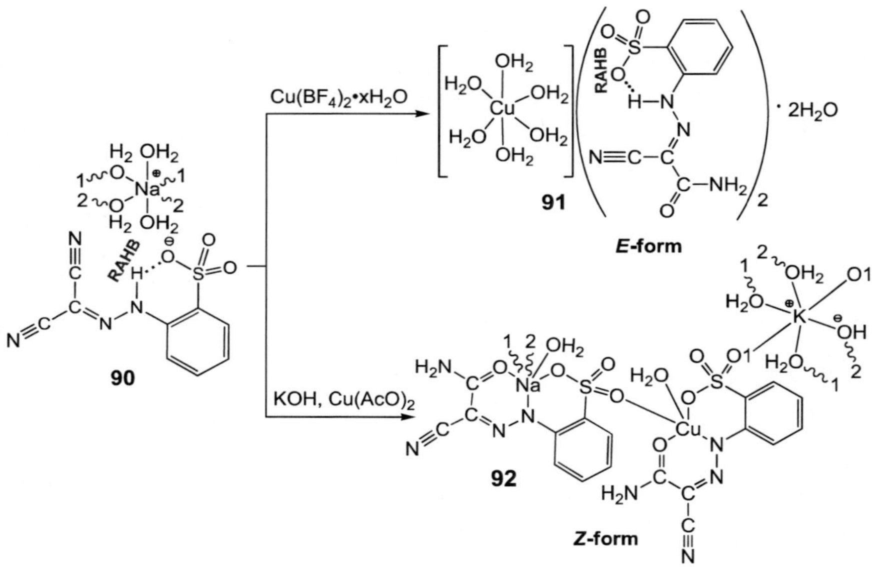

Scheme 15. Formation of *E* or *Z* isomers.

The RAHB assisted controllable *E*↔*Z* transition around a C=N bond has been described [15]. For that, the sodium salt (**90**) of **3** was reacted with copper tetrafluoroborate and copper acetate in the presence of KOH giving the *E* (**91**) and *Z* (**92**) isomers, respectively (Scheme 15). In the *E*-form, the hydrazone hydrogen creates a hydrogen bond with the negatively charged sulfonic group rather than with the neutral oxygen atom of the amido moiety. To switch **90** to the *Z* isomer, the RAHB system is firstly destroyed by deprotonation with KOH and then a Cu^{2+} ion is added. The Lewis acid Cu^{2+} accepts the free electron pairs of the sulfonyl and $C(NH_2)=O$ carboxyl groups thus stabilizing the *Z* configuration (Scheme 15, Route **90** → **92**). As it can be seen, the described *E*↔*Z* switch can be easily controlled by the sequence at which the acid and base are added.

CONCLUSION

The syntheses, organic transformations and coordination chemistry of AMANs are summarized and discussed herein. AMANs can be readily functionalized, providing rich opportunities to create multiple coordination modes. Thus, derivatives of AMANs are used for the preparation of complexes with Na^{I}, Ag^{I}, Ni^{II}, Cu^{II} and $Mn^{II,III}$. The role of different metal ions and the structure of AMANs in directing the preparative pathways and tuning particular properties of the resulting compounds are described. Several applications of AMANs, as well as of their complexes, are also indicated. Nevertheless, in spite of several attractive applications, the chemistry of this very promising class of ligands and starting materials for organic synthesis is only emerging and further interesting results are expected to appear in the near future.

ACKNOWLEDGMENTS

This work has been partially supported by the Foundation for Science and Technology (FCT), Portugal. K.T.M. and M.N.K. express gratitude to FCT for the post-doc fellowship and working contract.

REFERENCES

[1] Zollinger, H., *Diazo Chemistry I: Aromatic and Heteroaromatic Compounds*. New York: Wiley, 1994.

[2] Zollinger, H., *Diazo Chemistry II: Aliphatic, Inorganic and Organometallic Compounds*, VCH, Weinheim 1995.

[3] Roglans, A., Pla-Quintana, A., Moreno-Manas, M., *Chem. Rev.* 106 (2006) 4622–4643.

[4] Price, B.K., Hudson, J.L., Tour, J.M., *J. Am. Chem. Soc.* 127 (2005) 14867–14870.

[5] Mahouche-Chergui, S., Gam-Derouich, S., Mangeney, C., Chehimi, M.M., *Chem. Soc. Rev.* 40 (2011) 4143−4166.
[6] Mahmudov, K.T., Kopylovich, M.N., Pombeiro, A.J.L., *Coord. Chem. Rev.* 257 (2013) 1244−1281.
[7] Freeman, F., *Chem. Rev.* 69 (1969) 591−624.
[8] Freeman, F., *Chem. Rev.* 80 (1980) 329−350.
[9] Kukushkin, V.Y., Pombeiro, A.J.L., *Chem. Rev.* 102 (2002) 1771−1802.
[10] Shchegol'kov, E.V., Burgart, Y.V., Khudina, O.G., Saloutin, V.I., Chupakhin, O.N., *Rus. Chem. Rev.* 79 (2010) 31−61.
[11] Kandhavelu, M., Paturu, L., Mizar, A., Mahmudov, K.T., Kopylovich, M.N., Karp, M., Yli-Harja, O., Pombeiro, A.J.L., Ribeiro, A.S., *Pharmaceutical Chem. J.* 46 (2012) 157−164.
[12] Romualdo, L.L., Bessler, K.E., Deflon, V.M., Niquet, E., *Z. Anorg. Allg. Chem.* 628 (2002) 1098−1103.
[13] Kryštof, V., Cankař, P., Fryšová, I., Slouka, J., Kontopidis, G., Džubák, P., Hajdúch, M., Srovnal, J., de Azevedo Jr.W.F., Orság, M., Paprskářová, M., Rolčík, J., Látr, A.,Fischer, P.M., Strnad, M., *J. Med. Chem.* 49 (2006) 6500−6509.
[14] Anwar, M.U., Lan, Y., Beltran, L.M.C., Clerac, R., Pfirrmann, S., Anson, C.E.,Powell, A.K., *Inorg. Chem.* 48 (2009) 5177−5186.
[15] Kopylovich, M.N., Mizar, A., Guedes da Silva, M.F.C., Mac Leod, T.C.O.,Mahmudov, K.T., Pombeiro, A.J.L., *Chem. A Eur. J.* 19 (2013) 588−600.
[16] Kopylovich, M.N., Mahmudov, K.T., Mizar, A., Pombeiro, A.J.L., *Chem. Commun.* 47 (2011) 7248−7250.
[17] Han, L.-N., Zhang, M., Lu, R.-Z., Weia, W.-B., Wang, H.-B., *Acta Cryst.* (2010) E66 o196.
[18] Mitsuhashi, T., Yamamoto, G., Goto, M., Kondo, Y., *J. Chem. Soc. Perkin Trans.* 2 (1992) 563−566.
[19] Anwar, M.U., Tragl, S., Ziegler, T., Subramanian, L.R., *Synlett.* (2006) 627−629.
[20] Bertolasi, V., Gilli, P., Ferretti, V., Gilli, G., Vaughan, K., *New J. Chem.* 23 (1999) 1261−1267.
[21] Bertolasi, V., Gilli, P., Ferretti, V., Gilli, G., Vaughan, K., Jollimore, J.V., *Acta Cryst.* B55 (1999) 994−1004.
[22] Al-Mousawi, S.M., Moustafa, M.Sh., *Beilstein J. Org. Chem.* 3 (2007) 1−3.
[23] Al-Awadhi, H., Al-Omran, F., Elnagdi, M.H., *Tetrahedron* 51 (1995) 12145−12162

[24] Aziz, S.I., Anwar, H.F., Fleita, D.H., Elnagdi, M.H., *J. Heterocyclic Chem.* 44 (2007) 725–729.
[25] Koga, N., Koga, G., Springer, J.P., Arison, B.H., Anselme, J.-P., *J. Chem. Soc., Chem. Commun.* (1983) 610–612.
[26] Patel, H.M., *Euro. J. Chem.* 3 (2012) 44–50.
[27] Al-Sheikh, M.A., Elnagdi, M.H., *Molecules* 14 (2009) 4406–4413.
[28] Behbehani, H., Ibrahim, H.M., Makhseed, S., *Arkivoc* ii (2010) 267–282.
[29] Belskaya, N.P., Bakulev, V.A., Deryabina, T.G., Subbotina, J.O., Kodess, M.I.,Dehaen, W., Toppet, S., Robeyns, K., Meervelt, L.V., *Tetrahedron* 65 (2009) 7662–7672.
[30] Farag, A.M., Kheder, N.A., Dawood, K.M., *Int. J. Mod. Org. Chem.* 2 (2013) 26–39.
[31] Fun, H.-K., Hemamalini, M., Abdel-Aziz, H.A., Aboul-Fadl, T., *Acta Cryst.* E67 (2011) o2145–o2146.
[32] Wardakhan, W.W., El-Sayed, N.N.E., Mohareb, R.M., *Acta Pharm.* 63 (2013) 45–57.
[33] Uhde, M., Ziegler, T., *Synthetic Commun.* 40 (2010) 3046–3057.
[34] Elassar, A.-Z.A., Dib, H.H., Al-Awadi, N.A., Elnagdi, M.H., *Arkivoc* ii (2007) 272–315.
[35] Murahahaashi, S.-I., Takaya, H., *Acc. Chem. Res.* 33 (2000) 225–233.
[36] Turner, D.R., Chesman, A.S.R., Murray, K.S., Deacon, G.B., Batten, S.R., *Chem. Commun.* 47 (2011) 10189–10210.
[37] Kukushkin, V.Y., Pombeiro, A.J.L., *Inorg. Chim. Acta* 358 (2005) 1–21.
[38] Pombeiro, A.J.L., *New J. Chem.* 18 (1994) 163–174.
[39] Taylor, E.C.Jr., Barton, J.W., Osdene, T.S., *J. Am. Chem. Soc.* 80 (1958) 421–427.
[40] Shawali, A.S., Abdelkader, M.H., Attalbawy, F.M., *Tetrahedron* 58 (2002) 2875–2879.
[41] Elnagdi, M.H., Fleita, D.H., Elmoghayer, M.R.H., *Tetrahedron* 31 (1975) 63–67.
[42] Ibrahim, N.S., Galil, F.A., Abdel-Motaleb, R.M., Elnagdi, M.H., *Heterocycles* 24 (1986) 1219–1222.
[43] Sotelo, E., Mocelo, R., Suarez, M., Loupy, A., *Synth.Commun.* 27 (1997) 2419–2423.
[44] Kandeel, E.M., Sadek, K.U., Elnagdi, M.H., *Z. Naturforsch. B Anorg. Chem. Org. Chem.* 35 (1980) 91–94.
[45] Gewald, K., Rehwald, M., Mueller, H., Bellmann, P., Schaefer, H., *Monatsh. Chem.* 126 (1995) 341–347.

[46] Salah, A.M., Abdallah, T.A., Radwan, N.F., Hassaneen, H.M., *Z. Naturforsch., B: Chem. Sci.* 61b (2006) 1158−1161.
[47] Abdel-Motaleb, R.M., Makhloof, A.M.A., Ibrahim, H.M., Elnagdi, M.H., *J. Heterocycl. Chem.* 43 (2006) 931−934.
[48] Kurbatov, V.P., Antsyshkina, A.S., Kukharicheva, E.S., Garnovsky, A.D., *Russ. J. Inorg. Chem.* 41 (1996) 1315−1319.
[49] Zhao, H., Qu, Z.R., Ye, H.Y., Xiong, R.G., *Chem. Soc. Rev.* 37 (2008) 84−100.
[50] Gao, E.-Q., Liu, N., Cheng, A.-L., Gao, S., *Chem. Commun.* (2007) 2470−2472.
[51] *Molecular Switches*, Ed.; Feringa, B.L., Wiley-VCH: Weinheim, Germany, 2001.
[52] Landge, S.M., Tkatchouk, E., Benítez, D., Lanfranchi, D.A., Elhabiri, M., Goddard, W.A., Aprahamian, I., *J. Am. Chem. Soc.* 133 (2011) 9812−9823.
[53] Landge, S.M., Aprahamian, I., *J. Am. Chem. Soc.* 131 (2009) 18269−18271.
[54] Mahmudov, K.T., Kopylovich, M.N., Guedes da Silva, M.F.C., Pombeiro, A.J.L., *Chem. A Eur. J.* (2013) submitted.

In: Ligands
Editors: Paula Gawryszewska et al.
ISBN: 978-1-63117-143-7

Chapter 7

Powerful New Ligand Systems: Carbacylamidophosphates (CAPh) and Sulfonylamidophosphates (SAPh)

V. Amirkhanov[1*], V. Ovchynnikov[1], V. Trush[1], P. Gawryszewska[2,*], L. B. Jerzykiewicz[2]

[1]Taras Shevchenko National University of Kyiv, City of Kyiv, Ukraine
[2]Faculty of Chemistry, University of Wroclaw, Wroclaw, Poland

Abstract

In this chapter, in respect to searching for ligands allowing for the coordination of engineering towards biologically active metal-organic frameworks, we consider two types of organic ligands: carbacylamidophosphates (CAPh) – with the functional fragment -C(O)N(H)P(O)= and sulfonylamidophosphates (SAPh) - containing the -$S(O)_2N(H)P(O)=$ fragment, their sodium salts and their lanthanide complexes. The phosphoryl group, present in all ligands, has high affinity for the majority of metal ions, and especially for rare-earth elements. The possibility of being involved in the coordination sphere, both in

* Corresponding author: Taras Shevchenko National University of Kyiv, v_amirkhanov@yahoo.com
* Corresponding author (lanthanide complexes): Faculty of Chemistry, University of Wroclaw, paula.gawryszewska@chem.uni.wroc.pl

molecular and acidic forms, makes the coordination chemistry of these ligand systems extremely diverse and interesting.

In the first subsection some historical information about this of type of ligand, different methods of synthesis and, in more detail, the most effective new synthetic strategies are considered. Taking into account the weak acidic nature of the imide group (-NH-), present in all of the compounds under consideration, methods of obtaining sodium salts and lanthanide complexes are described.

The crystal structures of CAPh and SAPh compounds and their lanthanide complexes are also described. The influence of substituents' nature near -C(O)N(H)P(O)= and $-S(O)_2N(H)P(O)=$ functional fragments on structural parameters of compounds under consideration, and on the peculiarity of the intermolecular interaction in the crystal state of CAPh and SAPh compounds is discussed. The importance of the structural characteristics for the efficiency of the ligand-to-metal energy transfer in lanthanide complexes, and the efficiency of the metal luminescence, is shown.

IR, Raman, absorption and luminescence spectroscopic data, covering a wide range of temperatures from 4 to 295 K, are used for the identification of CAPh and SAPh compounds. It is demonstrated that the position of the ligand-to-metal charge transfer (LMCT) state is a prominent factor in the effectiveness of the sensitized Eu^{3+} luminescence.

The penultimate subsection is devoted to the different aspects of the biochemical activity of CAPh and SAPh compounds. The presence of the peptide (in the case of CAPh) or sulfonylamide group (for SAPh) and phosphoryl groups combined together in one molecule could be the main reason for these diverse biological activities. For example, some CAPh compounds, having ethylenimine bound to a phosphorus atom, are used in clinical practice as anticancer drugs. Some of these compounds are very effective insecticides. Complex formation can be considered as an instrument of influence on the biochemical activity of drugs, in particular their toxicity and selectivity.

In the last subsection of this chapter we briefly consider the general directions of CAPh and SAPh ligands coordination chemistry development.

Keywords: carbacylamido phosphates, sulfonylamido phosphates, organic synthesis, lanthanide, crystal structure, spectroscopy

INTRODUCTION

The chemistry of phosphor-containing compounds was described under the leadership of academician Aleksandr Kirsanov in the 1960s; nowadays, his so-called phosphazo-reaction [1] is an extensively used technique in modern phosphororganic chemistry. Many compounds of this type are already well-known as drugs, insecticides, acaricides, fungicides, herbicides, growth regulator for plants, defoliants, extragents, antioxidants, and anticorrosive additives [2-4]

In this short review we would like to present information concerning the synthesis and some properties of two types of phosphor-containing organic compounds: carbacylamidophosphates (CAPh) – substances with –C(O)N(H)P(O)= core – and sulfonylamidophosphates (SAPh) – compounds which contain -$S(O)_2N(H)P(O)$= fragment. The availability of donor groups in their structures makes these molecules behave as ligands, and therefore form coordination compounds with different metal ions.

Obtained for the first time at the beginning of the XX century, these organic compounds were for a long period difficult to synthesize, and were little-studied substances. However, our knowledge about CAPh and SAPh compounds has grown recently to fill a big new chapter in chemistry belonging to phosphor organic derivatives and the coordination chemistry of phosphoryl-containing ligands. Now, many representatives of this type of compound and their complexes have been synthesized and investigated [5,6].

The examination of the coordination chemical behavior of biologically active ligands is a serious problem, because in many cases the above-mentioned activities essentially differ for the "free" ligands and for corresponding complexes. Also, it is well known that transport and biochemical transmutations in living organisms are very often carried out with the participation of metal ions – the complex formation agents. It should be noted that CAPh and SAPh compounds are in fact structurally analogous with the widely-studied bidentate-chelating ligands - β-diketons, whose complexes have been practically applied for a long time [7-11].

Hetero substitution of carbon atoms of the chelating core often leads to useful new properties untypical for β-diketons. For example, replacing one carbonyl group in β-diketon's functional fragment to phosphoryl allows the discovery of a new class of effective extractants for trans plutonium elements [12] since the phosphoryl group has very high affinity to the majority of metal ions, and especially to the lanthanides and actinides [13]. The presence of the phosphoryl group, which possesses high affinity to lanthanides and actinides, allows the use of CAPh and SAPh in the coordination chemistry of these metals. Additionally, complexes of some carbacylamidophosphates can be formed by the neutral or deprotonated form of CAPh ligands because of the weak-acid nature of amide nitrogen [6, 14, 15].

For the CAPh compounds amide-imide prototropic tautomerism can be realized in solutions [16, 17]:

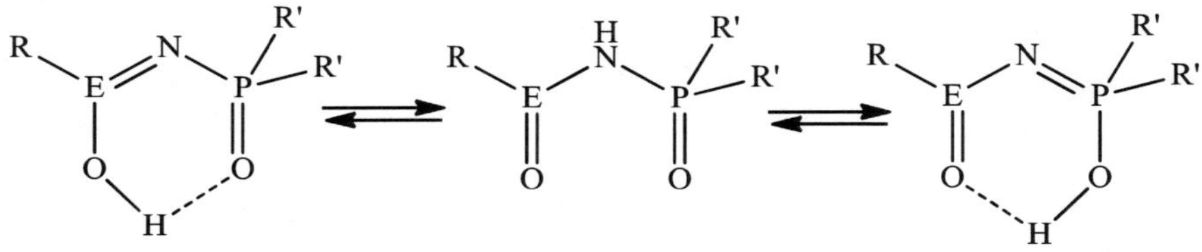

This balance strongly depends on the electrophility of substituents R and R'. In the case of the low value of electrophility the balance practically fully shifts to the form II. The last one always appears in the crystal structure [18-25].

1. Syntheses of CAPh and SAPh Compounds

The first synthetic step for both CAPh and SAPh compounds – the so called phosphoro-azo reaction – is interaction between the amide of the corresponding organic acid and phosphorus pentachloride [1, 26]. Initial compounds for this synthesis – amides of aliphatic, aromatic or heterocyclic carbonic or sulfonic acids – must be thoroughly purified and dried.

Amides of carbonic acids and sulfonamides can be synthesized from corresponding chloroanhydrides. Aliphatic amides can also be obtained via the reaction of esters with aqueous or alcoholic solutions of ammonia.

An important condition for carrying out such reactions is using absolute aprotic solvents, which must be inactive to the aggressive action of PCl_5. The boiling point of the solvent must be in the range 60-130°C. Typically, chloride alkanes (CCl_4 or $C_2H_4Cl_2$) or aromatic hydrocarbons: benzene (because of it's toxicity it is rarely used), toluene or xylenes are employed. Intermediate compounds ($RC(O)NPCl_3$ or $RS(O)_2NPCl_3$) are very active and can react with different nucleophiles, forming corresponding derivatives. The latter can be obtained by mixing equimolar amounts of the amide and phosphorus pentachloride with the practically quantitative yield.

It should be noted that there is a principal difference in the syntheses of trichlorophosphazo derivatives of carbonic and sulfonic acids and respective dichloroanhydrides, which are precursors for further syntheses of CAPh and SAPh compounds of both types: etheric and amidic. CAPh precursors ($RC(O)NPCl_3$ and $RC(O)NHP(O)Cl_2$) are much more active than SAPh ones ($RS(O)_2NPCl_3$ and $RS(O)_2NHP(O)Cl_2$). In fact, CAPh precursors are not stable, and even a trace amount of HCl catalyzes their decomposition to carbonitriles. Therefore, it is important to use the precursors immediately after obtaining them, and their storage should be avoided. In contrast, SAPh precursors can be stored without being kept in special conditions, such as inert atmosphere or low temperature for a long time.

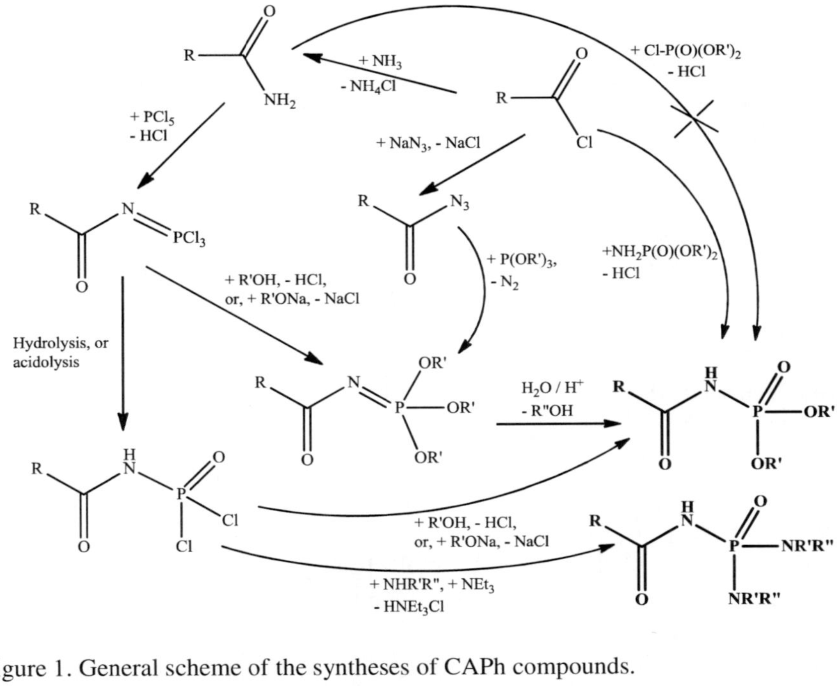

Figure 1. General scheme of the syntheses of CAPh compounds.

For the syntheses of the ester type ligands (*i.e.* ligands having two ester type substituents at the phosphorus atom) or the amide type (ligands having two amide substituents at the phosphorus atom), one- or polyatomic alcohols or phenols (and their derivatives), and primary or secondary amines are used along with an organic base as a hydrogen acceptor. The initial amine often plays the role of the base when added in more than double excess.

General schemes of the CAPh and SAPh compounds syntheses are presented (Figure 1 and Figure 2).

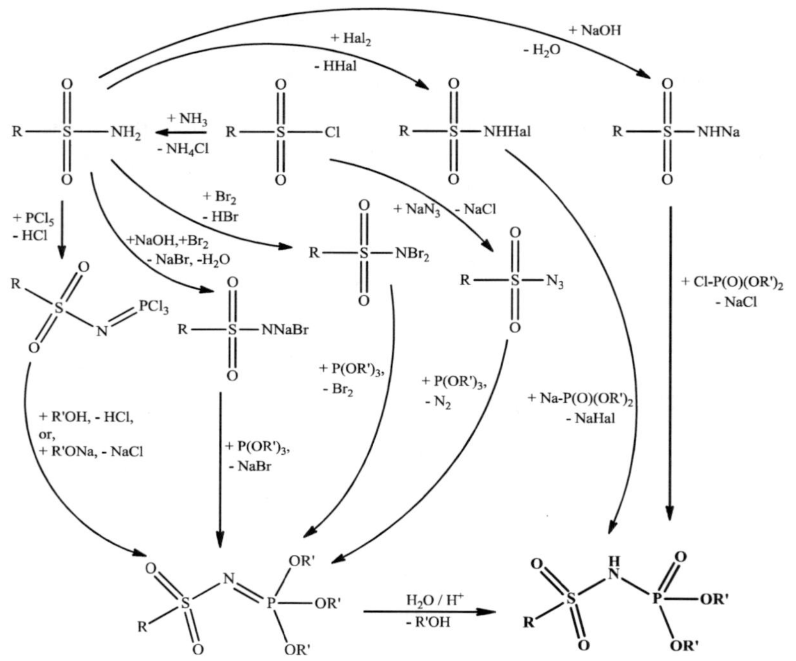

Figure 2. General scheme of the syntheses of SAPh compounds.

Some ligands based on phosphorylated carbamides, which contain 5- or 6-membered cycles (for example, 2-aminothiazoles, 2-aminopyridine) can be synthesized using phosphorus isocyanates (Figure 3):

It is possible to obtain the desirable final product with higher yield by employing the second approach presented in the scheme. Moreover, the reproducibility of this synthetic method is much higher.

Obtaining the dimethylether of isocyanatophosphate acid (4) (Figure 4) requires modification of the standard procedure [27]: instead of methyluretane an ethyluretane (1) was used.

This change leads to an increase of the boiling temperature of urethane phosphate acid ethers. As a result, the separation of the dimethylether of isocyanatophosphate acid becomes easier and the yield of the target products increases.

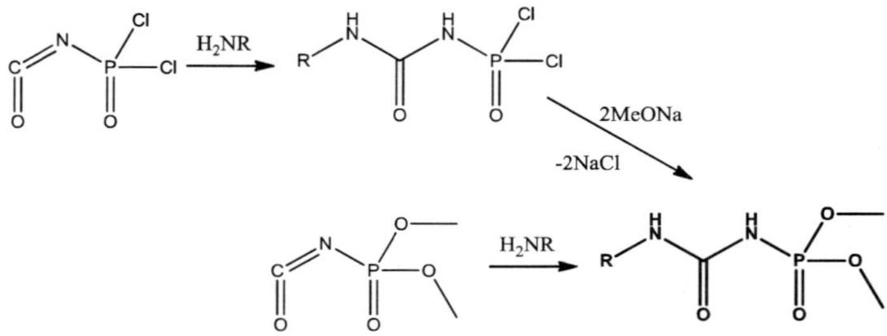

Figure 3. Scheme of the syntheses of CAPh ligands based on phosphorylated carbamides.

The syntheses of CAPh and SAPh ligands containing aromatic substituents at the phosphorus atom can be realized according to the scheme presented in Figure 5.

Syntheses of "Simple" Salts Based on Phosphoramide Ligands

The presence of the NH – proton in the structure of CAPh and SAPh ligands, situated between two donor groups C=O (O=S=O) and P=O, imparts to the compound the properties of weak acids. Some CAPh and SAPh derivatives of ether type have even higher acidity compared to acetic acid. This characteristic of ether type ligands in contrast to amide ones allows them to be dissolved in aqueous solutions of hydrocarbanates, carbonates or hydroxides of alkali metals. An exchange reaction between alcoholates in non-aqueous solutions is a simple method of obtaining alkali metal salts of CAPh or SAPh.

The above-mentioned methods have allowed the synthesis of these "simple" salts with lithium, sodium, potassium, rubidium and cesium cations. Also, some compounds with large cations such as tetraethylammonium, tetraphenylphosphonium, and tetraphenylstibonium have been prepared. Typically, the products precipitate out from the mother liquor by the slow evaporation of solvents. Most "simple" salts, based on CAPh and SAPh compounds, are quite stable in the air for a long time. They are soluble in many polar organic solvents, but these compounds quickly hydrolyze in the aqueous solutions precipitating CAPh or SAPh "free" ligands.

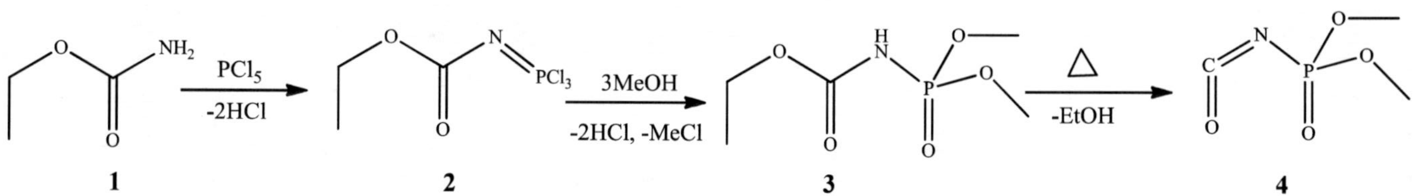

Figure 4. The obtaining of the dimethylether of isocyanato phosphate acid.

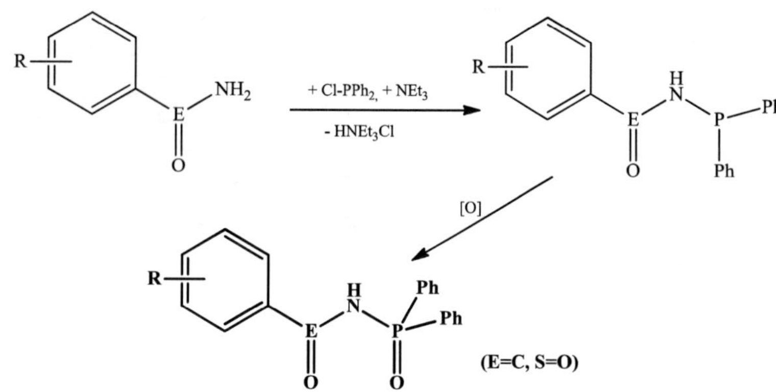

Figure 5. Syntheses of CAPh and SAPh ligands containing aromatic substituents at the phosphorus atom.

Representatives of CAPh and SAPh Ligands

The Cambridge Structural Database contains over 2,000 structures with the -C(O)N(H)P(O)= system (ligands CAPh) and about 10 structures with the -S(O)2N(H)P(O)= fragment (ligands SAPh). Most representatives of the CAPh and SAPh ligands which have been already studied are listed in Table 1.

Table 1. Some representatives of CAPh and SAPh ligands

Ligand	Formula	Title	Ref.
HL1	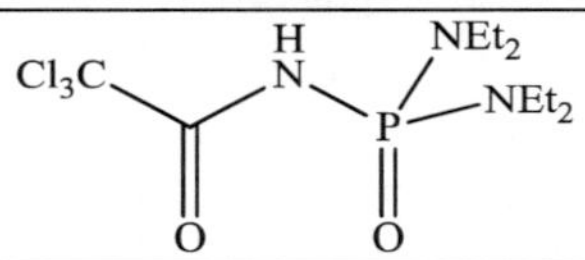	*N*-[bis(diethylamino)phosphoryl] -2,2,2-trichloroacetamide	[20]
HL2	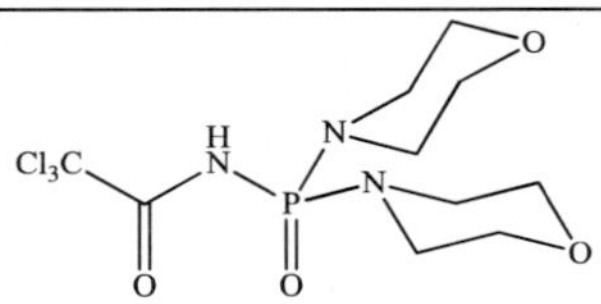	2,2,2-trichloro-*N*- (dimorpholin-4-ylphosphoryl)acetamide	[28, 21]
HL3	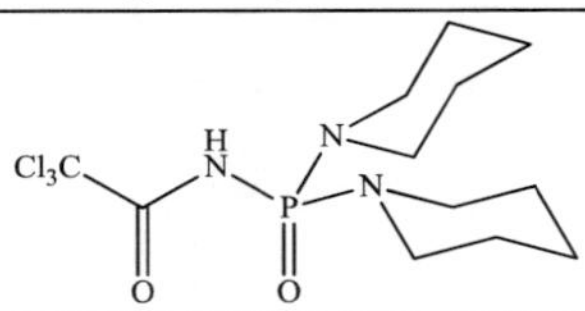	2,2,2-trichloro-*N*- (dipiperidin-1-ylphosphoryl)acetamide	[29]
HL4	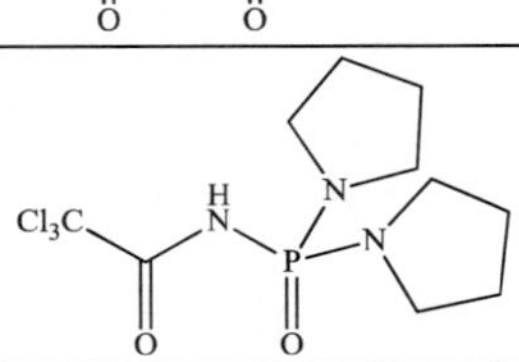	2,2,2-trichloro-*N*- (dipyrrolidin-1-ylphosphoryl)acetamide	[30, 31]
HL5	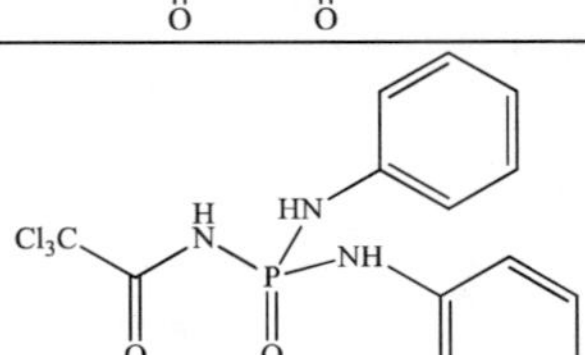	N,N'-diphenyl-N''-trichloroacetylphosphorylamide	[32, 22]
HL6	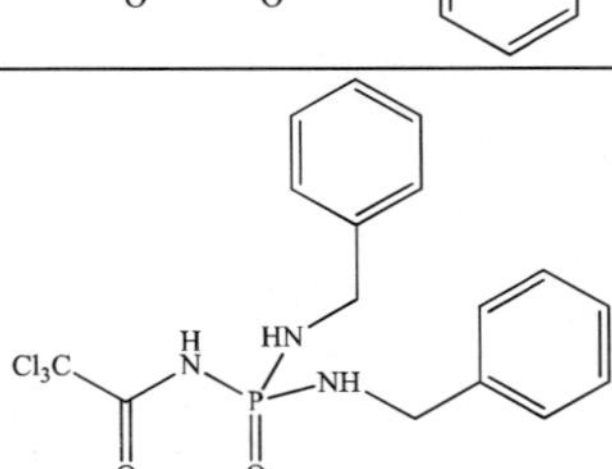	N,N'-dibenzyl-N''-trichloroacetylphosphorylamide	[33, 34]

Table 1. (Continued)

Ligand	Formula	Title	Ref.
HL7	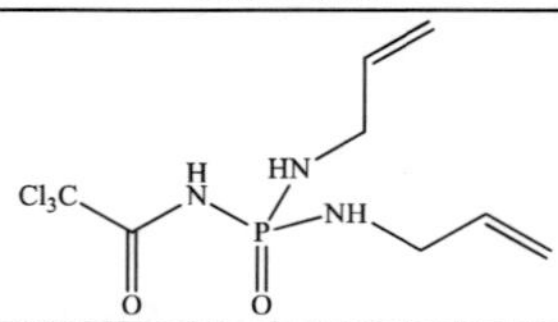	N,N'-diallyl-N''-trichloroacetylphosphorylamide	[35]
HL8	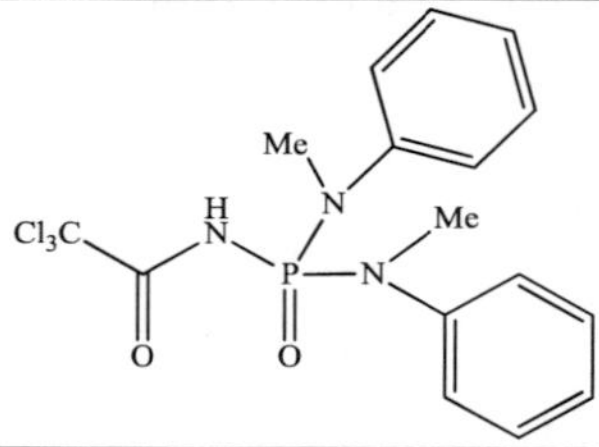	*N*-{bis[methyl(phenyl)amino] phosphoryl}-2,2,2-trichloroacetamide	[36]
HL9	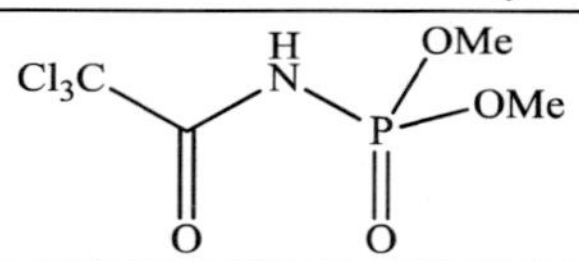	Dimethyl-N-(trichloroacetyl)amidophodphate	[32, 37, 38]
HL10	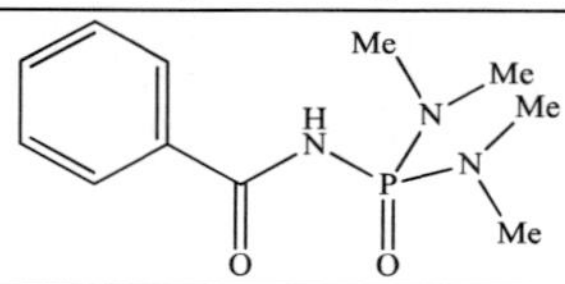	*N*- [bis(dimethylamino) phosphoryl]benzamide	[24]
HL11	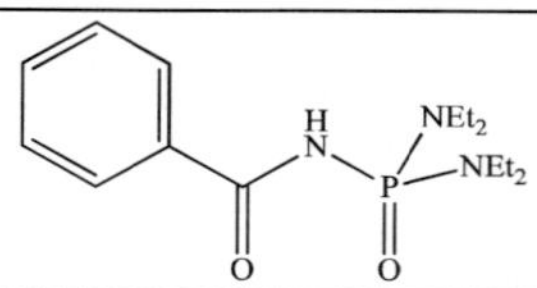	*N*-[bis(diethylamino)phosphoryl] benzamide	[39]
HL12	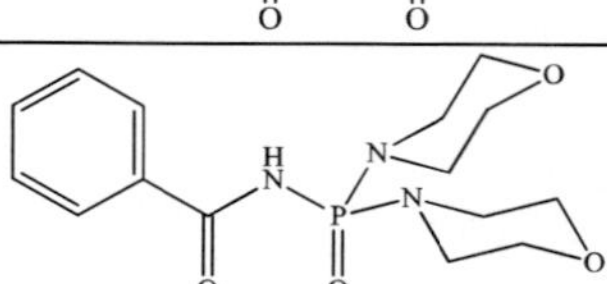	*N*- (dimorpholin-4-ylphosphoryl)benzamide	[24]
HL13	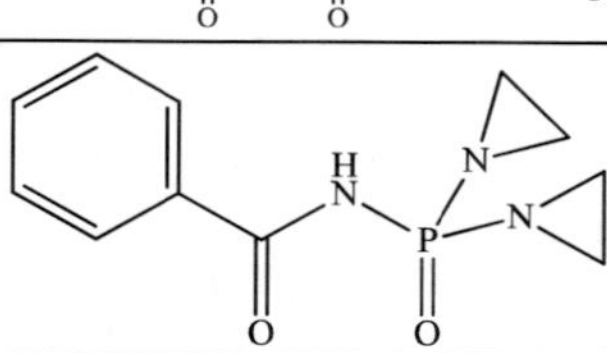	*N*- [bis(aziridin-1-yl)phosphoryl] benzamide	[40, 41]
HL14	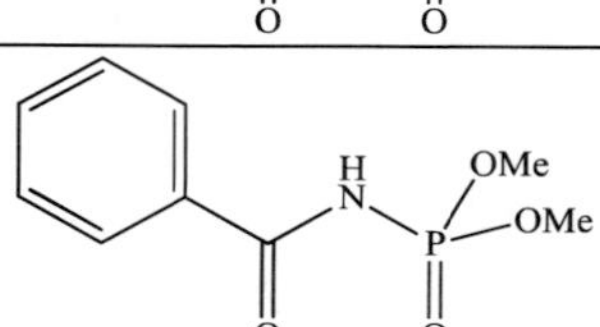	dimethyl (trichloroacetyl) amido-phosphate	[38, 18]

Ligand	Formula	Title	Ref.
HL15	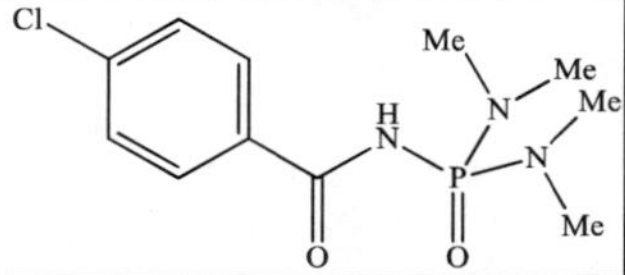	bis(dimethylamino)phosphoryl] -4-chlorobenzamide	[25]
HL16	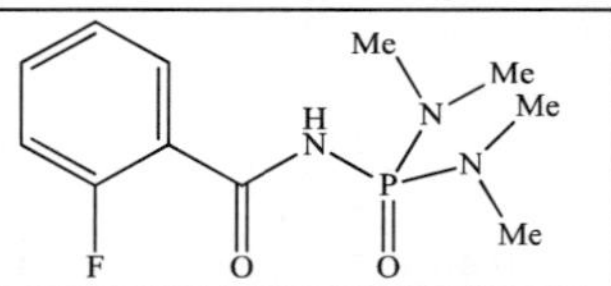	bis(dimethylamino)phosphoryl] -2-fluorobenzamide	[25]
HL17	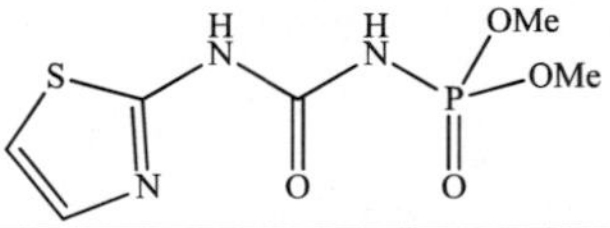	dimethyl [(1,3-thiazol-2-ylamino) carbonyl]amidophosphate	[42]
HL18	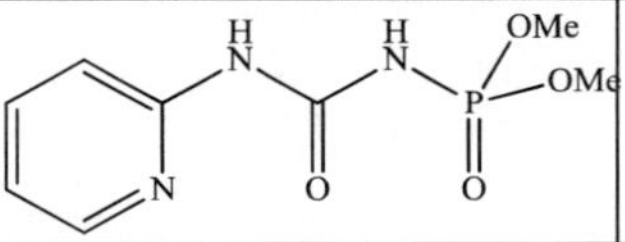	dimethyl[(pyridin-2-ylamino) carbonyl] amidophosphate	[42]
HL19	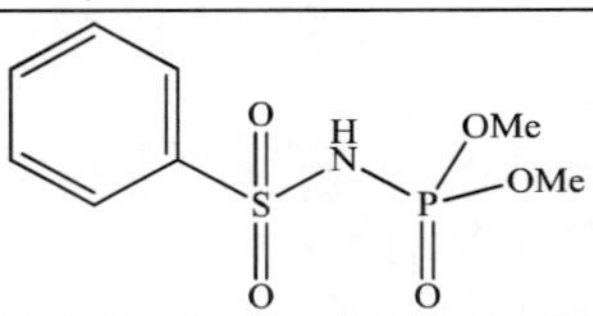	dimethyl (phenylsulfonyl)amido-phosphate	[43-47]
HL20	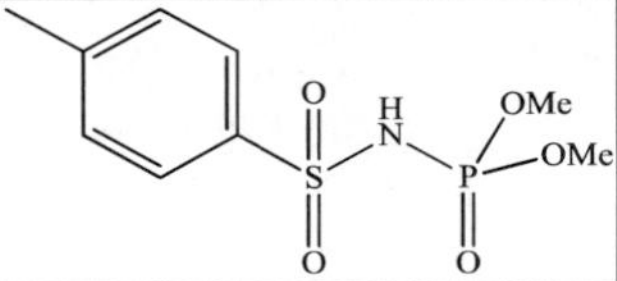	dimethyl [(4-methylphenyl) sulfonyl]amidophosphate	[44, 48, 49]
HL21	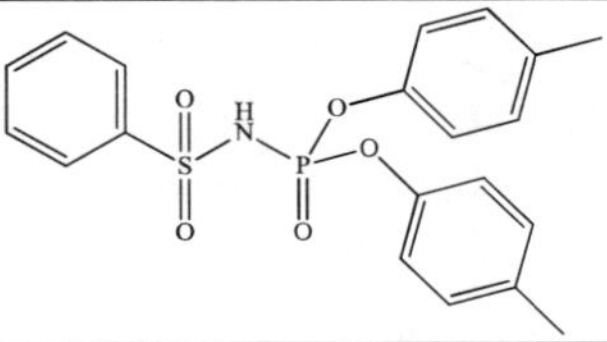	bis(4-methylphenyl) (phenylsulfonyl)amidophosphate	[50, 51]
HL22	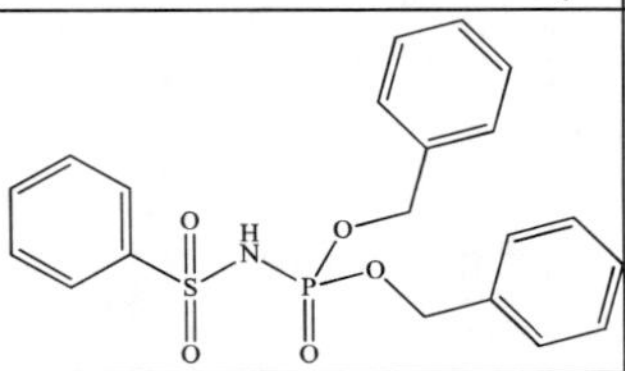	dibenzyl (phenylsulfonyl) amidophosphate	[50]

Table 1. (Continued)

Ligand	Formula	Title	Ref.
HL23	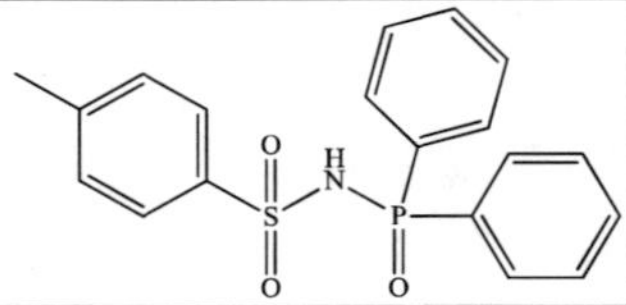	*N*- (diphenylphosphoryl) -4-methylbenzenesulfonamide	[52-56]
HL24	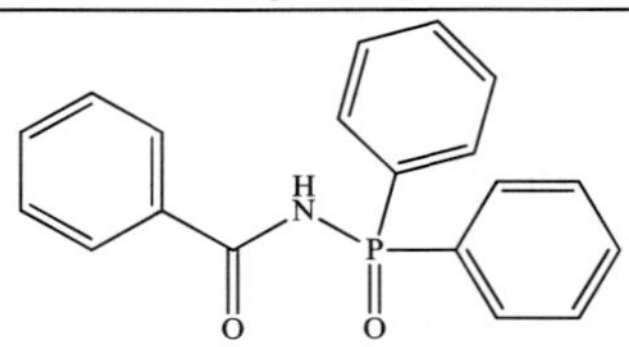	*N*-(diphenylphosphoryl)benzamide	[57-61]
HL25	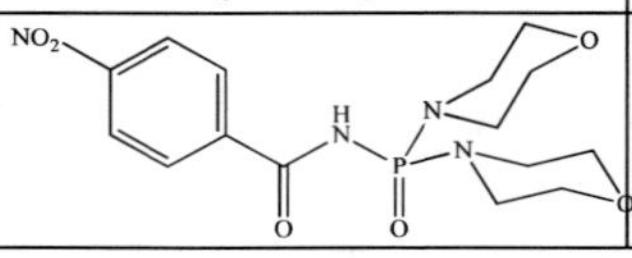	*N*- (dimorpholin-4-ylphosphoryl) -4-nitrobenzamide	[62]

2. X-RAY CHARACTERIZATION OF CAPH AND SAPH COMPOUNDS

The crystal structures of some CAPh and SAPh compounds were estimated by means of X-ray analysis. Selected structural parameters of the chelating core ($R_1E(O)NHP(O)R'_2$ where E= C or SO) and IR data are presented in Table 2.

The X-ray diffraction studies show that CAPh ligands in the crystal state usually form dimeric species: two molecules are bounded through two hydrogen -P=O…H-N- bonds. Also, they can form unlimited chains in the crystal state by means of the above-mentioned hydrogen bonds. The disposition of C=O and P=O groups differs from the ideal *anti* – configuration. Generally, according to the structural data, all CAPh compounds could be divided into two groups. The first one contains molecules with a planar functional fragment–C(O)NHP(O)= where C=O and P=O groups are located in *anti-* positions to each other. The value of the O-P-N-C torsion angle in this case is close to zero. The second group contains molecules with inhibited conformation (corresponding O-P-N-C torsion angle is near 60º). Only the latter conformation facilitates the formation of unlimited chains through hydrogen bonds. It is possible that in the –C(O)NHP(O)= core the potential barrier of rotation P(O) group around the P-N bond is very low, which is the reason for conformation lability. However, the -C-C(O)N- unit is practically planar for all the investigated compounds.

First of all, while analyzing bond lengths it is necessary to notice the difference in the structure of compounds with alkyl and aryl substituents in the carbonyl group. The length of the C-C(O) bond in the case of aryl derivatives is shorter than for alkyl ones. This could be explained by the π-conjugation of the double bond C=O with the aromatic system of the aryl substituent. The slightly longer bond lengths C=O in the case of aromatic substituents also confirm this assumption.

Usually, the angle between the planes formed by the aromatic ring and the C=O group has a value in the range 5-37°[63], which allows the above-mentioned π-conjugation to exist, and the degree of conjugation increases by decreases of this angle.

The C-N bond length for alkyl derivatives in the compounds under consideration is closer to the standard value in the amides of carbonic acids (1.32 Å), than to the single C-N bond (1.47 Å). We observed an average value for the aryl derivatives. It can be explained by the higher degree of the π-conjugation of the C-N bond with the carbonyl one in the case of alkyl derivatives. The existence of the dπ -pπ coupling between phosphor and nitrogen atoms explains the shortening of P-N bond length in comparison with the usual P-N single bond ($d(P\text{-}N)_{single} \approx 1.77$ Å). In the case of the amide type of CAPh ligands the nitrogen atoms belonging to substituents at the PO group and N atom of the chelate core C(O)NHP(O) take part in the π-conjugation. It is necessary to note that P-N bond lengths of terminal groups are always shorter than P-N distance in the C(O)NHP(O) core. This means that the degree of π-conjugation with terminal N atoms is higher than in the chelate core. The phosphor atoms have a slightly distorted tetrahedral configuration. Due to steric effects, the values of the OPX angles in the derivatives of phosphoric acid are usually larger than those of the ideal tetrahedron.

3. IR Spectra of CAPh or SAPh Compounds

The positions of ν(CO) or ν(SO) and ν(PO) absorption bands in the IR spectra of the investigated compounds are given in Table 2.

Table 2. Selected interatomic distances (Å) and IR parameters (ν(CO) or ν(SO) and ν(PO) – in cm^{-1}) for the phosphorylic compounds under consideration

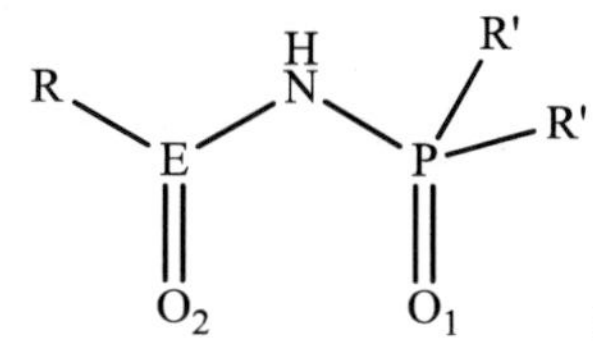

(E = C or SO)

	R	R'	R-E	$E=O_2$	E-N	$P=O_1$	P-N	P-R'*	ν(EO)	ν(PO)
CAPh, E=C										
HL1	CCl_3	NEt_2	1.549	1.207	1.338	1.475	1.696	1.625	1725	1235
HL2	CCl_3	Mrf	1.558	1.202	1.346	1.477	1.697	1.629	1728	1201
HL3	CCl_3	Pip	1.561	1.200	1.350	1.474	1.702	1,633	1729	1194
HL4	CCl_3	Pyr	1.552*	1.199*	1.345*	1.471*	1.697*	1.620	1728	1188
HL5	CCl_3	N(H)Ph	1.559	1.217	1.321	1.449	1.685	1.620	1730	1220
HL6	CCl_3	BnNH	1.574	1.212	1.353	1.480	1.709	1.630	1710	1212
HL7	CCl_3	AllNH	1.543*	1.207*	1.338*	1.475*	1.696*	1.622	1700	1215
HL8	CCl_3	N(Me)Ph	1.562*	1.203*	1.357*	1.480*	1.691*	1,642	1730	1210
HL9	CCl_3	OMe	1.566	1.202	1.347	1.459	1.676	1.550	1732	1212
HL10	Ph	NMe_2	1.519	1.235	1.387	1.498	1.709	1.656	1680	1280
HL11	Ph	NEt_2	1.495	1.216	1.373	1.478	1.684	1.632	1675	1280
HL12	Ph	Mrf	1.514	1.239	1.387	1.500	1.710	1.663	1685	1200
HL13	Ph	NC_2H_4	1.502	1.211	1.376	1.475	1.671	1.644	1682	1206

HL14	Ph	OMe	1.473	1.219	1.393	1.461	1.667	1.548	1682	1242
HL15	*p*-ClPh	NMe_2	1.496*	1.217*	1.369*	1.472*	1.678*	1.616	1685	1270
	R	**R’**	**R-E**	**$E=O_2$**	**E-N**	**$P=O_1$**	**P-N**	**P-R’***	**ν(EO)**	**ν(PO)**
HL16	*o*-FPh	NMe_2	1.499	1.204	1.368	1.475	1.687	1.631	1680	1250
HL24	Ph	Ph	1.491*	1.226*	1.37*	1.479*	1.686*	1.789	1670	1199
HL25	*p*-NO_2-Ph	Mrf	1.510	1.220	1.377	1.481	1.688	1.634	1672	1196
SAPh, E=SO										
HL19	Ph	OMe	1.760	1.423	1.6421	1.4562	1.6617	1.549	1320 1185	1260
HL22	Ph	OCH_2Ph	1.760	1.423	1.6430	1.4589	1.6512	1.565	1345 1180	1255

where BnNH = benzylamine, Mrf = morpholine, Pip = Piperidine, Pyr = Pyrrolidine, AllNH = allylamine.

* - average value.

For the majority of CAPh compounds the most typical in the IR spectra is the absorption band of the carbonyl group ν(CO). The region 1670-1730 cm-^{1} is characteristic for the vibrations of the double C=O bond. The position of this band strongly depends on the nature of the substituents at the carbonyl atom. For example, in the case of compound **1** (alkyl substituent at the carbonyl group) the ν(CO) band is situated at 1725 cm^{-1}. The substitution of the trichloromethyl group for a more electronegative phenyl group (compound **11**) leads to a low frequency shift to 1675 cm^{-1}.

Assignment of the absorption bands of the phosphoryl groups is more difficult due to the complicated character of the spectra. Nevertheless, the comparative analysis of the IR spectra of the compounds under consideration helps us to estimate more or less reliable data.

The intense absorption observed around 3060 cm^{-1} for HL belongs to the ν(NH) stretching vibrations. The low frequency value of this band explains the fact that hydrogen atoms of the amide group participate in the formation of NH···OP type hydrogen bonds. Surely, this band disappears in the spectra while deprotonating of HL. All the ν(NH) bands overlap with each other in the IR spectra of the HL.

According to the IR spectra of sodium salts after deprotonation, the order of the CO and PO bonds decreases for all CAPh compounds due to an efficient π-conjugation in the OCNPO fragment.

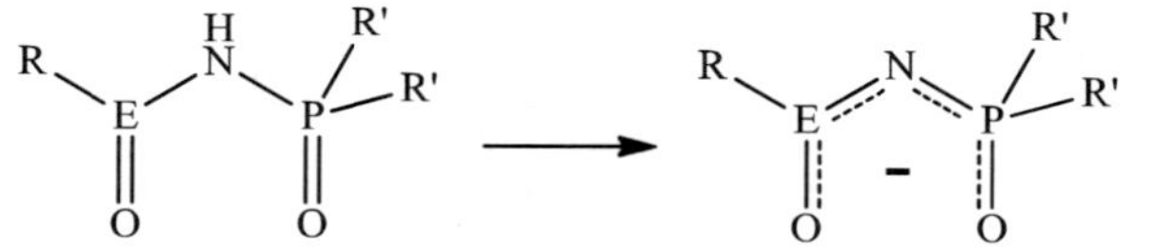

The PO vibrations are less sensitive because of P-O and P-N vibration modes coupling. Infrared spectroscopic investigations revealed a bathochromic shift of valence vibrations of the CAPh ligands (HL) in comparison with their sodium salts NaL by values of $\Delta\nu(C{=}O) = 15\text{-}40\ cm^{-1}$ and $\Delta\nu(P{=}O) = 30\text{-}90\ cm^{-1}$.

Assignment of the characteristic absorption bands in the IR spectra of SAPh compounds (ν_{as}(SO), ν_s(SO) and ν(PO)) becomes a more complex task because many other bands are presented in this region. The data for several SAPh compounds with different substituents at the functional fragment OS(O)NPO should be considered for the reliable assignments [64-66]. In addition, the results of quantum chemistry calculations should be taken into account [67-69]. Decreasing of ν_{as}(SO), ν_s(SO) and ν(PO) vibrations is observed by deprotonation of the SAPh ligands, which can be explained by reduction of the orders of double bonds because of electron density delocalization in the OS(O)NPO fragment.

4. Bioactivity of Representatives of CAPh Compounds

The presence of the peptide and phosphoryl groups combined together in one molecule could be the main reason for the interesting biological activities. Some of the CAPh compounds show a wide range of biological activities, especially their ethylenimine derivatives, which are applied in diagnostics and cancer therapy [41]. Complex formation with such ligands can be considered as an instrument to influence on the biochemical activity of drugs, in particular their toxicity and selectivity [70], as demonstrated *in vitro* for lanthanide complexes with N-benzoyl-N',N',N",N"-di-ethylenephosphoryltriamide ("Benzotef").

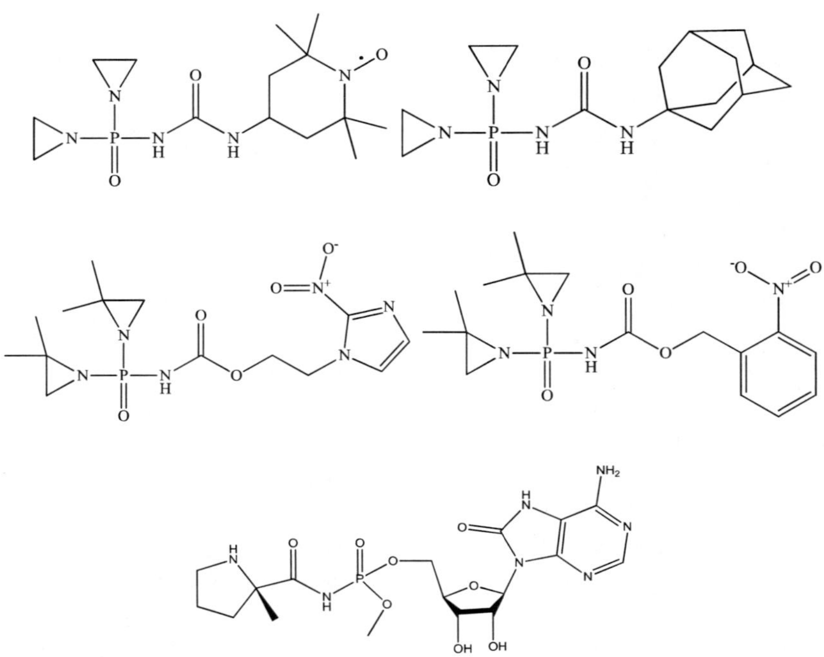

Figure 6. Some anticancer drugs, which can be attributed to CAPh compounds.

Several anticancer drugs [71, 72, 73], have been introduced into clinical oncology and applied against different kinds of carcinoma, and can also be attributed to the CAPh compounds.

5. Lanthanide Complex with CAPh and SAPh

5.1. General Information about Lanthanide Complexes

The lanthanide chelates with strong luminescence properties continue to be of great interest because of their potential to be exploited in applications ranging from biomedical to sensing areas and luminescence imaging; this led to most interesting developments in the coordination chemistry of these ions. The continued development and refinement of the lanthanide agents relies on the ability to construct complexes that meet fairly rigorous requirements of stability in aqueous media containing relatively large concentrations of competing cations, high-luminescence intensity and accessible chemistry to design an overall structure that is highly selective to the system of interest [74].

So far, Eu^{3+} and Tb^{3+} complexes have been used most often as luminescent markers and probes. The luminescence of these ions is characterized by: large Stokes shift upon excitation to the ligand state, narrow emission bands and long luminescence decay time, which makes them ideal for application in the hetero- and homogeneous fluoroimmuno assay methods [75, 76, 77], as labels of proteins and nucleic acids [78], and as various luminescent probes [79, 80, 81, 82, 83, 84]. For the applications, for example, in quadruple-plex compounds, fluoroimmuno assay next to Eu^{3+} and Tb^{3+} compounds studies of Dy^{3+} and Sm^{3+} complexes are also carried out [85,86]. Lanthanide ions, such as Er^{3+}, Nd^{3+} and Yb^{3+}, emitting in the near infra-red (NIR) range, can be excited through an organic ligand or d-elements chromophors using lower energy than for excitation of Eu^{3+} and Tb^{3+}. This is beneficial for medical imaging methods due to the fact that lower energy radiation does not damage tissues, and penetrates deeper into the tissues at the same time. The biological material has the so-called spectral window (no absorption in range about 0.8 – 1.0 μm) within the range of emission of Nd^{3+}, Er^{3+} or Yb^{3+} ions [87, 88,89].

The Ln^{3+} centers cannot efficiently absorb light (the extinction coefficient for such trivalent cations is $\varepsilon \approx 1–10 M^{-1}\ cm^{-1}$) [52]; the way of choice to sensitize their luminescence is to employ chromophores. The strong sensitized photoluminescence of lanthanide is achieved by excitation of a chelating chromophore to its singlet state and subsequent effective transfer of the energy absorbed to the metal ion excited state – the so-called antenna effect. In this way it is possible to overcome the low value of the molar absorption coefficient, associated with intra-configurational 4f-4f transitions of lanthanide ions. The antenna effect makes it possible to create luminophores with various application capabilities, depending on the choice of lanthanide ion as well as the organic ligands. The choice of an organic ligand and its modification by, e.g., bonding various functional groups, enables control of the process of converting energy in the very broad range of UV-Vis. Due to a large Stokes shift, this type of compound falls to the light converting molecular devices (LCMDs). In the light conversion process the values that influence emission intensity include: molar absorption coefficient of chromophore, ligand-to-metal energy transfer efficiency, emission quantum yield of the lanthanide ion.

A review that describes the basic principles for the design of organic ligands for alkali and alkali-earth cations was reported by Lehn in 1973 [90] and later by other Authors [91,92]. The design of efficient luminescent lanthanide-based probes for application in biological systems also requires detailed knowledge concerning solution structure and dynamics, and the influence that these have on the important photochemical and photophysical properties. However, exact information about the structure of a complex can only be obtained from the solid state through diffraction techniques. Furthermore, the investigation of the solid-state structure allows the probing of interactions between the complex and its environment in solutions, and specifically the interaction with solvent water molecules, which are potentially important.

5.2. X-Ray Data of Lanthanide Complexes with SAPh and CAPh

The different types of coordination compounds were found in lanthanide complexes with CAPh and SAPh ligands.

Structures of some representatives of Ln complexes with CAPh and SAPh ligands with indication of type coordination of ligands are given in Table 3.

In the first type of coordination mode was observed only for complexes with neutral ligands, where the chelate ligand coordinates as a monodentate through the oxygen atom of the phosphoryl group (**C1 – C3**).

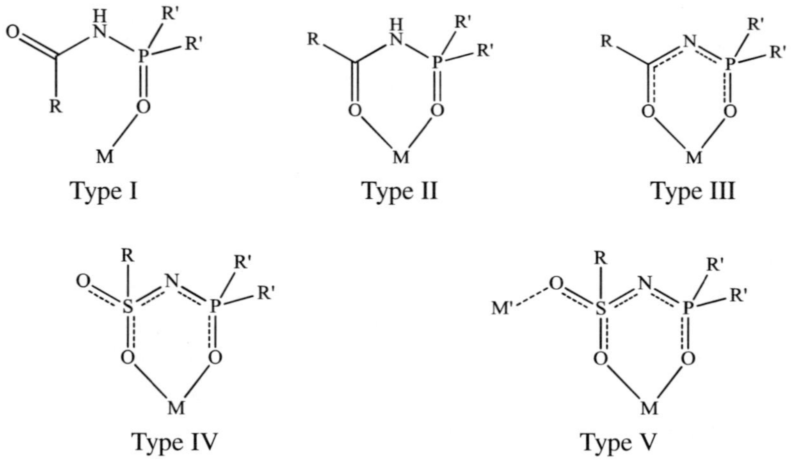

Figure 7. Types of CAPh and SAPh ligands coordination in lanthanide complexes.

In the second type the ligand coordinates in molecular form as a chelate through both donor oxygen atoms (O=P and O=C) of ligands (**C4**).

The third type of coordination is the most common. This bidentate mode is observed for the complexes with deprotonated ligands (**C5**-**C8**).

A few crystal structures of lanthanide complexes with SAPh ligands have been reported in the literature up to date (**C9**-**C15**). The first crystallographic data of Nd^{3+} complex with dimethyl(phenylsulfonyl)amidophosphate (**C15**) was published in 2007 [93]. A series of monomeric Yb^{3+} complexes with this ligand was published in 2011 [43]. A second series of heterometallic one-dimensional coordination polymers, assembled from deprotonated SAPh ligands, have been investigated [94]. The lanthanide ions in all these compounds are eight coordinated. The most common coordination polyhedra, in complexes with SAPh ligands, are slightly distorted square antiprisms. Only in the structure of $Na[Tb(\mathbf{L22})_4]_n$ (**C11**) coordination polyhedron around the eight-coordinated metal center is a slightly distorted dodecahedron. In all cases the geometrical arrangements around the metal center are imposed by the small chelating O–M–O angle of the SAPh ligands (av. over value of these chelate O-M-O angles is 73°). In structures of coordination polymers, the π–π stacking interactions play an important role in crystal packing.

Table 3. Some representative complexes with CAPh and SAPh ligands

Complex	View of structure	Name	Ref.
C1-HL11 (type I)	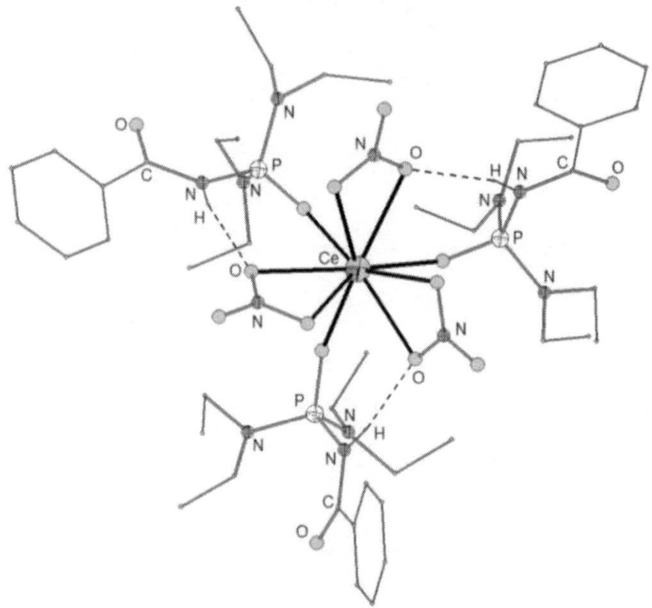	*tris*(N,N,N',N'-Tetraethyl-N''-benzoylphosphoryltriamide-O) -*tris*(nitrato-O,O') -cerium(III), [Ce$(NO_3)_3$(**HL11**)$_3$]	[95]

Table 3. (Continued)

Complex	View of structure	Name	Ref.
C2-HL1 (type I)	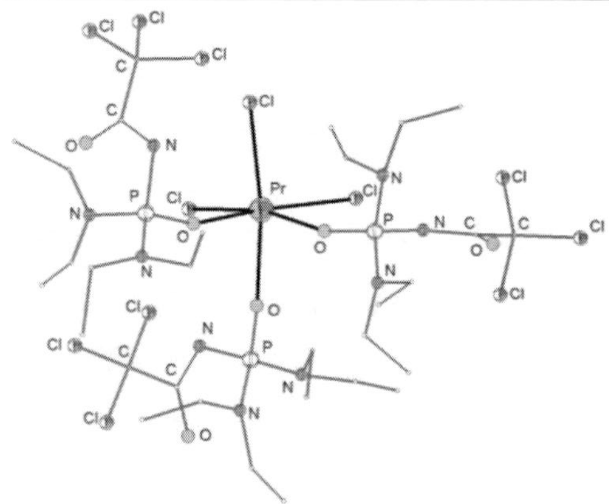	Trichloro-*tris*(*bis*(diethylamido) -trichloroacetylamido-phosphoric acid) -praseodymium(III), $[PrCl_3(\mathbf{HL1})_3]$	[96]
C3-HL25 (type I)	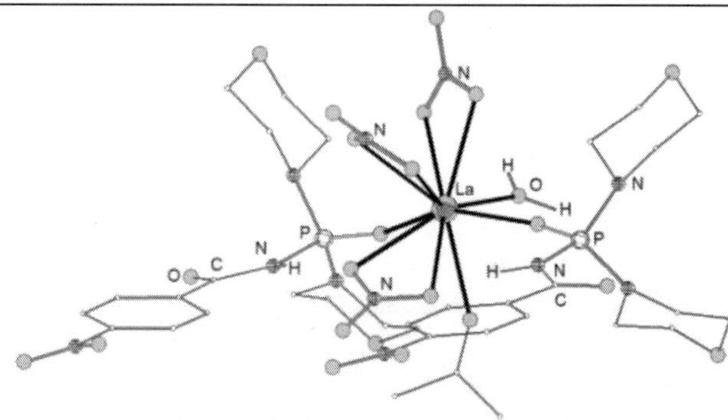	Acetone-aqua-*bis*(N-(*bis*(morpholin-4-yl) phosphoryl) -4-nitro benzamide-O) -tris(nitrato-O,O') -lanthanum(III), $[La(NO_3)_3(\mathbf{HL25})_3(H_2O)((CH_3)_2CO)]$	[62]

Complex	View of structure	Name	Ref.
C4-HL10 (type II)	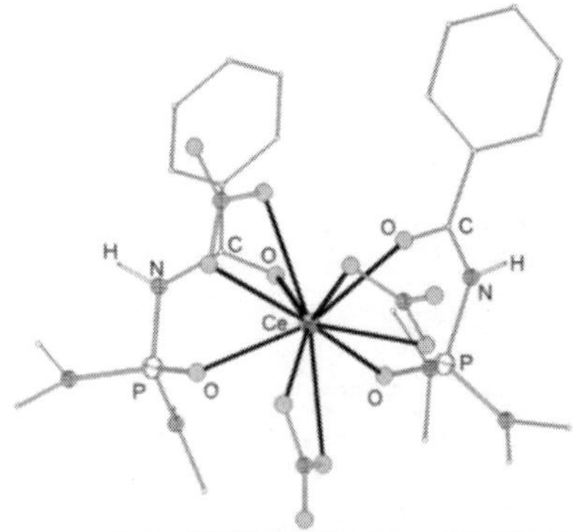	*bis*(N-benzoyl-N',N',N",N"-tetramethylphosphoryltriamide-O,O') -*tris*-(nitrato-O,O') -cerium(III), $[Ce(NO_3)_3(\mathbf{HL10})_2]$	[97]
C5-L1 (type III)	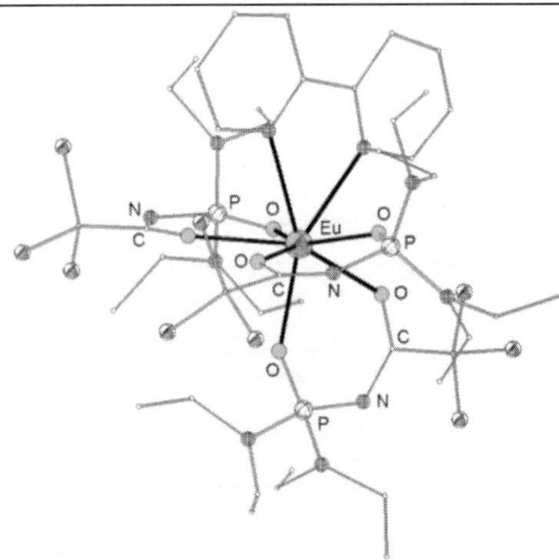	*tris*(N,N,N',N'-Tetraethyl-N"-trichloroacetylphosphoramido) - (2,2'-bipyridyl) - europium(III), $[Eu(\mathbf{L1})_3bpy]$	[98]

Table 3. (Continued)

Complex	View of structure	Name	Ref.
C6-L9 (type III)	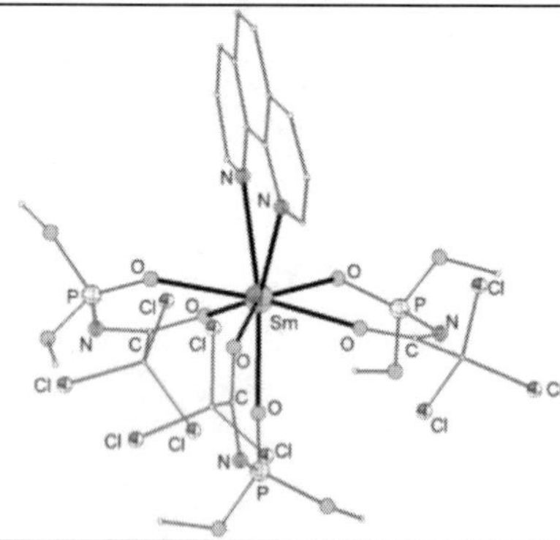	*tris*(Dimethyl-N-trichloroacetylamidophos phate) - (1,10-phenanthroline) - samarium(III), [Sm(**L9**)$_3$phen]	[99]
C7-L9 (type III)	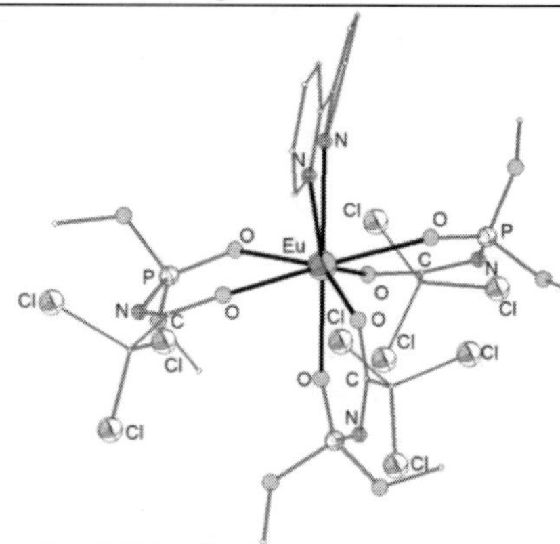	*bis*(*tris*(Dimethyl N-(trichloroacetyl)amidoph osphate-O,O') - (2,2'-bipyridyl) - europium(III)) 2,2'-bipyridyl, [Eu(**L9**)$_3$bpy]	[14]

Complex	View of structure	Name	Ref.
C8-L9 (type III)	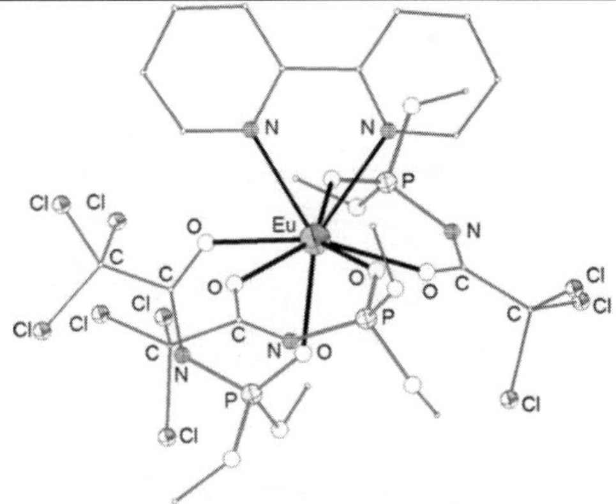	*tris*(Dimethyl N-(trichloroacetyl) amidophosphate-O,O') - (2,2'-bipyridyl) - europium(III), [Eu(**L9**)$_3$bpy]·bpy	[100]
C9-L19 (type IV)	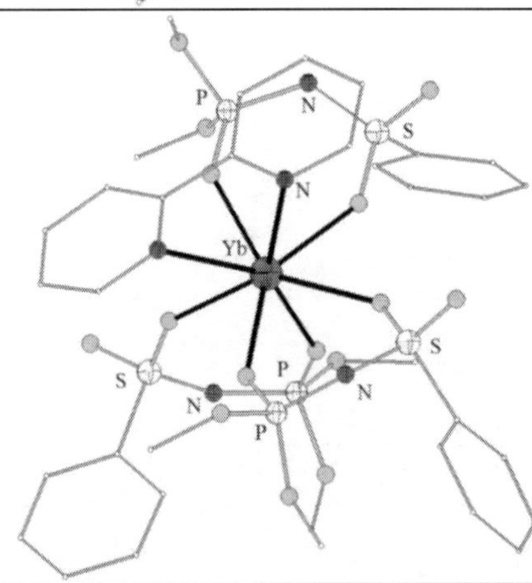	(2,2'-Bipyridine-N,N') - *tris*(dimethyl (phenylsulfonyl)phosphoramidato-O,O') - ytterbium(III) monohydrate [Yb(**L19**)$_3$bpy]·H_2O	[43]

Table 3. (Continued)

Complex	View of structure	Name	Ref.
C10-L19 (type IV)	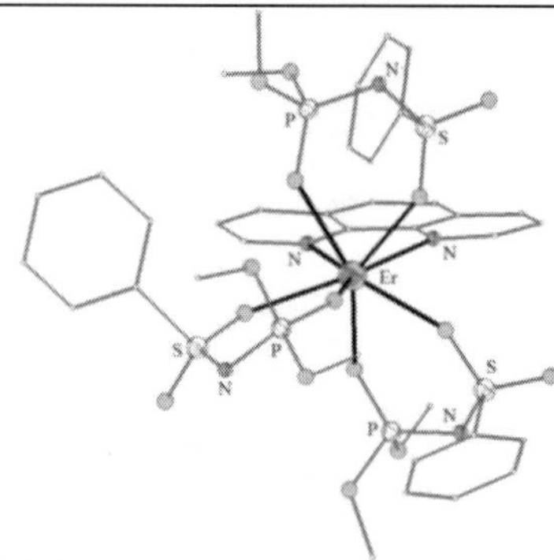	*tris*(Dimethyl (phenylsulfonyl) phosphoramidato-O,O') - (1,10-phenanthroline-N,N') -erbium(III), [Er(**L19**)$_3$phen]	[43]
C11-L22 (type IV)	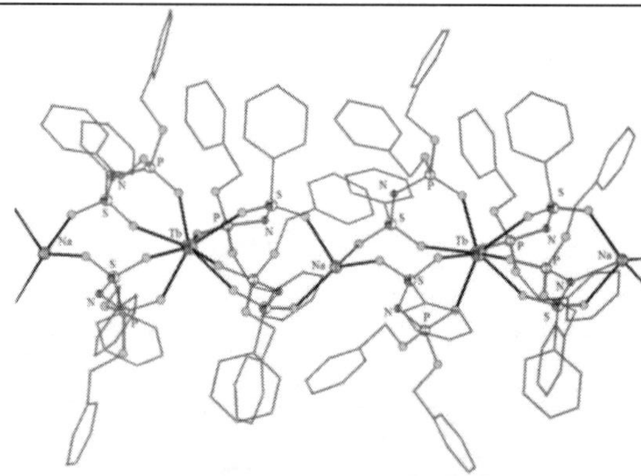	sodium *tetrakis*((*bis*(benzyloxy) phosphoryl)(phenylsulfonyl)azanide) – terbium(III), {Na[Tb(**L22**)$_4$]}$_n$	[94]

Complex	View of structure	Name	Ref.
C12-L23 (type IV)	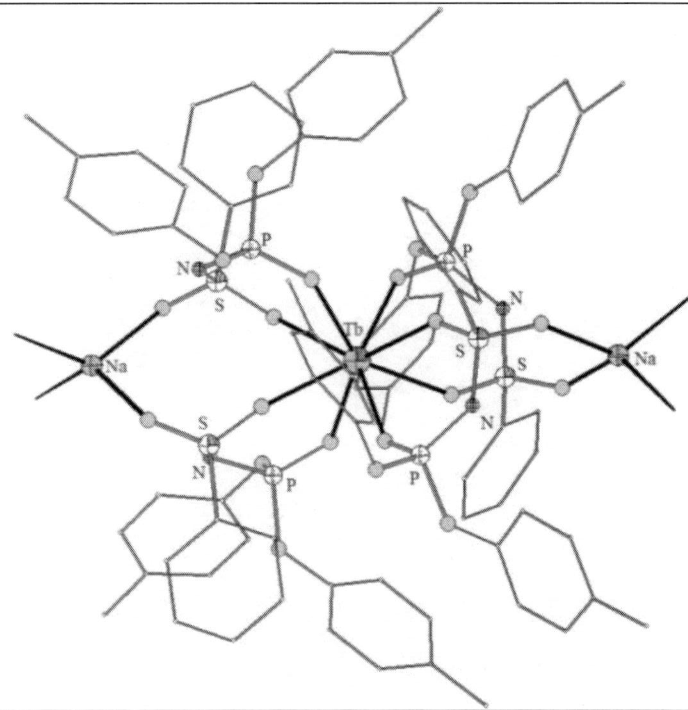	sodium *tetrakis*((*bis*(4-methylphenoxy)phosphoryl)(phenylsulfonyl)azanide) -terbium(III), $\{Na[Tb(\mathbf{L21})_4]\}_n$	[94]

Table 3. (Continued)

Complex	View of structure	Name	Ref.
C13-L22 (type IV)	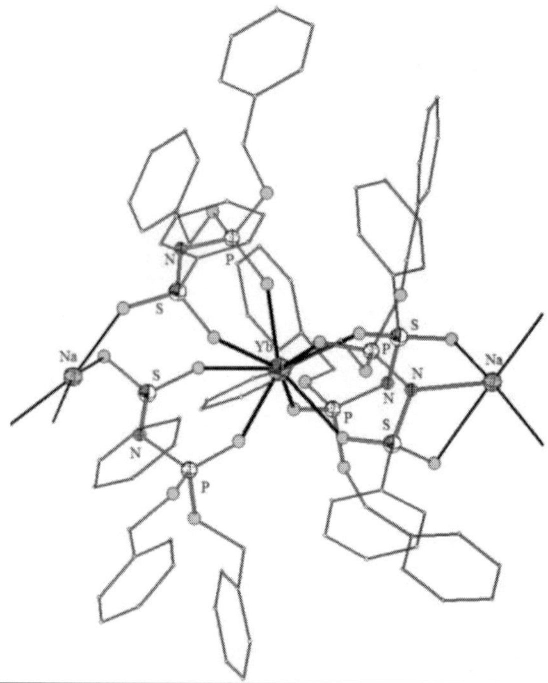	sodium *tetrakis*((*bis*(benzyloxy) phosphoryl)(phenylsulfonyl)azanide) - ytterbium(III), $\{Na[Yb(\mathbf{L22})_4]\}_n$	[94]

Complex	View of structure	Name	Ref.
C14-L23 (type IV)	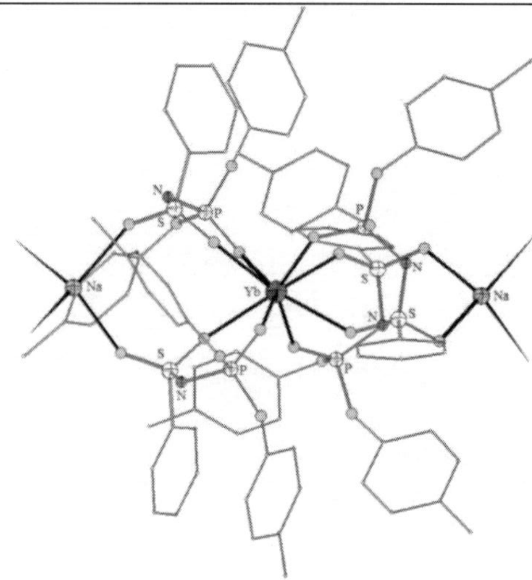	sodium *tetrakis*((*bis*(4-methylphenoxy) phosphoryl)(phenylsulfonyl)azanide) - ytterbium(III), $\{Na[Yb(\mathbf{L21})_4]\}_n$	[94]
C15-L19 (type IV)	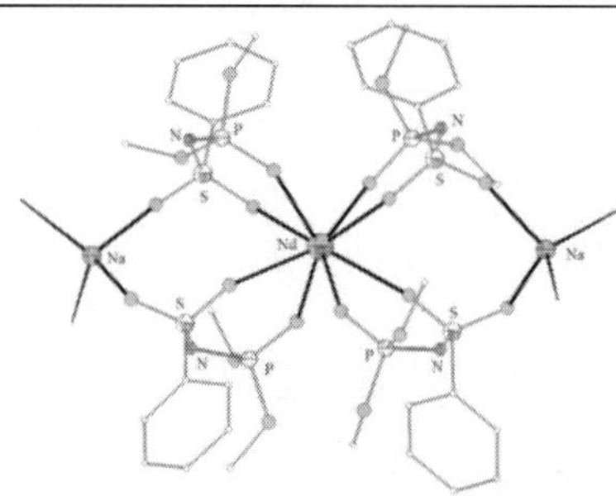	*catena*-Poly[neodymate(III)*bis*[μ-dimethyl (phenylsulfonyl) amidophosphato]sodium (I) *bis*[μ-dimethyl (phenylsulfonyl)amidophosphato]], $\{Na[Nd(\mathbf{L19})_4]\}_n$	[93]

Table 4. Selected geometric parameters (Å, °) for some representative complexes with CAPh and SAPh ligands (the average values)

No	M	P-O1	E-O2	M-O_1	M-O_2	O_1-M-O_2	P-N-E	O1-P-N	O2-E-N	O_1-P-N-E	M-O_1- P-N	M-O_2-E-N
Type I E = C												
C1	Ce	1.485	1.215	2.402	-	-	126.69	104.15	121.61	52.27	-2.51	-
C2	Pr	1.470	1.204	2.357	-	-	122.59	102.92	124.22	-174.16	-39.38	-
C3	La	1,491	1.215	2.417	6.490	25.05	128.30	103,71	121,34	-178.95	-22.64	-2.76
Type II E = C												
C4	Ce	1.494	1.244	2.444	2.544	69.08	125.81	109.63	121.61	25.58	-17.99	-34.31
Type III E = C												
C5	Eu	1.496	1.257	2.328	2.419	72.85	122.22	117.01	131.98	3.64	-3.62	-0.70
C6	Sm	1.473	1.243	2.352	2.427	72.52	122.16	118.47	130.95	-5.49	-3.97	11.62
C7	Eu	1.464	1.244	2.332	2.413	73.22	122.57	118.37	131.59	6.10	2.97	-6.76
C8	Eu	1.474	1.253	2.324	2.407	73.57	123.13	118.33	131.16	-0.16	-8.40	3.72
Type IV E = S												
C9	Yb	1.490	1.471	2.247	2.36	74.76	124.11	117.38	112.78	-10.35	-12.65	-8.88
C10	Er	1.491	1.471	2.269	2.382	73.79	122.48	117.38	112.53	-17.35	-3.00	-0.21
Type V E = S												
C11	Tb	1.491	1.476	2.315	2.458	73.20	123.56	116.93	113.26	-31.38	-9.72	-1.20
C12	Tb	1.492	1.467	2.315	2.466	72.80	123.54	117.92	113.30	-11.11	-14.16	-4.07
C13	Yb	1.489	1.464	2.226	2.406	75.44	122.58	117.72	113.43	1.51	-4.12	-1.36
C14	Yb	1.489	1.463	2.256	2.428	73.43	122.90	117.71	113.33	-10.97	-13.54	-4.15
C15	Nd	1,487	1,461	2,382	2,496	72,67	126,84	117,04	113,96	9,11	5,50	5,32

5.3. Spectroscopic Properties of Lanthanide Complexes with SAPh and CAPh Ligands

The complexes with SAPhs ligands constitute a new class of complexes with lanthanide ions; until 2007 no crystallographic data [93], and until 2010 – no spectroscopic data [101] had been available on complexes of this type. There are also only a few publications concerning spectroscopic characterization of the rare-earth compounds with CAPh

Carbacylamidophosphates and sulfonylamidophosphates are the structural analogues of *β*-diketones, which are known to be excellent emission sensitizers in complexes with lanthanide ions [102]. In the CAPhs and SAPhs ligands, C=O vibrations (~1600 cm^{-1}), present in *β*- diketones, have been reduced (CAPh) or eliminated (SAPh). They have been replaced with the lower energetic vibrations P=O (~1250 cm^{-1}) and S=O (~1350 cm^{-1}). Replacing the carbon atom at the same time, which is in *β*-diketones, with the nitrogen atom eliminates the high-power vibrations of C-H from the six-membered chelate ring, which is created as a result of coordination with the lanthanide ion. Therefore, the structure of CAPhs and SAPhs ligands favors decreasing of the multiphonon quenching of the lanthanide emission to a greater extent than the structure of *β*- diketone ligands. This has particular significance in the case of the lanthanide ions with a small energy gap (Sm^{3+}, Dy^{3+}, Nd^{3+}, Er^{3+}) between the energy levels, among which the emission process takes place. It is also essential in the case of Yb^{3+}, whose energy gap is approximate to Eu^{3+} (Eu-circa 12000 cm^{-1}, Yb-circa 10000 cm^{-1}) due to the fact that the emission of Yb^{3+} falls in the infrared range (960 – 1050 nm), where the detection sensitivity is much smaller than it is for the emission in the visible range. Moreover, the presence of the phosphorus atom in these ligands enables the possibility of bonding the additional chromophore in relation to the carbon atom, which enables stronger sensitizing of the emission of the lanthanide ions. The amidophosphates have created thermodynamically stable complexes, which do not undergo degradation when exposed to UV laser radiation or even synchrotron radiation.

5.3.1. Spectroscopic characterization of lanthanide complexes with CAPHs

The following types of complexes were obtained and investigated: $Na[Ln(CAPh)_4]\cdot xH_2O$ with **5**, **9**, **10** ligands [70, 103,104] (where Ln = Pr^{3+}, Nd^{3+}, Eu^{3+}, x = 1 or 2); $[Ln(\mathbf{L1})_3Cl_3]$ and $[Ln(\mathbf{L1})_3(NO_3)_3]$ [105,106] (where Ln = Pr^{3+}, Nd^{3+}, Eu^{3+}); $[Ln(\mathbf{L9})_3bpy](1/2bpy)$ and $[Ln(\mathbf{L9})_3bpy]$ [14, 100]

(where bpy – 1,2-bipyridine, Ln = Eu^{3+}, Gd^{3+}, Tb^{3+}) as well as [Ln(**L9**)$_3$phen] [99] (where phen – 1,10- phenantroline, Ln = Sm^{3+}, Eu^{3+}, Gd^{3+}, Tb^{3+}); Ln(**HL11**)$_3$(NO_3)$_3$ (where Ln = La^{3+}, Ce^{3+}, Nd^{3+}, Sm^{3+}, Lu^{3+}, Y^{3+}) [107].

The correlation between structure and spectroscopy was investigated for all the complexes. The strong interaction of the ligands with the lanthanides was confirmed by IR, Raman and electronic spectra.

For the Na[Ln(CAPh)$_4$]·xH_2O complexes with **5**, **9**, **10** ligands normal coordinate analysis (NCA) was performed and IR and Raman spectra were analyzed. The complex motions of the chelate rings are characterized by three types of vibrations: chelate ring breathing $\nu(\phi_n)$, in-plane bending $\delta(\phi n)$ and out-of-plane bending $\Pi(\phi n)$. These modes are observed in the 625–840, 420–540 and 180–395 cm^{-1} regions. Some of these bands are split due to coupling with ν(Ln-O) and δ(OLnO) vibrations. It was shown that the ν(Ln-O) stretching vibrations are strongly coupled with chelate ring deformations. Therefore, the bands sensitive to changes in the metal could be observed, even above 400 cm^{-1}. These bands are observed in the broad frequency region 630–320 cm^{-1}. The broad and middle intensity bands observed in the IR and Raman spectra in the 50–280 cm^{-1} region are described as δ(OLnO) and γ(OLnO) motions. The NCA calculations performed allow the description of the weak satellite lines observed in the absorption spectra at a high-energy range of electronic transitions, as well as in excitation and emission spectra. Several of these lines correspond to the ν(CO), ν(PO), ν(PNC), ν(Ln-O) and δ(OLnO) vibrations. The participation of these modes in the electron energy transfer confirms the existence of the strong coupling between the lanthanide ion and chelate ring, as well as inside the chelate ring alone. One can observe very intensive vibronic components in the excitation spectra in the high energy range of $^7F_0 \rightarrow {}^5D_2$ transition, corresponding to ν_S(P-O) and ν_S(C-O) stretching vibrations. This confirms very strong M-OP bonding with the lanthanide ions. [104] Additionally, it was demonstrated that substitution of the -NP(O)(OCH_3)$_2$ ligand group by NP(O)(N(CH_3)$_2$)$_2$ one in the Na[Ln(**L10**)$_4$]·H_2O complex changes the biological function of the Na[Ln(**L9**)$_4$]·H_2O compound. In vitro cytotoxicity experiments with lanthanide compound exhibit the anticancer effect of complex and free ligands on the cell strains tested [70].

In the complexes with the L:M molar ratio 3 (types [Ln(**L1**)$_3$$Cl_3$] and [Ln(**L1**)$_3$($NO_3$)$_3$]) it is possible to design the symmetry of the coordination environment by the suitable choice of an anion. The complex with chloride anions has coordination number (CN) 6, while the complex with nitrate anions has CN 9 (nitrate anions are bidentate ligands). Analysis of the electronic components in the low temperature spectra point to the D_3 symmetry site in [Ln(**L1**)$_3$Cl$_3$] and most probably C_{2v} in [Ln(**L1**)$_3$(NO$_3$)$_3$]. The changes in oscillator strength values versus temperature for [Pr(**L1**)$_3$Cl$_3$] and comparison of these for both types of praseodymium compounds: [Pr(**L1**)$_3$Cl$_3$] and [Pr(**L1**)$_3$(NO$_3$)$_3$] indicate the contribution of the dynamic coupling mechanism as well as the role of the polarizability mechanism in f–f transition intensities. Vibronic components are related to the internal ligand modes and to the localized Ln–L modes: νM–O(NO$_3$), νM–Cl and M–O(HX) as well as internal ligand modes. The P–O mode promotes the strongest vibronic components, thus confirming the coordination mode of ligand molecules, which corresponds well to the X-ray analysis. The resonance effect, in vibronic coupling with the δ(PN) mode, manifests as the additional splitting (of the order of 6 cm^{-1}) of two electronic components of the $^7F_1$ level in the emission spectra of [Eu(**L1**)$_3$Cl$_3$] at low temperature. [105, 106]

In the [Eu(**L9**)$_3$bpy](1/2bpy), [Eu(**L9**)$_3$bpy] and [Ln(**L9**)$_3$phen] complexes, where Ln = Sm^{3+}, Eu^{3+}, Tb^{3+} the photophysical properties were analyzed. The ligand-to-metal energy transfer was proved for europium and terbium complexes. The LMCT (ligand – to – metal charge transfer) the state of Eu^{3+} was localized by comparison of Eu^{3+} and Tb^{3+} excitation spectra. The LMCT states for [Ln(**L9**)$_3$phen] and [Eu(**L9**)$_3$bpy](1/2bpy) are at approximately: 27 397 cm^{-1} (overlapped by ligand transitions) and 29 620 cm^{-1}. In the latter case the localization of the LMCT state between the singlet and triplet states of the ligand suggests its important role in the energy transfer processes. The lowest energy of the ligand triplet state in the [Eu(**L9**)$_3$bpy](1/2bpy) complex was localized at 23 810 cm^{-1} based on the phosphorescence spectra of [Lu(**L9**)bpy(1/2bpy)] at 77 K. The efficiency of the energy transfer process at 298 K as well as at 77 K seems to be larger in compound [Eu(**L9**)$_3$bpy](1/2bpy) than [Eu(**L9**)$_3$bpy], for which the higher relative intensity of the ligand band compared with the narrow Eu^{3+} transitions is observed. This is most probably caused by the contribution of intermolecular energy transfer via non-coordinated 2,2-bipyridine. The luminescence decay times of [Eu(**L9**)$_3$bpy](1/2bpy) and [Eu(**L9**)$_3$bpy] compounds in the solid state are almost identical and relatively large (~1.8 ms) at 298 K, and they were found to be temperature-independent. Based on the comparison of the IR and emission spectra, a resonance effect was discovered in all [Eu(**L9**)$_3$bpy](1/2bpy), [Eu(**L9**)$_3$bpy] and [Eu(**L9**)$_3$phen] compounds, and δ(OPO) or ν(OLnO)*π (φ); δ(CCl$_3$)* δ(ϕ_n) modes were determined to be responsible for this phenomenon. [14, 99, 100]

5.3.2. Photophysical properties of lanthanide complexes with SAPh ligands

Important parameters characterizing the emission of Ln^{3+} ions are: (1) the decay time of the excited state $\tau_{obs} = 1/k_{obs}$ with k_{obs} being the constant rate (in s^{-1}) of the de-population of the excited state; (2) the quantum yield Q, which is equal to the ratio between the number of emitted photons divided by the number of absorbed photons. If the metal ion is directly excited into a 4f level, these two quantities are related by:

$$Q_{Ln}^{Ln} = \tau_{obs}/\tau_{rad} = k_{rad}/k_{obs}$$

where k_{rad} is the radiative rate constant. The value of Q_{Ln}^{Ln}, the intrinsic quantum yield, reflects the extent of non-radiative deactivation processes occurring both in the inner- and outer-coordination spheres of the metal ion. The so-called overall quantum yields have been obtained by excitation into the metal–ion surroundings, with η_{sen}, the sensitization efficiency, defined as the efficacy with which energy is transferred from the feeding levels of the metal-ion surroundings onto the Ln^{3+}excited states [91].

$$Q_{Ln}^{L} = \eta_{sen} Q_{Ln}^{Ln}$$

There are many different factors influencing the overall quantum yield of the metal-centered luminescence, e.g., the rate of the intersystem crossing and the rate of the ligand-to-metal energy transfer, the presence of high-energy vibrations deactivating an excited state of the metal ion. The localization of the low-lying ligand-to-metal charge transfer state (LMCT) in relation to the excited states of Eu^{3+} is very important, since the influence of the LMCT on quenching of the Eu^{3+} emission is well documented [108, 109]. Also, the localization of the ligand triplet state and ligand excited singlet state in relation to each other and in relation to the Ln^{3+} excited states are very important.

The following types of complexes were obtained and investigated: $Na[Ln(SAPh)_4]$ with **19 21** and **22** ligands (where Ln = Nd^{3+}, Eu^{3+}, Gd^{3+}, Tb^{3+}, Yb^{3+}, Lu^{3+}) and [Ln(L**19**)$_3$L’] with additional bidentate chelate ligands (L’): 2, 2’ - bipirydine (bpy) or 1, 10 - phenantroline (phen).

Yb^{3+} and Nd^{3+} complexes

To determine the energy of the ligand excited singlet state the reflection spectra of Na[Ln(**L19**)$_4$], Na[Ln(**L21**)$_4$], Na[Ln(**L22**)$_4$] and [Ln(**L19**)$_3$L] type complexes were recorded. They are shown in Figure 8. Creating six-membered chelating rings with a π-electron system in the complexes is manifested by a hypsochromic shift of absorption bands of free ligands for about 10 nm. The energy of the ligand excited triplet states was determined on the basis of the phosphorescence spectra of Gd^{3+} or Lu^{3+} complexes. Table 5 summarizes the energies of both states.

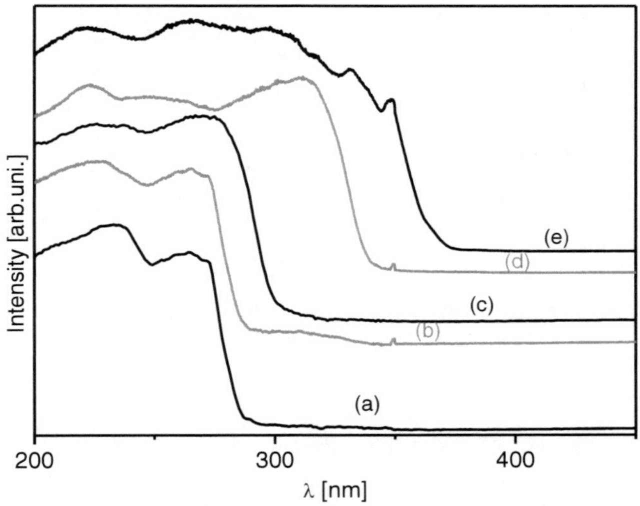

Figure 8. Reflection spectra of (a) Na[Gd(**L22**)$_4$], (b) Na[Gd(**L19**)$_4$], (c) Na[Lu(**L21**)$_4$], (d) [Gd(**L19**)$_3$bpy], (e) [Gd(**L19**)$_3$phen] at 295.

Table 5. The lowest energy of ligand excited singlet states and triplet states in the complexes

Complexes	**The lowest energy of ligand excited singlet state [cm^{-1}]**	**The lowest energy of ligand excited triplet state [cm^{-1}]**
Na[Gd (**L19**)$_4$]		25920
Na[Lu(**L21**)$_4$]	33110	25350
Na[Lu(**L22**)$_4$]	35210	26330
[Gd(**L19**)$_3$bpy]	29480	23310
[Gd(**L19**)$_3$phen]	26810	22222

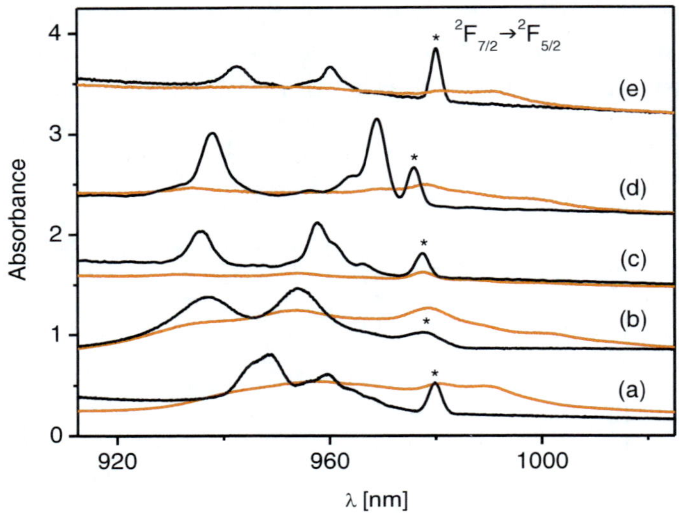

Figure 9. The absorption spectra of (a) Na[Yb(**L19**)$_4$], (b) [Yb(**L19**)$_3$phen], (c) [Yb(**L19**)$_3$bpy], (d) Na[Yb(**L21**)$_4$], (e) Na[Yb(**L22**)$_4$] at 295 (yellow line) and 4 K (black line).

The absorption and emission spectra of Yb^{3+} complexes with bands assigned to the $^2F_{7/2} \rightarrow {}^2F_{5/2}$ and $^2F_{5/2} \rightarrow {}^2F_{7/2}$ transitions are presented in Figures 9 and 10. All investigated complexes exhibit sensitized luminescence of Yb^{3+} in the temperature range from 295 to 77 K. The electronic emission and absorption lines at low temperature are accompanied by a large number of weaker lines of mostly vibronic origin (Figs. 9, 10). It is known that for Yb^{3+} strong electron-phonon mixes of both electronic and vibronic transitions occur. The 4 K absorption spectra consist of three main components corresponding to the $^2F_{7/2}(0) \rightarrow {}^2F_{5/2}(0', 1', 2')$ transitions. At 295 K additional components corresponding to the transitions from thermally populated ligand-field levels of the ground state appear at wavelengths longer than the lowest component of the $^2F_{5/2}$ state. The lowest components of the $^2F_{5/2}$ state (zero-phonon line) are marked with an asterisk (Figure 9). The singlet zero-phonon line and number of electronic components in the absorption spectra at 4 K clearly indicate the presence of only one crystallographic position of the ytterbium ion in the crystal structures. The ligand-field sublevels and total ligand-field splitting of $^2F_{5/2}$ state are shown in Table 6. The total ligand-field splitting of the excited and ground states of [Yb(**L19**)$_3$bpy] and [Yb(**L19**)$_3$phen] is very similar. This is due to the similar crystal structure of both complexes, which differ in the co-ligand molecule (phen or bpy). The higher site symmetry of Yb^{3+} of Na[Yb (**L19**)$_4$] (close D_{2d} [43, 101]) results in the smaller total ligand-field splitting of the $^2F_{7/2}$ and $^2F_{5/2}$ states, as compared to [Yb(**L19**)$_3$bpy] and [Yb(**L19**)$_3$bpy] with site symmetry close to C_{2v}.

Table 6. The ligand-field sublevels and total ligand-field splitting of $^2F_{5/2}$ state

Complex	The ligand-field sublevels of $^2F_{5/2}$ state [cm-^{1}]	The total ligand-field splitting of Yb^{3+} states [cm-^{1}]
Na[Yb (**L19**)$_4$]	10540; 10422; 10206	334 ($^2F_{5/2}$) 414 ($^2F_{7/2}$)
Na[Yb(**L21**)$_4$]	10662, 10321, 10248	414 ($^2F_{5/2}$) 524 ($^2F_{7/2}$)
Na[Yb(**L22**)$_4$]	10607, 10413, 10202	405 ($^2F_{5/2}$) 482 ($^2F_{7/2}$)
[Yb(**L19**)$_3$bpy]	10687; 10442; 10229	458 ($^2F_{5/2}$) 544 ($^2F_{7/2}$)
[Yb(**L19**)$_3$phen]	10673; 10484, 10227	446 ($^2F_{5/2}$) 549 ($^2F_{7/2}$)

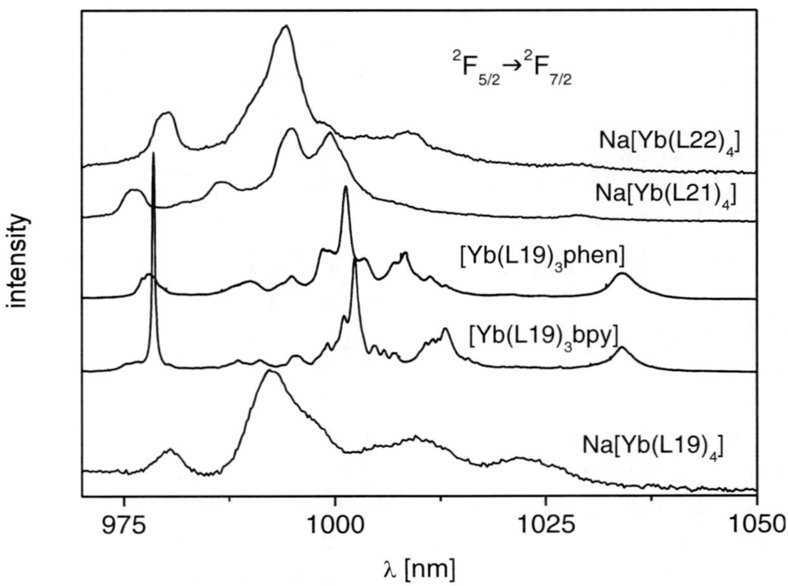

Figure 10. The emission spectra at 77 K, λ_{exc} = 266 nm.

Table 7. Photophysical data for Yb^{3+} complexes

Complexes	ΔE (E_T-$E_{2F5/2}$) [cm^{-1}]	τ_{77K}, τ_{295K} [μs]	τ_{rad} [μs]	Q_{Ln}^{Ln} [%]
Na[Yb (**L19**)$_4$]	15335	35, 40	1100	3.1
Na[Yb(**L21**)$_4$]		70, 66		
Na[Yb(**L22**)$_4$]		48, 55		
[Yb(**L19**)$_3$bpy]	12620	15.5, 16.5	914	1.6
[Yb(**L19**)$_3$phen]	11550	20, 18	943	2.2

The decay time values were estimated with an error of 10 %. $\frac{1}{\tau_{rad}} = 2303 \times \frac{8\pi c n^2 \tilde{\nu}^2 (2J+1)}{N_A (2J'+1)} \int \varepsilon(\tilde{\nu}) d\tilde{\nu}$ [91] where c is the speed of light in cms^{-1}, N_A is Avogadro's number, and J and J' are the quantum numbers for the ground and excited states, respectively, $\int \varepsilon(\tilde{\nu}) d\tilde{\nu}$ is the integrated spectrum of the f-f transition, $\tilde{\nu}$ is the barycentre of the transition and n is the refractive index.

The photophysical data for the Yb^{3+} complexes are summarized in Table 7. The longest emission decay time of the $^2F_{5/2}$ level is for Na[Yb(**L21**)$_4$]. In turn, replacing one molecule of the **19** ligand with the phen or bpy molecule caused a reduction in the emission decay time at 295 K from 35 μs for the complex Na[Yb(**L19**)$_4$] to 20 and 15.5 μs for the complexes [Yb(**L19**)$_3$phen] and [Yb(**L19**)$_3$bpy] respectively. An important factor, which reduces the decay time and the intrinsic emission quantum yield (Q_{Ln}^{Ln}) for complexes as a result of introduction of the co-ligand, is the increased contribution of the multiphonon process of emission quenching. In the quenching mechanism the high energetic C-H vibration stretching (≈3000 cm^{-1}) and deformation (≈1600 cm^{-1}) of heteroaromatic rings are involved. Moreover, the presence of the one water molecule in the outer coordination sphere of [Yb(**L19**)$_3$bpy] is an additional factor of the emission quenching.

The facts that: the sensitized Yb^{3+} emission at room and 77 K was easily recordable for all the complexes using a 400 W xenon lamp as an excitation source, which is known to have a very low emission intensity at the UV range; the lack or weak ligand phosphorescence accompanying Yb emission prove to be a relatively efficient ligand-to-metal energy transfer. An additional element unfavorable to the UV excitation is an energy gap between the triplet excited state of the ligands and the $^2F_{5/2}$ excited level of the Yb^{3+} ion (see Tables 5 and 6). Consequently, the exchange or dipol-dipol mechanism cannot be accepted for such a large energy gap. It has to be associated with a mechanism of energy transfer from the ligand to the metal ion, where the excitation of the Yb^{3+} ion may occur through a double electron transfer [110] in addition to a common phonon-assisted transfer mechanism [111].

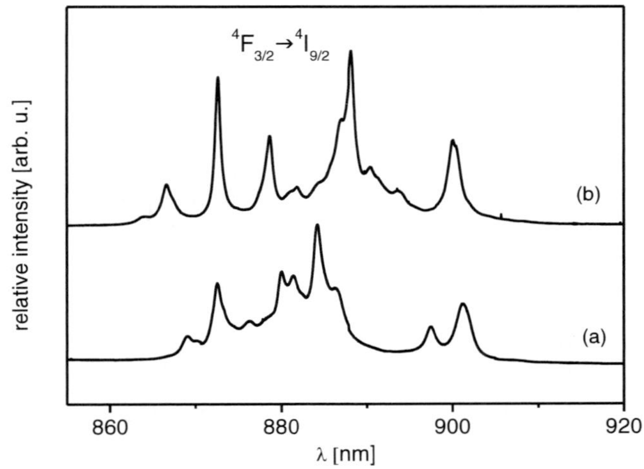

Figure 11. The luminescence spectra of (a) Na[Nd(**L22**)$_4$], (b) Na[Nd(**L21**)$_4$] at 77 K.

The sensitized lanthanide emission was also proved for Nd^{3+} complexes (Na[Nd(**L19**)$_4$], Na[Nd(**L21**)$_4$], Na[Nd(**L22**)$_4$]) [94, 101]. The luminescence spectra are presented in Figure 11. The Nd^{3+} energy gaps between the lowest luminescence state and the highest non-luminescence state are smaller in comparison with the Yb^{3+} energy gap. The contribution of the multiphonon process of emission quenching in comparison with Yb^{3+} complexes is larger in Nd^{3+} complexes. Their emission is the most sensitive towards high vibration quenching, which is reflected in the short emission decay times of the $^4F_{3/2}$ level, which are 3.6 (Na[Nd(**L21**)$_4$]), 5.0 (Na[Nd(**L22**)$_4$]) and 2.4 μs (Na[Nd(**L19**)$_4$]) at 295 and 77 K. In the quenching mechanism the high energetic ν(PO) (≈1270 cm-^{1}) and ν(SO) (≈1450 cm-^{1}) vibration, as well as the ν(CH) (≈2900 cm-^{1}) vibration of CH_3 groups and aromatic rings are involved. Additionally, spectroscopic studies of these neodymium complexes demonstrate that the use of high resolution absorption spectroscopy at low temperatures makes it possible to determine subtle differences in the crystallographic positions of lanthanide ions, and ligand-field splitting of the states of lanthanide ions. As shown in Figure 12, a double number of electronic lines (2(J+1/2)) is observed in these transitions of Na[Nd(**L22**)$_4$] in comparison to Na[Nd(**L21**)$_4$], clearly indicating the presence of two non-equivalent crystallographic positions of the lanthanide ion in the structure of Na[Nd(**L22**)$_4$], and one equivalent crystallographic position of the lanthanide ion in the structure of Na[Nd(**L21**)$_4$].

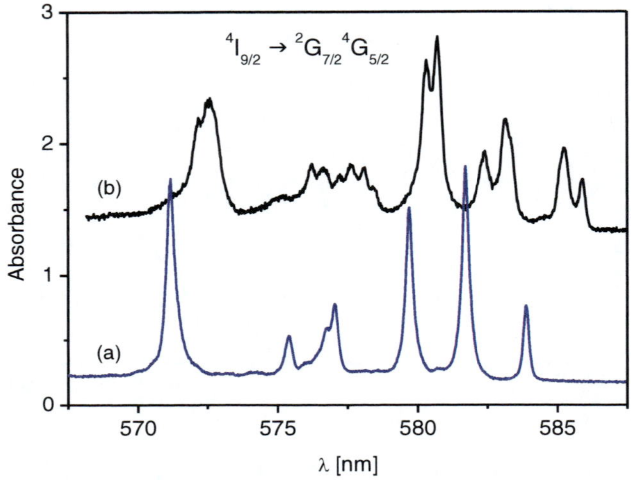

Figure 12. The absorption spectra of (a) Na[Nd(**L21**)$_4$], (b) Na[Nd(**L22**)$_4$] at 4 K.

Eu and Tb complexes

The Eu^{3+} and Tb^{3+} complexes with SAPhs ligands show very intensive emission when excited directly into the lanthanide ion, which is in agreement with the long decay times of the $^5D_0$ state of Eu^{3+} and the $^5D_4$ state of Tb^{3+} and the large value of Q_{Ln}^{Ln} (where $Q_{Ln}^{Ln} = \tau_{obs}/\tau_{rad}$). The radiative lifetime (τ_{rad}) of the $^5D_0$ excited state of Eu^{3+} was calculated directly from its corrected emission spectrum using the following equation [112].

$$1/\tau_R = A_{MD,0} n^3 (I_{tot}/I_{MD})$$

In this formula, n is the refractive index of the crystal (1.5 - Na[Eu(**L21**)$_4$]); 1.5 - Na[Eu(**L22**)$_4$]), $A_{MD,0}$ is the spontaneous emission probability for the $^5D_0 \rightarrow {}^7F_1$ transition in vacuum (14.65 s^{-1}), and I_{tot}/I_{MD} is the ratio of the total area of the corrected Eu^{3+} emission spectrum to the area of the $^5D_0 \rightarrow {}^7F_1$ band. The Q_{Ln}^{Ln} of Eu^{3+} complexes are very high (71% for Na[Eu(**L21**)$_4$] and 80% for Na[Eu(**L22**)$_4$]), and reflect the tiny participation of the multiphonon process in Eu^{3+} emission quenching. The sample spectra of the emissions are presented in Figure 13. Photophysical data are presented in Table 8. Taking into account that the energy gap between the $^5D_4$ emitting level of Tb^{3+} and the $^7F_0$ ground level is bigger as compared to the related energy gap in Eu^{3+}, participation of the multiphonon process in emission quenching must be even smaller in Tb^{3+} complexes than in Eu^{3+} complexes.

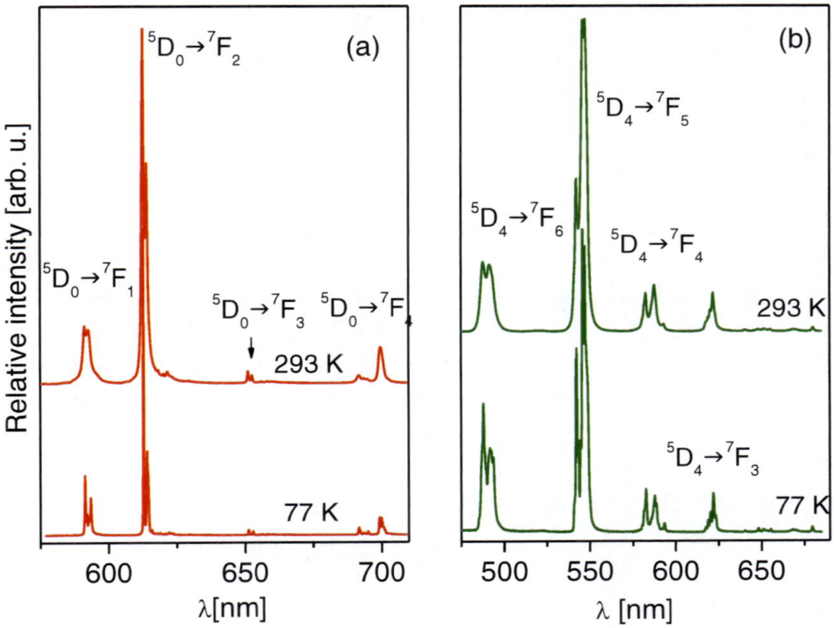

Figure 13. The emission spectra of (a) Na[Eu(**L22**)$_4$] and Na[Tb(**L22**)$_4$] at 295 and 77 K.

Table 8. Photophysical data for Tb^{3+} and Eu^{3+} complexes

complex	τ_{295K} [μs]	τ_{77K} [μs]	τ_{rad} [μs]	Q_{Ln}^{Ln} [%]	Q_{Ln}^{L} [%]
Na[Eu(**L21**)$_4$]	1620	1610	2280	71	0.42
Na[Eu(**L22**)$_4$]	2320	2520	2900	80	3.7
Na[Tb(**L21**)$_4$]	1700	1710			20
Na[Tb(**L22**)$_4$]	2360	2440			2.4

Comparison of the physicochemical properties of the four complexes: Na[Eu(**L21**)$_4$], Na[Tb(**L21**)$_4$], Na[Eu(**L22**)$_4$], Na[Tb(**L22**)$_4$] and some others, not yet published, allowed us to draw some preliminary conclusions about the ability of SAPh to sensitize the metal emission, which of course must be confirmed on a larger number of test compounds. There is a big difference in luminescence intensity between europium and terbium complexes under excitation into the ligand singlet state, which is seen particularly well on the excitation spectra of Na[Eu(**L21**)$_4$] and Na[Tb(**L21**)$_4$] complexes at 295 and 77 K (see Figure 14). It is also reflected in the value of the overall emission quantum yield (Q_{Ln}^{L}) (see Table 8). The Na[Eu(**L21**)$_4$] and Na[Tb(**L21**)$_4$] complexes are isostructural, so the influence of the geometric factor on the efficiency of the ligand-to-metal energy transfer is the same for both complexes. The difference only depends on the energetic factor of Eu^{3+}. It may be concluded that the LMCT state takes part in sensitized luminescence quenching in the europium complex. In this case the observation of the LMCT state is impossible due to overlap with the more intense absorption ligand band. Sometimes the energy of the LMCT state can be determined by comparing the Tb^{3+} excitation spectrum with Eu^{3+} excitation spectrum as well as the Tb^{3+} reflection spectrum with Eu^{3+} reflection spectrum, when it turns out that the band in the Eu^{3+} spectrum is widened at the lower energy side. This suggests the presence of an additional band that can be assigned to the ligand-to-metal charge transfer (LMCT) transition. In this way, the position of the LMCT state in Na[Eu(**L22**)$_4$] was localized, which is shown in Figure 15. In this complex the LMCT state is in resonance with the ligand excited singlet state (S_1) and the bottom of the LMCT state lies slightly below the S_1 state. It is also supported by the reflection spectra of Na[Eu(**L22**)$_4$] and Na[Tb(**L22**)$_4$].

An interesting example is the pair of Na [Tb(**L21**)$_4$] and Na [Tb(**L22**)$_4$] complexes, which differ in their abilities to convert UV light to Vis at room temperature. For Na[Tb(**L22**)$_4$] the intensity of sensitized Tb^{3+} emission is much lower at room temperature than at 77 K (Figure 15), and thus the additional nonradiative process contributes in this case. The positions of the singlet and triplet states for these two Tb^{3+} complexes seem to be too similar to explain the observed differences. This phenomenon can be explained by energy migration in Na[Tb(**L22**)$_4$] between the anions, [Tb1(**L22**)$_4$]$^-$ and [Tb2(**L22**)$_4$]$^-$, which are interconnected in chains by the π-π stacking interactions. This shows the importance of non-covalent interaction influence on the photophysical properties of lanthanide complexes.

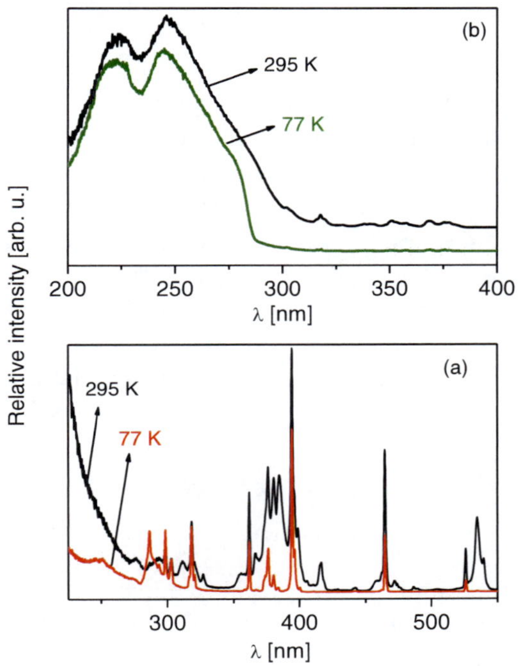

Figure 14. The excitation spectra of (a) Na[Eu(**L21**)$_4$] and (b) Na[Tb(**L21**)$_4$] at 295 and 77 K.

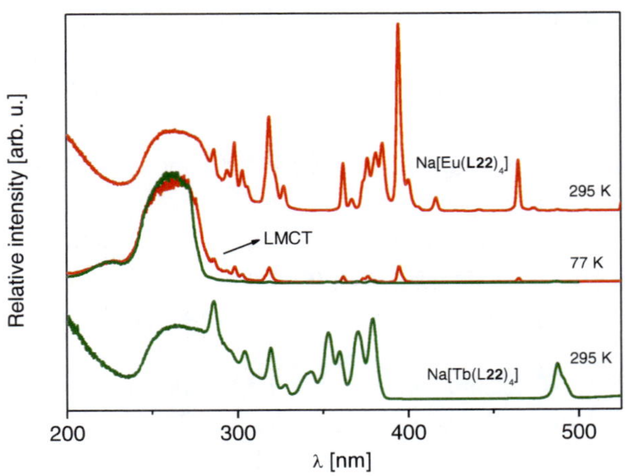

Figure 15. The excitation spectra of Na[Eu(**L22**)$_4$] and Na[Tb(**L22**)$_4$] at (a) 295 K and (b) 77 K.

CONCLUSION

The lanthanide complexes with carbacylamidophosphates and Tb^{3+} and Yb^{3+} complexes with sulfonylamidophosphates, exhibiting relatively efficient ligand-to-metal energy transfer and strong metal-centered emissions, are promising candidates for effective UV-to-visible energy converters. The high stability of the complexes, their resistance to UV laser radiation or even synchrotron radiation, the ability to create complexes in protonated and deprotonated forms, and the ability of coordination with d-elements and f-elements make the ligands described here very useful for coordination engineering towards biologically active metal-organic frameworks.

REFERENCES

[1] Derkatsch, G., Zmurova, I., Kirsanow, A., et al. (1965). *Phospazosoedinenia*. – Kiev.: Naukova dumka, -284.

[2] Cherkasov, R., Zimin, M., Kutyrev, G., et al. (1981). *Chemistry and application of phosphororganic compounds*. Kiev.: Naukova dumka,- 146.

[3] Shokol, V. & Kozhushko, B. (1992). *Phosphorus isocyanates*. – Kiev.: Naukova dumka, -134.

[4] Zabirov, N., Shamsevaleev, F. & Cherkasov, R. (1991). *N*-Phosphorylated amides and thioamides. *Russ. Chem. Rev. 60*, 10, 1128–1144.

[5] Polovinko, V., Rudzevich, V. & Amirkhanov, V. (1994), *Zh. Neorg. Khim.* (Russ). *39*, 4, 640-643.

[6] Skopenko, V., Amirkhanov, V., Sliva, T., Vasilchenko, I., Anpilova, E. & Garnovskii, A. (2004). *Russian Chemical Reviews*, *73*, 8, 737-752.

[7] Wenzel, N. J. (1987). NMR Shift Reagents. - CRC Press, Boca Raton, FL., – 286 p.

[8] Moril, T. (1986). *Lantanide Shift Reagents in Stereochemical Analysis*. - VCH Publishers. Deerfield Beach, FL.,– 193 p.

[9] Buenzli, J. C. G. & Chopin, G. R. (Eds.) (1989). *Lanthanide probe in Life, Chemical and Earth Sciences, Theory and Practice*. – Elsevier. – Amsterdam,– 234 p.

[10] Morton, R. & Diamandis, E. (1990). *Anal. chem.*, *62*, 17. 1841-1845.

[11] Silva, F. R., Goncalves, de, Longo, R., Malta, O. L., Piguet, C. & Buenzli, J. C. G. (2000). *Phys. Chem. Chem. Phys.*, *2*, 23, 5400 – 5403.

[12] Chmutova, M., Kochetkova, N., Koiro, O., Myasoedov, B., Medved, T., Nesterova, N. & Kabachnik, M. (1983), *J. Radioanal. Chem.*, *80*, 1, 63-69.

[13] Normant, H. (1970), *Russ. Chem. Rev.*, *39*, 6, 457–484.

[14] Borzechowska, M., Trush, V., Turowska-Tyrk, I., Amirkhanov, V. & Legendziewicz, J. (2002). *J. All. and Comp.*, *341*. 98-106.

[15] Trush, V., Swiatek–Kozlowska, J., Skopenko, V. & Amirkhanov, V. Z. (2001), *Naturforsch.*, *56b*, 249-254.

[16] Kabachnik, M., Gilyarov, V. & Popov, E. (1961). *Russian Chemical Bulletin*, 1022-1030.

[17] Kabachnik, M., Gilyarov, V., Chjan, Chjen & Matrosov, E. (1962). *Russian Chemical Bulletin*. 1589-1599.

[18] Mizrahi, V. & Modro, T. (1982). *Cryst. Struct. Commun.*, *11*, 627-631.

[19] Amirkhanov, V. & Trush, V. (1995). *Russian Journal of General Chemistry*, *65*, No.7. Part 1. 1021-1024.

[20] Amirkhanov, V., Ovchinnikov, V., Trush, V. & Skopenko, V. (1996). *Russian Journal of Organic Chemistry*, *32* (3), 356-360.

[21] Ovchynnikov, V., Amirkhanov, V., Timoshenko, T., Glowiak, T. & Kozlowski, H. Z. (1998). *Naturforschung.*, *53b*, 481-484.

[22] Amirkhanov, V., Ovchynnikov, V., Glowiak, T. & Kozlowski, H. Z. (1997). *Naturforsch., B: Chem.Sci.*, *52*, 1331-1336.

[23] Gubina, K., Ovchynnikov, V., Amirkhanov, V., Sliva, T., Skopenko, V., Glowiak, T. & Kozlowski H. Z. (1999). *Naturforschung.*, *54b*, 1357-1359.

[24] Gubina, K., Ovchynnikov, V., Amirkhanov, V., Skopenko, V. & Shishkin O. (2000). *Zeitschrift fuer Naturforschung, B: Chemical Sciences.*, *55* (6), 495-498.

[25] Gubina, K., Amirkhanov, V. Z. (2000). *Naturforschung.*, *55b*, 1015-1019.

[26] Kirsanow, A. & Makitra, R. (1955). *Russian Journal of General Chemistry*, *36*, 2134-2137.

[27] Kirsanow, A. & Marenets, M. (1959). *Russian Journal of General Chemistry*, *29*, (7). 2256-2262.

[28] Gholivand, K., Mojahed, F. & Alizadehgan, A. (2007). *Polish Journal of Chemistry*, *81*, 3, 393 – 402.

[29] Gholivand, K., Alizadehgan, A., Mojahed, F. & Soleimani, P., (2008). *Polyhedron*. *27*. 1639-1649.

[30] Gholivand, K., Alizadehgan, A., Arshadi, S. & Firooz, A. (2006). *Journal of Molecular Structure.*, *791*, No 1-3, 193 – 200.

[31] Znovjyak, K., Moroz, O., Ovchynnikov, V., Sliva, T., Shishkina, S. & Amirkhanov, V. (2009). *Polyhedron*, *28*, 13, 3731-3738.

[32] Steinkopf, *Chemische Berichte*. 41, 3585. (1910) *Journal fuer Praktische Chemie (Leipzig)*. (1908). *81*. 241.

[33] Rudavskii, V. & Kucherova, M. (1972). *Russian Journal of General Chemistry*, *42*, 797- 801. (Engl. Transl.) 788-791.

[34] Ovchynnikov, V. (2010). *Acta Crystallogr. Sect.E* ., *66*, o1425.

[35] Gubina, K., Maslov, O., Trush, E., Trush, V., Ovchynnikov, V., Shishkina, S. & Amirkhanov, V. (2009). *Polyhedron*, *28*, 2661-2666.

[36] Znovjyak, K., Ovchynnikov, V., Sliva, T., Shishkina, S. & Amirkhanov, V. (2009). *Acta Crystallogr., Sect. E:* ., *65*, o2812.

[37] Kirsanow, A. & Derkatsch, G. (1956). *Russian Journal of General Chemistry*, *26*, 2631-2638. (Engl. Transl.). 2935 -2941.

[38] Derkatsch, G., Dregval, G. & Kirsanow A. (1960). *Russian Journal of General Chemistry*, *30*, 3393-3397. (Engl. Transl.) 3359-3363.

[39] Derkatsch, G., Dregval, G. & Kirsanow, A. (1961). *Russian Journal of General Chemistry*, vol. *31*, 2385-2390. (Engl. Transl.) 2223-2228.

[40] Prozenko, L. & Kornew, K. (1958). *Ukrainskii Khimicheskii Zhurnal (Russian Edition)*, *24*, 636-638.

[41] Rebrova, O., Byushkin, V., Malinovskii, T., Protsenko, L. & Dneprova, T. (1982). *Dokl.Akad.NaukSSSR(Russ.). (Proc.Nat.Acad.Sci. USSR).*, *266*, 1391.

[42] Kirsanow, A. & Lewtschenko, E. (1957). *Russian Journal of General Chemistry*, *27*, 2585- 2590. (Engl. Transl.). 2642-2645.

[43] Gawryszewska, P., Kulesza, D., Moroz, O., Trush, V. & Amirkhanov, V. (2011). *Journal of Photochemistry and Photobiology A: Chemistry*, *217*, 1, 1 – 9.

[44] Kirsanow, A. & Schewtschenko, A. (1954). *Russian Journal of General Chemistry*, *24*, 474-484. (Engl. Transl.). 483-492.

[45] Goldstein, J. (1977). *Journal of Organometallic Chemistry*, *42*, 2466-2469.

[46] Kozlov, E., Kolesnik, N. & Shokol V. (1978). *J. Gen. Chem. USSR*. (Engl. Transl.). *48*, 783-787, 716-720.

[47] Iwanowa, Zh., Levchenko, E. & Kirsanow, A. J. (1965). *Gen. Chem. USSR*. (Engl. Transl.), *35*, 1607-1612, 1611-1616.

[48] Kirsanow, A. & Makitra, T. (1957). *Zhurnal Obshchei Khimii*, *27*, 245 -248, engl.Ausg. 273 – 276.

[49] Glidewell, G. (1976). *Journal of Organometallic Chemistry, 108*, 335-344.
[50] Moroz, O., Trush, V., Znovjyak, K., Konovalova, I., Omelchenko, I., Sliva, T., Shishkin, O. & Amirkhanov, V. (2012). *Journal of Molecular Structure, 1017*, 109–114.
[51] Kirsanow, A. & Schewtschenko, V. (1954). *Zhurnal Obshchei Khimii, 24*, 1980-1993; engl. Ausg. 1949 – 1962.
[52] Pietraszkiewicz, M., Karpiuk, J. & Staniszewski, K. (2002). *Journal of Alloys and Compounds., 341*, 267 – 271.
[53] Regitz, M. & Anschuetz, W. (1969). *Chemische Berichte., 102*, 2216-2229.
[54] Regitz, M., Bennyarto, F. & Heydt, H. (1981). *Liebigs Annalen der Chemie., 6*, 1044 – 1051.
[55] Schewtschenko, I., Stratienko, V. & Pinchuk, A. (1960). *J. Gen. Chem. USSR.*, (Engl. Transl.). *30*, 1566 -1570.
[56] Bock, H. & Wiegraebe, W. (1966). *Chemische Berichte., 99*, 1068-1076.
[57] Derkach, G., Gubnizkaja, E., Shokol, V. & Kirsanow, A. (1962). *J. Gen. Chem. USSR.*, (Engl. Transl.). *32*, 1874 - 1878, 1853 – 1857.
[58] Derkach, G. & Gubnizkaja, E. (1964). *J. Gen. Chem. USSR.*, (Engl. Transl.). *34*, 604 -609, 605 -610.
[59] Derkach, G., Gubnizkaja, E. & Kirsanow, A. (1961). *J. Gen. Chem. USSR.*, (Engl. Transl.). *31*, 3679 – 3684, 3434 – 3439.
[60] Tomaschewski, T. & Kuehn, G. (1968). *Zeitschrift fuer Chemie*, 8, 461.
[61] Birdsall, D., Green, J., Ly, T., Novosad, J., Necas, M., Slawin, A., Woollins, J. & Zak, Z. (1999). *European Journal of Inorganic Chemistry*, 9, 1445 – 1452.
[62] Gholivand, K., Mostaanzadeh, H., Koval, T., Dusek, M., Erben, M., Stoeckli-Evans, H. & Della Védova, C. O. (2010). *Acta Cryst. B., 66*, 441-450.
[63] Gadekar, M. & Ross, E. Phosphorus analogs of Carbromal. *J. Org. Chem., 26*. 606-607.
[64] Zhu, S., Zhang, J. & Xu B. (1996) *J. of Fluorine Chemistry.*, (1961) *78*, 183-185.
[65] Boudjabi, S., Dewynter, G. & Montero J.-L. (2000). *Synlett., 5*, 716-718.
[66] Cromer, D., Ryan, R., Karthikeyan, S. & Paine, R. (1990). *Inorganica Chimica Acta, 172*, 165-172.
[67] Iriarte, A., Erben, M., Gholivand, K., Jios, J., Ulic, S. & Vedova, C. (2008). *Journal of Molecular Structure, 886*, 66–71.

[68] Gholivand, K., Vedova, C., Erben, M., Mahzouni, H., Shariatinia, Z. Amiri S. (2008). *Journal of Molecular Structure*, *874*, 178–186.

[69] Gholivand, K., Vedova, C., Erben, M., Mojahed, F. & Alizadehgan, A. (2007). *Journal of Molecular Structure*, *840*, 66–70.

[70] Amirkhanov, V., Ovchynnikov, V., Legendziewicz, J., Graczyk, A., Hanuza, J. & Macalik, L. (1996). *Acta Physica Polonica A.*, *90*, 2, 455-460.

[71] Sosnovsky, G., Uma Maheswara Rao, N. & Shu, Wen Li. (1986). *J. Med. Chem.*, *29*, 2225- 2230.

[72] Perlman, M., Dunn, J., Piscitelli, T., Earle, J., Rose, W., Wampler, G., MacDiarmid, J. & Bardos, T. (1991). *J. Med. Chem.*, *34*, 1400-1407.

[73] MacDiarmid, J., Rose, W., Biddle, W., Perlman, M., Breiner, R., Ambrus, J. & Bardos, T. (1985). *J. Med. Chem.*, *28*, 1685-1691.

[74] Kido, J. & Okamoto, Y. (2002). *Chem. Rev.*, *102*, 2357.

[75] Hemmila, I. & Laitala, V. (2005). *J. Fluoresc.*, *15*, 529-542.

[76] Hemmila, I. Mukkala, V.-M. & Takalo, H. (2005). *J. All. Comp.*, *249*, 158-162.

[77] Bazin, H., Trinquet, E. & Mathis, G. (2002). *Rev. Mol. Biotechnol.*, *82*, 233.

[78] Yam, V. W. W. & Lo, K. K. W. (1999). *Coord. Chem. Rev.*, *184*, 157.

[79] Bünzli, J.-C. G. (1989). Luminescent probes. In *Lanthanide Probes in Life, Chemical and Earth Sciences. Theory and Practice*; 1 ed.; Bünzli, J.-C. G., Choppin, G. R., Eds.; Elsevier Science Publ. B.V.: Amsterdam, pp 219.

[80] Parker, D. (2000). *Coord. Chem. Rev.*, *205*, 109-130

[81] Butler Stephen, J. & Parker David. (2013). *Chem. Soc. Rev.*, *42*, 1652.

[82] Gawryszewska, P., Sokolnicki, J. & Legendziewicz, J. (2005). *Coord Chem. Reviews*, *249*, 2489.

[83] Legendziewicz, J. & Huskowska, E. (1989). Application of Luminescence in Studies of Ln(III) Ions Interaction with Amino Acids. In *Excited States of Transition Elements*, Legendziewicz, J., Strek, W., Eds., World Scientific Publishing Co. Pte. Ltd., Singapore, pp 222. D.

[84] Brites, C. D. S., Lima, P. P., Silva, N. J. O., Millán, A., Amaral, V. S., Palacio, F. & Carlos, L. D. (2013). *J. Lum.*, 133230.

[85] Hemmilä, I. & Mukkala, V.-M. (2001) Time-resolution in fluorometry, technologies, labels, and applications in bioanalytical assays. *Crit Rew Clin Lab Sci*, *38*, 441–519.

[86] Blomberg, K. R., Mukkala, V.-M, Hakala, H. H. O., Mäkinen, P. H., Suonpää, M. U. & Hemmilä, I. A. (2011). *Anal. Bioannal. Chem., 399,* 1677.

[87] Wers, M. H. V., Wondenberg, R. H., EMMerink, P. G., Van Gassel, R., Hofstraat, J. W. & Vorhoeven, J. W. (2000). *Angew. Chem. Int. Ed., 39,* 4542-4544.

[88] Comby, S. & Bünzli, J.-C. (2007). G *In Handbook on the Physics and Chemistry of Rare Earth*; K. A. Gschneidner, Jr., J.-C. G. Bünzli V. K. Pecharsky, Eds. Elsevier Science BV; Amsterdam, Vol. *37*, chapter 234, p 217.

[89] Svetlana, V., Eliseeva & Jean-Claude G., Bunzli. (2010). *Chem. Soc. Rev., 39*(1) 189.

[90] Lehn, J.-M. *Design of Organic Complexing Agents Strategies Towards Properties*, Springer, Berlin, Heidelberg, 1973.

[91] "Lanthanide Luminescence Photophysical, Analytical and Biological Aspects" Volume Editors: Pekka Hänninen Harri Härm Springer-Verlag Berlin Heidelberg 2011.

[92] Armelao, L., Quici, S., Barigelletti, F., Accorsi, G., Bottaro, G., Cavazzini, M. & Tondello, E. (2010). *Coor. Chem. Rev., 254*, 487–505.

[93] Moroz, O., Shishkina, S., Trush, V., Sliva, T. & Amirkhanov, V. (2007). *Acta Cryst., E63*, m3175-m3176.

[94] Gawryszewska, P., Moroz, O. V., Trush, V, Amirkhanov, V. M., Lis, T., Sobczyk, M. & Siczek, M. (2012). *ChemPlusChem, 77*, 482-496.

[95] Gubina, K., Shatrava, J., Ovchynnikov, V. & Amirkhanov, V. (2000). *Polyhedron, 19*, 2203–2209.

[96] Amirkhanov, V., Ovchynnikov, V., Kapshuk, A. & Skopenko, V. (1995). *Russ. Journal of Inorg. Chem., 40*, No 11, 1869-1873.

[97] Gubina, K., Ovchynnikov, V., Amirkhanov, V., Fischer, H., Stumpf, R. & Skopenko, V., 2000. *Zeitschrift fuer Naturforschung, 55b*, 576-582.

[98] Ovchynnikov, V., Amirkhanov, V., Domasevich, K., Sieler, I. & Skopenko, V. (2001). *Russ. Journal of Inorg. Chem., 46*, No.4, 615-622.

[99] Oczko, G., Legendziewicz, J., Trush, V. & Amirkhanov V. (2003). *New J. Chem. 27*, 948 –956.

[100] Puchalska, M., Turowska-Tyrk, I., Trush, V. & Legendziewicz, J. (2008). *Journal of Alloys and Compounds, 451* (1-2). 264-269.

[101] Kulesza, D., Sobczyk, M., Legendziewicz, J., Moroz, O. & Amirkhanov, V. (2010). *Struct. Chem., 21*, 425-438.

[102] Binnemans, Koen, (2005). "Rare-earth beta-diketonates", in Handbook on the Physics and Chemistry of Rare Earth, Elsevier Science B. V., *Amsterdam*, vol. *35*, chapter 225

[103] Legendziewicz, J., Amirkhanov, V., Janczak, C., Macalik, L. & Hanuza, J. (1995). *J. Appl.Spectr.*, *62*, 5-17.

[104] Sokolnicki, J., Legendziewicz, J., Amirkhanov, V., Ovchynnikov, V., Macalik, L. & Hanuza, J. (1999). *Spectrochim. Acta. Part A.*, *55*, 349-367.

[105] Legendziewicz, J., Oczko, G., Amirkhanov, V., Wiglusz, R. & Ovchynnikov, V. (2000). *J. Alloys Compd.*, 300–301, 360-369.

[106] Legendziewicz, J., Oczko, G., Wiglusz, R. & Amirkhanov, V. (2001). *J. Alloys Compd.*, 323–324, 792.

[107] Gubina Katerina, E., Shatrava, Julia, A. & Ovchynnikov, Vladimir, A., Amirkhanov Vladimir M. (2000). *Polyhedron*, *19*, 2203–2209.

[108] Sabbatini, N., Guardigli, M. & Lehn, J.-M. (1993). *Coord. Chem. Rev.*, *123*, 201

[109] de Sá, G. F., Malta, O., de Mello Donega, C., Simas, A. M., Longo, R. L., Santa-Cruz, P. A. & da Silva, E. F. (2000). *Coord. Chem. Rev.*, *196*, 165.

[110] Horrocks, Jr W. D., Bolender, J. P., Smith, W. D. & Supkowski, R. (1997). *J. Am. Chem. Soc.*, *119*, 5972-5973.

[111] Reinhard, Ch. & Güdel, H. U. (2002). *Inorg. Chem.*, *41*, 1048-1055.

[112] van der Tol, E. B., van Ramesdonk, H. J., Verhoeven, J. W., Steemers, F. J., Kerver, E. G., Verboom, W. & Reinhoudt, D. N. (1998). *Chem. Eur. J.*, *4*, 2315; Werts M. H. V., Jukes R. T. F. and Verhoeven J. W. (2002). *Phys. Chem. Chem. Phys.*, *4*, 1542.

In: Ligands
Editors: Paula Gawryszewska et al.
ISBN: 978-1-63117-143-7

Chapter 8

SYNTHESIS, APPLICATIONS AND COORDINATION CHEMISTRY OF FORMAZANS

Yauhen Yu. Karabach* and Maximilian N. Kopylovich*

Centro de Química Estrutural, Instituto Superior Técnico,
Universidade de Lisboa, Lisbon, Portugal

ABSTRACT

This chapter represents an overview of the existing information on the chemistry and some applications of formazans, compounds with the $R_1NH\text{-}N{=}CR_2\text{-}N{=}NR_3$ azohydrazone conjugated system. The main method of formazan synthesis involves the interaction of bis-diazo salts with methylene-active and methyne compounds, typical yields being within 10-50%. Representatives of the farmazan family are applied for the synthesis of different heterocycles, while the "formazan – tetrazolium salt" system is widely used as a redox-based staining component for different biochemical studies. Moreover, formazans are valuable spectrofotometric analytical agents and ionophors; they can be used for the simultaneous selective determination of several metal cations. Polymers with formazan moieties are known as selective complexing sorbents for the concentration of micro quantities of elements. The high

* Corresponding Authors' address, Centro de Química Estrutural, Complexo I, Instituto Superior Técnico, Universidade de Lisboa, Av. Rovisco Pais, 1049–001 Lisbon, Portugal; Email: karabach@ist.utl.pt (Ya.Yu. Karabach), kopylovich@yahoo.com (M.N. Kopylovich).

photo- and thermo-resistance of some formazan complexes allow their application as registering components in optical data storage devices. However, the coordination chemistry of formazans remains essentially undeveloped; only a very limited number of metal complexes with formazan ligands have been fully characterized. Several representatives of formazan complexes are described in this chapter, and some conclusions on the typical coordination modes and other structural details are provided.

Keywords: Formazans, redox staining, coordination compounds, *E*/*Z* isomerisation

1. INTRODUCTION

Recent coordination chemistry has mainly been directed towards the search for new materials with bioactive, magnetic, catalytic, sorption and other useful properties. The preparation of such compounds is closely related to the synthesis of new ligands that can arrange metal ions into certain assemblies with necessary properties. β-diketones and their analogues, such as β-diimines or triazapentadienes, provide rich opportunities for the coordination of metal ions in bridging and/or chelating modes and thus creating immense possibilities for their arrangement. Formazans also belong to this isoelectronic family; these are compounds which contain the usually coplanar R_1NH-$N{=}CR_2$-$N{=}NR_3$ azohydrazone conjugated system; thus, formally they can be considered as ketohydrazones.

The electronic structure of formazans ensure their high tautomeric and conformational flexibility (Scheme 1) [1,2], the configuration of particular compounds being dependent on the position and nature of substituents [3]. Thus, the formazan chain can exhibit several distinctive isomeric configurations (Scheme 1a), although the bulky substituent R_2 will hamper the formation of *EEZ* and *EEE* isomers (herein the sequence corresponds to the [N(R_1)-N=C(R_2)-N] formazan skeleton fragment). The inclusion of heterocyclic substituents, such as azoles, can provide additional tautomeric forms due to proton migration to the heteroatom of the substituent (Scheme 1b) [3].

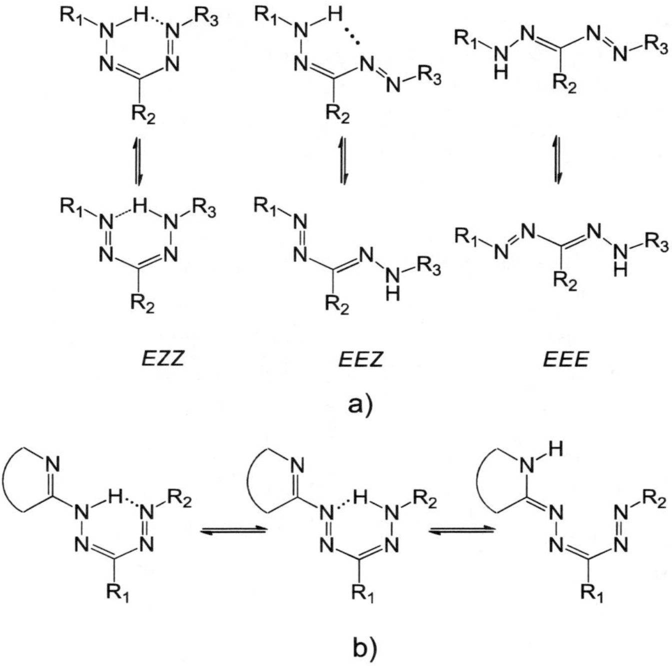

Scheme 1. Examples of some formazan isomeric and tautomeric configurations.

As can be seen, formazans possess rich possibilities for conformational and tautomeric isomery. The EZZ-, EEZ-, and EEE- isomers were structurally characterized, while the EEZ-configuration was shown to be more stable in the gas phase [2]. In solution, 1,3,5-triaryl formazans were known to exist in "closed" (red) and "open" (yellow) forms. However, more detailed investigation allowed the distinguishing of EZZ- (red), EEZ- (orange) and EEE (yellow) isomers. Some other studies relate the color of formazans to electronic distribution and interaction with solvents rather than to the specific configuration [2-5].

The preferential existence of a particular isomer depends on substituents (their volume; the possibility to form hydrogen bonding and π-interactions), and some external conditions, e.g., type of crystallization. This property, together with the system of extensive conjugated double bonds, creates possibities to use formazans as adjustable dyes, molecular switches and chromophoric agents, e.g., in analytical or biological chemistry. Formazans are also used for selective extraction, metal ion detection and diagnostics. As can be seen, the applications of formazans are related to their electronic structure, the possibility to exist in different isomeric and tautomeris forms, and also to their ability to form coordination compounds.

Thus, in this chapter we introduce this interesting class of ligands and give a short overview of their synthetic potential in coordination chemistry. Some known and potential applications of these compounds are also discussed.

2. Synthesis of Formazans

Up to the present time there are a number of synthetic methods that lead to formazans, and the most important ones involve: 1) reaction of diazonium salts and a) arylhydrazones, b) compounds with an active methylene group; 2) reaction of arylhydrazines with imino esters [1].

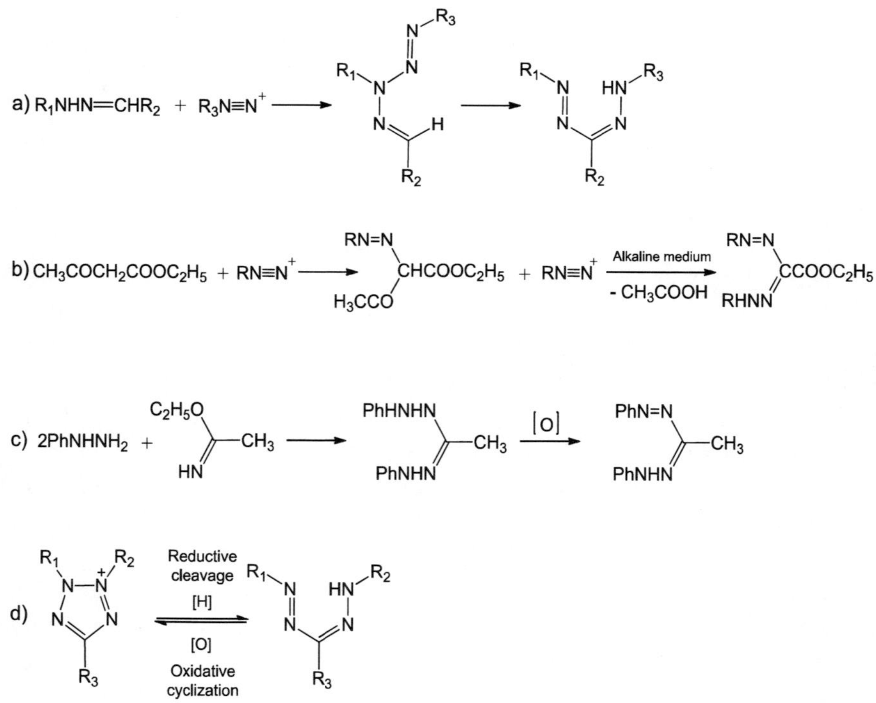

Scheme 2. Common synthetic procedures towards formazans.

It was demonstrated that the above reactions proceed in a few steps. In the case of 1a) the coupling between diazonium salt and imide nitrogen occurs at the first stage, then the formed azo compound undergoes rearrangement to form formazan (Scheme 2a). In the case of 1b), the arylhydrazone is formed at the first stage and then reacts with the second diazonium ion, yielding the corresponding formazan (Scheme 2) [1]. The second method leads to hydrazidines that are readily oxidized to the corresponding formazans. However, in some cases the formation of a mixture of products occurs (Scheme 2c) [1]. Apart from the above common synthetic procedures, formazans can be obtained via reductive cleavage of the respective tetrazolium salts (Scheme 2d). Although this method is much less synthetically significant, it is widely used for specific staining in molecular biology, e.g., for the preparation of colorimetric assays [4].

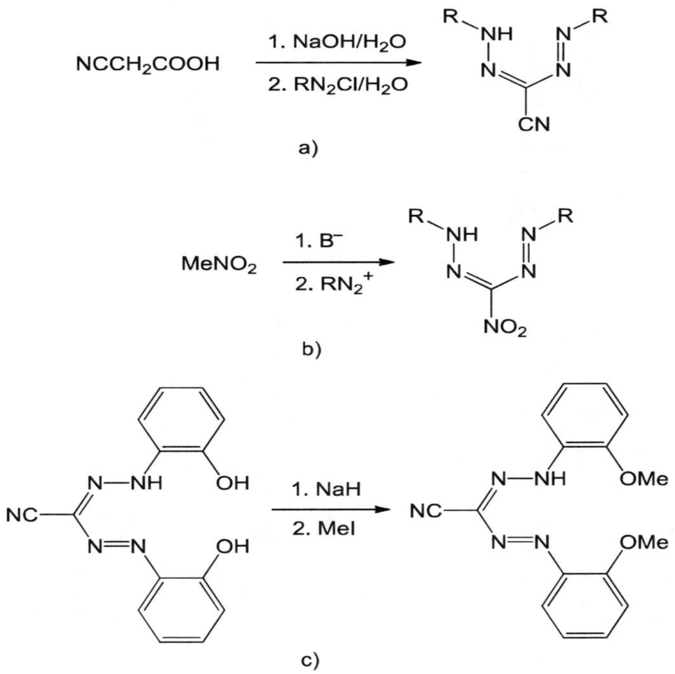

Scheme 3. Synthesis of 3-cyano- (ac) and 3-nitro- (b) formazans.

The ligands discussed below are mainly obtained through the modified method 1b), i.e., upon the interaction of diazonium with methylene active compounds. Consequently, 3-cyano- and 3-nitroformazans can be readily synthesized by reacting cyanoacetic acid or nitromethane, respectively, with diazonium salts (Scheme 3a,b) [5]. In the first case, diazonium cation attacks the *in situ* generated carbanion, leading to the hydrazone-type intermediate, which, being deprotonated, reacts with the second diazonium cation to give 3-cyanoformazan. In the case of 3-nitroformazans, it was found that tetrafluoroborate salts of diazonium cations are stable enough in nonaqueous conditions to couple with deprotonated nitromethane to lead directly to 3-nitroformazans with bulky aryl groups [5]. The thus prepared 3-cyano- and 3-nitroformazans can be modified further using common synthetic procedures. Thus, the tetradentate methoxy substituted 3-cyanoformazan can be easily prepared from the hydroxy one by reaction with methyl iodide (Scheme 3c).

Benzothiazolylformazans can be synthesized by coupling diazonium salts with benzothiazolylhydrazones (Scheme 4), while the latter were obtained by reaction of the corresponding aldehyde with 2-hydrazinobenzothiazole [6].

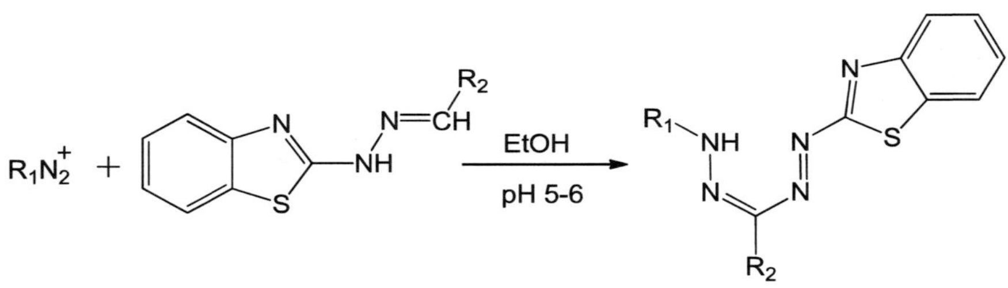

Scheme 4. Synthesis of benzathiazolylformazans.

The synthesis of 3-pyridylformazans is based on the coupling of phenylhydrazones of the corresponding pyridinecarbaldehydes with diazonium salts (Scheme 5) [7].

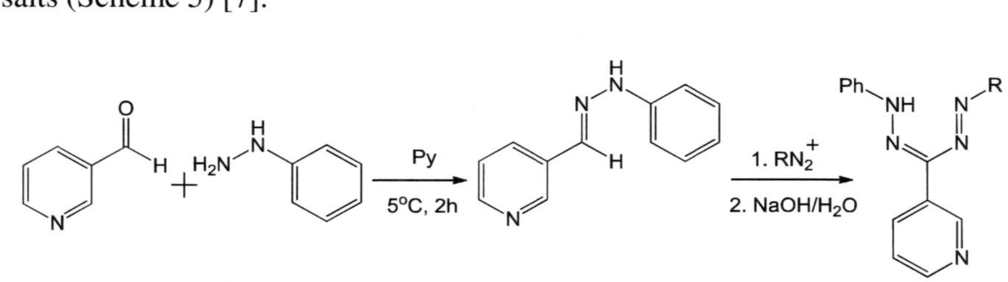

Scheme 5. Synthesis of 3-pyridylformazans.

There are a few methods that lead to macrocyclic crown formazans [8]; the most convenient one is based on the sequential azo-couplings of diazonium salt with methylene active compounds. Thus, bisdiazotized bis(2-aminophenoxy)-1,3-propane reacts with cyanoacetic, acetoacetic or malonic acids, yielding corresponding crown formazans in the presence of copper ions and pyridine (Scheme 6, type I) [9]. In the absence of copper ions the corresponding arylhydrozones are the main products (Scheme 6, type III) [9]. However, it was shown that formazan type II is also formed (the yield is quite low, *ca.* 5%), and can be isolated from the reaction mixture [10].

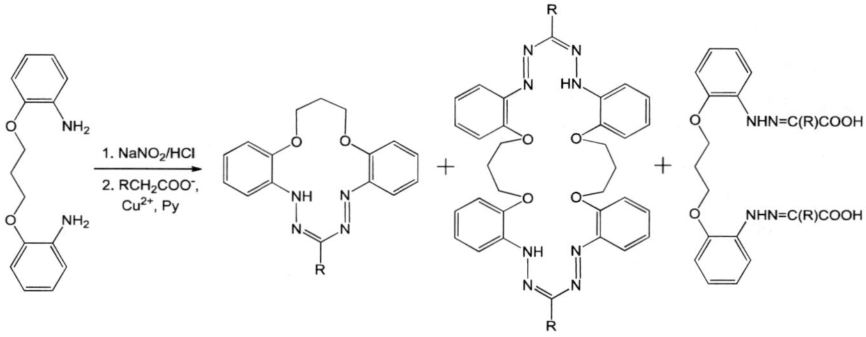

I II III

Scheme 6. Synthesis of macrocyclic crown formazans (for I and III, R=CN, $COCH_3$ or H, for II, R=CN).

In general, the synthetic procedures towards formazans are quite flexible and allow the preparation of ligands with various functional groups. This flexibility, along with formazan chelating abilities, lead to numerous applications in bio- and analytical chemistry. Moreover, formazans and their complexes exhibit attractive photo- and thermochromic properties [11,12].

3. Applications of Formazans in Medicine and Molecular Biology

As mentioned above, due to the easy reduction of colorless or slightly colored tetrazolium salts into bright formazans, this reaction is widely used in molecular biology, medicine, and the analysis of organic and inorganic compounds, among many other applications. Thus, the "formazan – tetrazolium salt" systems play an important role in cell biology as a crucial part of colorimetric assays to detect superoxide radicals [13,14], cell proliferation and viability [4,15], etc. The specific properties of such systems can also be used for drug cytoxicity determination [16], diagnosis of neurocysticercosis [17], bactericidal activity [18,19], and other related applications [20]. A number of tetrazolium salts are used, depending on the object under study, e.g. , MTT (3-(4,5-dimethylthiazol-2-yl)-2,5-diphenyltetrazolium bromide) [4,21,22], MTS (3-(4,5-dimethylthiazol-2-yl)-5-(3-carboxymethoxyphenyl)-2-(4-sulfophenyl)-2H-tetrazolium, inner salt) [23], XTT (2,3-bis(2-methoxy-4-nitro-5-sulphophenyl)-5-carboxanilide-2H-tetrazolium, monosodium salt) [13] and WST-1 (2-(4-iodophenyl)-3-(4-nitrophenyl)-5-(2,4-disulfophenyl)-2H-tetrazolium, monosodium salt) [24].

The MTT assay of cell viability is based on the reductive cleavage of the tetrazilum salt within living cells to the corresponding formazan, and involves pyridine nucleotide cofactors NADH and NADPH [25]. The formed formazan (Scheme 7) can be determined spectrophotometrically [15], i.e. the value of optic density is related to the amount of living cells. However, the MTT assay has some limitations [26-29]. For instance, the formazan, formed upon reduction of MTT, is not soluble in aqueous media and readily crystallizes; this can cause some difficulties in the extraction step [4]. Another method of formazan detection – surface enhanced resonance Raman scattering (performed with the addition of Au nanoparticles) – was proposed recently [15]. This method allows the exclusion of the extraction step and a decrease in the time of the procedure.

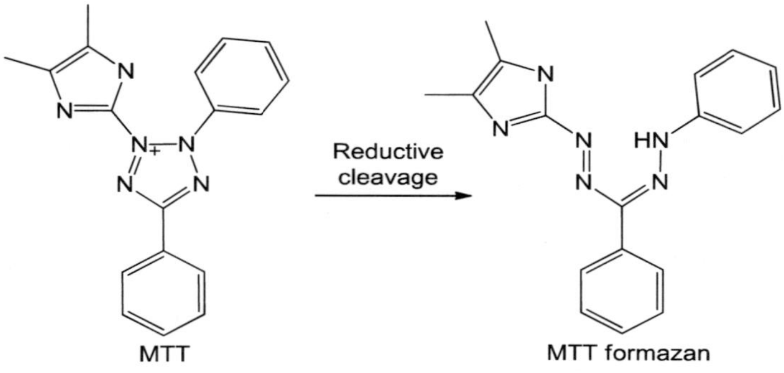

Scheme 7. Reductive cleavage of MTT within living cells [15].

To improve the availability of assays, other tetrazolium salts, which form water soluble formazans upon reduction, were introduced, e.g., MTS [20,23,30], XTT [20,31,32] and WST-1 [14,20,24,33,34]. The formation of water soluble formazan eliminates the solubilisation step that is required for the MTT assay. However, it is necessary to use an intermediate electron acceptor – phenazine methosulphate (PMS) (Scheme 8). It was reported that XTT assays show unstable results, which could be caused by the formation of charge transfer complexes between XTT and PMS, while the MTS/PMS system does not follow this trend and shows higher precision and stability [32,35].

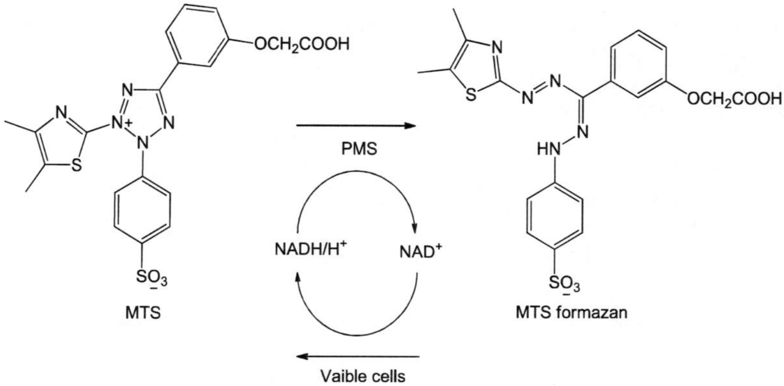

Scheme 8. Reductive cleavage of MTS within living cells [35].

The WST-1 assay for superoxide anion radicals is based on the ability of tetrazolium salt to react with radicals yielding the corresponding water-soluble formazan (Scheme 9) [14,33,34].

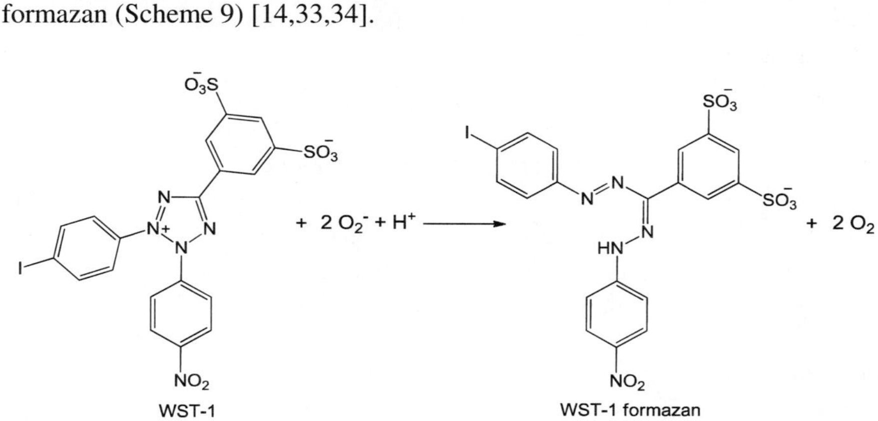

Scheme 9. Reductive cleavage of WST-1 by superoxide anion radicals [33].

Recently a method for the detection of trace copper(II) by the quantum photoelectric effect using nitro blue tetrazolium (NBT, 2H-Tetrazolium, 2,2'-(3,3'-dimethoxy[1,1'-biphenyl]-4,4'-diyl)bis[3-(4-nitrophenyl)-5-phenyl-, dichloride) was proposed [36]. The method is based on the interaction of NBT with quantum dots (QDs) under illumination [37]. The escaped photoelectrons can be captured by NBT and lead to conversion of NBT to the corresponding insoluble formazan, which can be detected as a purple precipitate (Scheme 10). The interaction of copper ion with QDs produces trapping sites that block the escape of photoelectrons and thus eliminate the response. The performed studies show good anti-interference of copper ions with other metal ions, with the exception of Ag^+ [36].

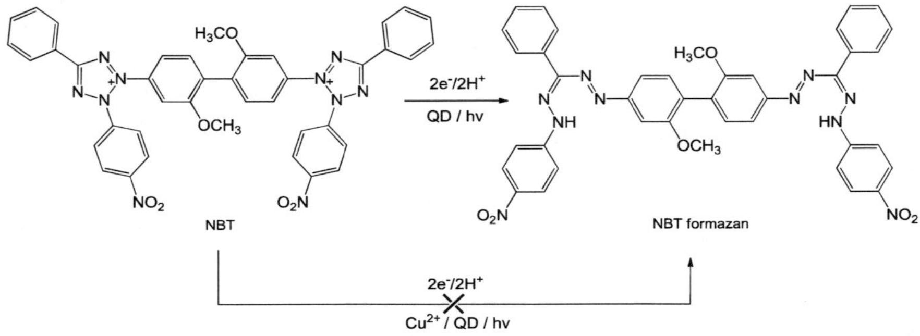

Scheme 10. Reduction of NBT to the corresponding formazan.

Apart from the above-mentioned applications in cell biology, the derivatives of formazans are known to possess antibacterial, antiviral and antifungial activity [38].

4. Formazans as Dyes

The photochemical properties of formazans allow their use as analytical agents and resistant dyes. Many formazan dyes involve transition metals, e.g., Cu(II), to increase the stability and photochromic properties of chromophores. The color of formazan dyes and their intensity strongly depend on the substituents. Thus, aryl derivatives are deeply colored in contrast to those with aliphatic substituents; moreover, presence of a nitro group within the aryl moiety leads to the color deepening if compared with the methyl group [6].

Along with their photochromic behavior, formazan dyes with heterocyclic substituents can also exhibit thermochromic properties [11,12]. Thus, the chloroform solutions of benzimidazolyl formazans change color from violet or blue-violet to bright blue or blue-black, respectively, upon cooling down to -10 °C [11]. This phenomenon is caused by the formation of intramolecular hydrogen bonds (Scheme 11) [11].

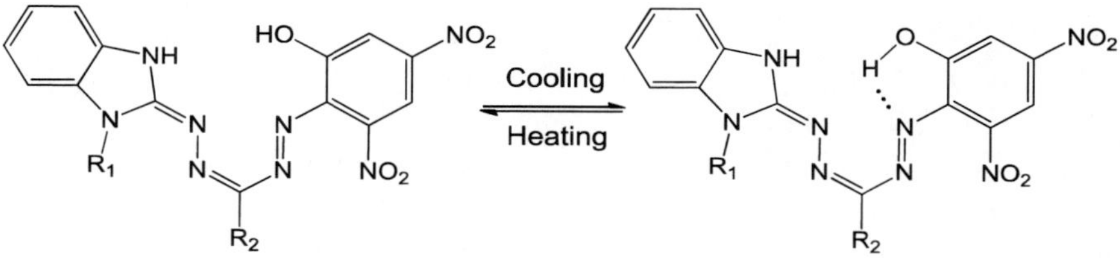

Scheme 11. The formation of intermolecular hydrogen bond in some thermochromic formazans (R_1 = aliphatic, R_2 = aliphatic/aromatic).

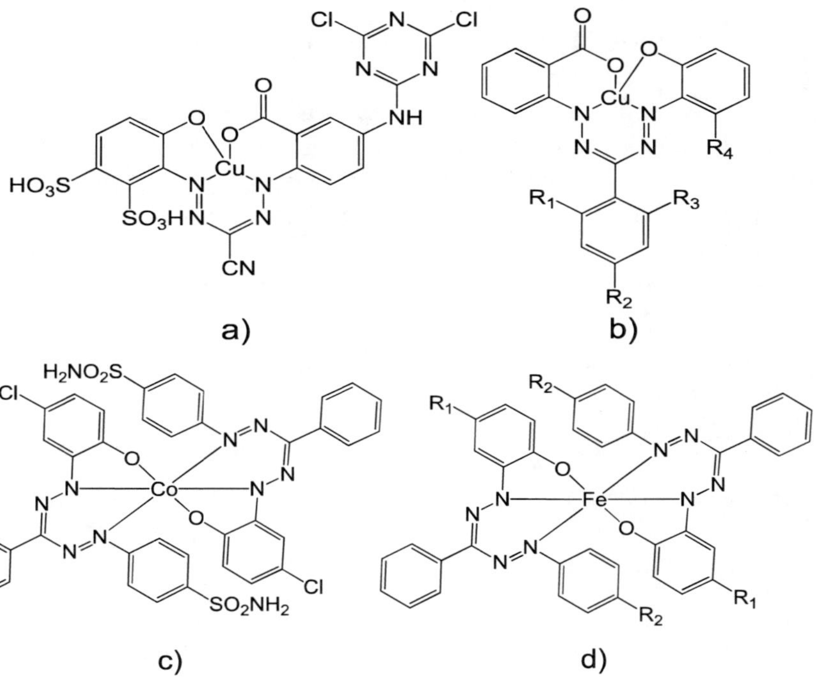

Scheme 12. Representatives of some formazan dyes [39,42]. For b): 1) R1=R2=R3=H, R4=SO_2NH_2; 2) R1=R2=R3=H, R4=SO_3H; 3) R1=H, R2=R3=Cl, R4= SO_2NH_2; 4) R1=R3=Cl, R2=H, R4= SO_2NH_2; 5) R1=R2=H, R3=SO_3H, R4= NO_2; 6) R1=H, R2=R3=Cl, R4=SO_3H [39]; for d): 1) R1=SO_2NH_2, R2=H; 2) R1=Cl, R2= SO_2NH_2; 3) R1=NO_2, R2= SO_2NH_2; 4) R1=H, R2= SO_2NH_2 [40].

Many known formazan dyes are 1:1 and 1:2 metal complexes of polydentate formazans [mainly of copper(II), cobalt(II), chromium(III) and iron(III))] (Scheme 12) [39-43]. Their technical properties, as well as aquatic toxicity, depend on the nature of the formazan and metal ion used. The 1:1 copper(II) complexes of tetradentate formazans have a blue color and are used as colorants of cellulose fibers (Scheme 12a) or wool (Scheme 12b) [39]. The 1:2 cobalt(III) complexes of tridentate formazan acid dyes, such as those depicted on Scheme 12c, are used for the coloration of polyamide and protein fibers [42]. The 1:2 iron complexes of tridentate formazans (Scheme 12d) [40,44] are also applied as prospective dyes for wool and nylon, due to their high photostability, wet fastness and relatively low chronic toxicity [40,42].

5. Formazans for the Selective Sorption and Determination of Metal Ions

Along with the above-mentioned applications, formazans are prospective materials for selective extraction, metal ion detection and diagnostics [45]. Thus, 1-(2-chlorophenyl)-2-(2-chlor-5-sulfophenyl)-3-phenylcarboxamide formazan and 1,5-bis-(2-chloro-5-sulfophenyl)-3-phenylcarboxamide formazan (

Scheme 13), supported on silica and polystyrene respectively, were tested as ion exchangers [46]. It was assumed that their interaction with various metal ions differs crucially and depends on the coordination mode of formazans.

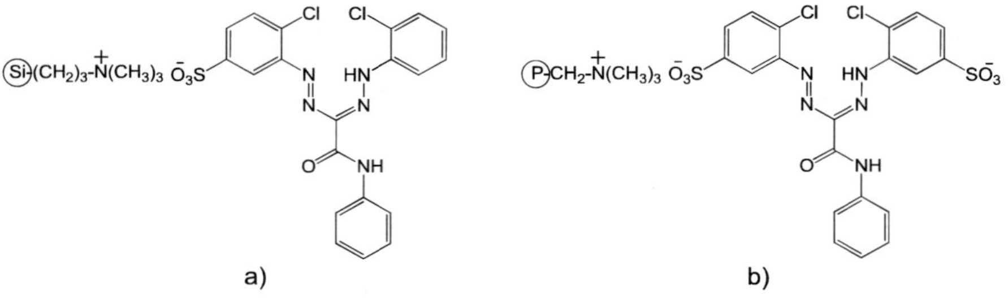

Scheme 13. Formazan-modified silica (a) and polystyrene (b) [46].

Indeed, various metal ions can be chelated by polydentate formazans differently, i.e. via *N,N*- or *N,O*- coordination modes. In spite of several possible tautomeric forms (Scheme 14) [47], performed spectroscopic studies show that the trans-anti form is preferable (Scheme 14b) [47]. Hence, in order to chelate a particular metal, the polydentate formazans must undergo tautomeric transformation from trans-anti to trans-syn form, and vice versa (Scheme 14a) [47]. Such behavior, together with the stability of the formed complexes, favors the selective sorption of some metal ions and hence can be used for specific separation and analytical determination. For example, 1-(o-chlorophenyl)-3-phenyl-5-(6-methyl-4-oxo-pyrimidinyl-2-)formazan (Scheme 15) was used as an ionophor in modified thick-film electrodes, and demonstrated high sensitivity towards the simultaneous determination of lead, copper, cadmium and zink ions [48,49]. A similar formazan system was used for the determination of manganese in natural water [50].

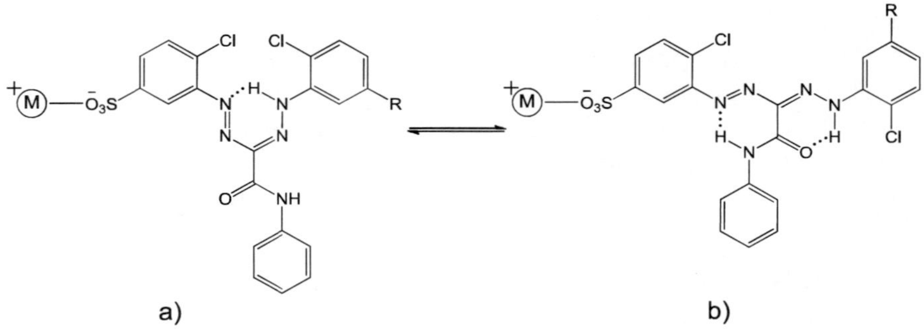

Scheme 14. Isomers of immobilized formazans; trans-syn (a) and trans-anti (b). R=H or SO_3^-, M^+ – polystyrene or silica gel [47].

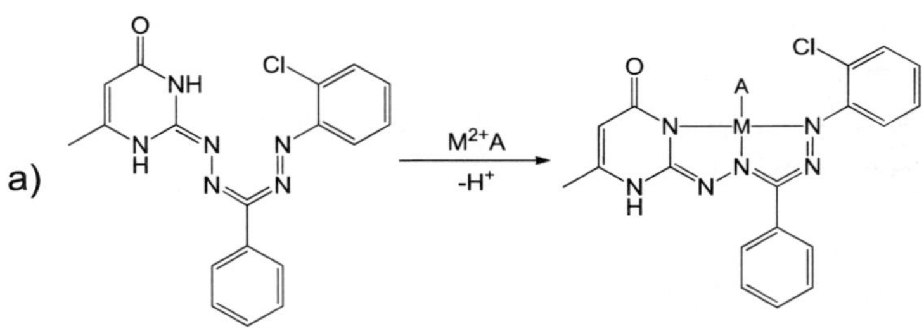

b) $M^{2+}R_{electr.} + 2e^- \longrightarrow M^0{}_{electr.} + R_{electr.}$

c) $M^0{}_{electr.} \longrightarrow M^{2+}{}_{sol.} + 2e^-$

Scheme 15. Formation of metal formazanate on electrode surface (a) and reactions used as response for determination of Cu(II), Pb(II), Cd(II) and Zn(II) (b) and (c). M – metal ion, A – counter ion, $R_{electr.}$ – formazan modifier [49].

A series of tetradentate formazans with *N,N,N,N*-, *N,N,N,O*- and *N,N,O,O*-donor sites were used as precolumn chelating reagents for the determination of ultratrace metal ions in high-performance liquid chromatography [51]. An acetonitrile-water mixture was used as an eluent. Moreover, it was shown that the addition of tetrabutylammonium bromide positively influences the retention and resolution of formazan chelates. The *N,N,O,O*-formazans were found to be good for the determination of zinc(II), copper(II), aluminum(III), gallium(III) and manganese(III) ions at low concentrations (down to nanomoles per litre level). However, to surpass the overlapping of peaks for Mn(III) and Al(III), the masking of the latter is required. The *N,N,N,N*-chelating formazan was found to be a good chelating reagent for the determination of nickel ions [51].

As can be seen from this and other above examples, the understanding of the mechanisms of formation and structure of formazan complexes with metal ions is crucial for the development of many important applications. Thus, some aspects of the coordination chemistry of formazans are discussed below.

6. Coordination Chemistry of Formazans

The known syntheses of formazan complexes are mainly based on the direct interaction of metal salts with formazan ligands, while the reaction conditions and structures of the formed complexes depend on the nature of the metal ion and/or ligand used. Curiously, there is a rather limited number of fully characterized formazanates; usually they involve Pd(II), Cu(II), Ni(II), Co(II), and Fe(II) metal ions. Some of the structurally characterized complexes are listed below.

6.1. Complexes of Bidentate Formazans

Palladium(II) complexes of nitroformazans **1**-**3** were obtained by reacting bis(hexafluoroacetylacetonato)palladium(II) with 3-p-tolyl-, 3-mesityl-, 3-di-iso-propylphenylformazans under reflux in methanol (Scheme 16) [52]. The structures of complexes (Figure 1) are similar and contain both hexafluoroacetylacetonate and formazan ligands that chelate the metal ion forming two six-membered metallacycle rings. Palladium(II) center exhibits the square planar coordination sphere filled with two nitrogen atoms of formazan ligand and two oxygen atoms of hexafluoroacetylacetonate. Due to steric hindrance between the substituents, the structures are significantly non-planar. The N=N=C bonds within formazan ligands are highly delocalized, which is typical for formazan complexes [52].

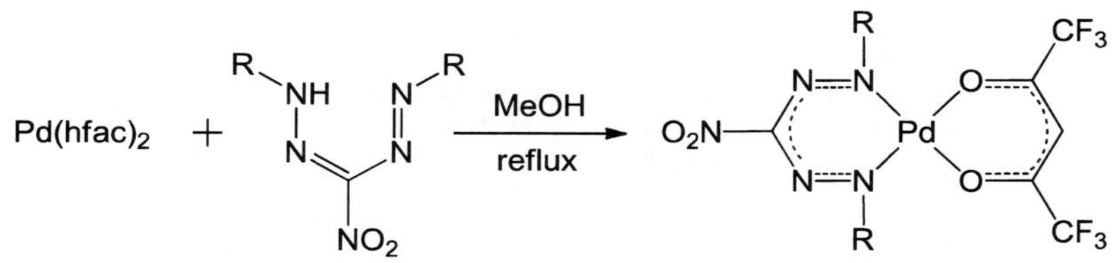

Scheme 16. Synthesis of palladium(II) complexes of nitroformazans (**1**-**3**). R = 3-p-tolyl-, 3-mesityl-, 3-di-iso-propylphenyl.

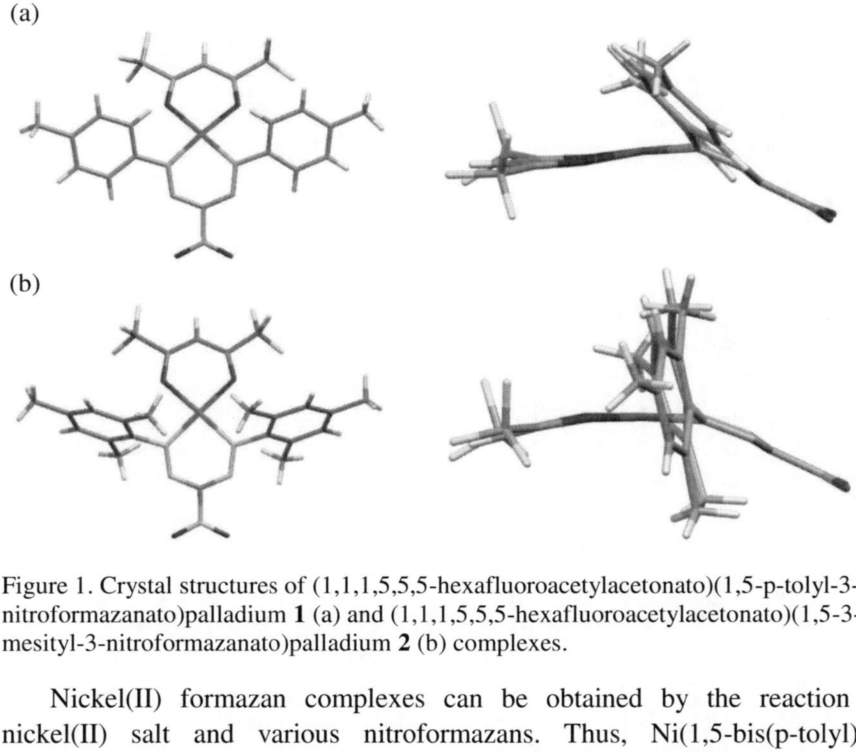

Figure 1. Crystal structures of (1,1,1,5,5,5-hexafluoroacetylacetonato)(1,5-p-tolyl-3-nitroformazanato)palladium **1** (a) and (1,1,1,5,5,5-hexafluoroacetylacetonato)(1,5-3-mesityl-3-nitroformazanato)palladium **2** (b) complexes.

Nickel(II) formazan complexes can be obtained by the reaction of nickel(II) salt and various nitroformazans. Thus, Ni(1,5-bis(p-tolyl)-3-nitroformazan)$_2$ **4** was synthesized via combination of Ni(AcO)$_2$ with the corresponding 3-nitroformazan under reflux in methanol (Scheme 17) [53]. Similarly to the previous case, the π system within the ligand is highly delocalized. The nickel ion exhibits square planar coordination geometry and is chelated by four nitrogen atoms of the two ligands (Figure 2). However, due to the steric hindrance between the bulky groups of formazan ligands the whole structure is not planar.

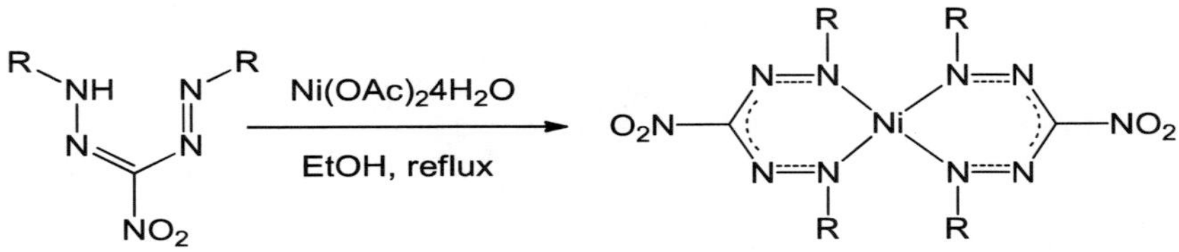

Scheme 17. Synthesis of Ni(1,5-bis(*p*-tolyl)-3-nitroformazan)$_2$ complex **4**.

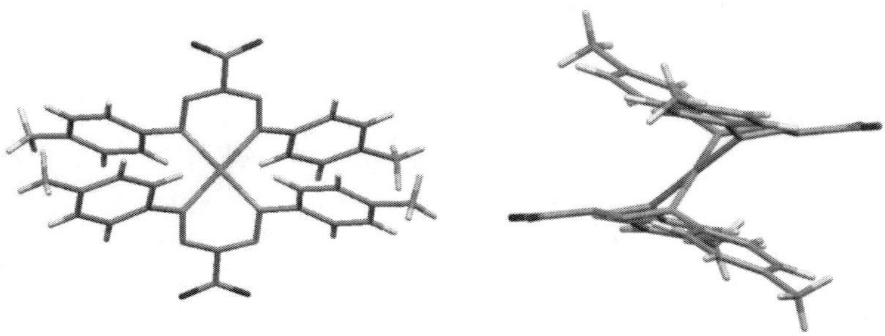

Figure 2. Crystal structure of complex **4**.

In this aspect the bulky groups of the formazan ligand can prevent the formation of the diligand mononuclear formazan complex ML_2 and support the formation of "half sandwich" or polynuclear complexes. Thus, the reaction of nickel(II) acetate hydrate and 1,5-bis(mesityl)-3-nitroformazan or 3-cyano-1,5-bis(2,6-methylphenyl)formazan under reflux in ethanol leads to dinuclear complexes $Ni_2(OH)_2$(1,5-bis(mesityl)-3-nitroformazan)$_2$ **5** or $Ni_2(OH)_2$(3-cyano-1,5-bis(2,6-methylphenyl)formazan)$_2$ **6**, respectively (Scheme 18, Figure 3) [53]. Complexes **5** and **6** are less bent if compared with the mononuclear metal formazan complex **4,** and consist of two nickel(II)(formazan) fragments interconnected via two hydroxyl bridges. Consequently, the nickel ions exhibit a square planar coordination arrangement filled with two formazan backbone nitrogen atoms and two oxygen atoms of two bridging hydroxy groups, thus forming the MN_2O_2 core.

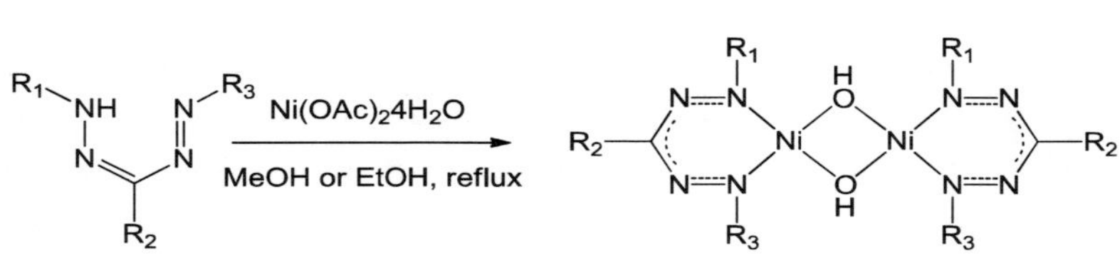

Scheme 18. Synthesis of the $Ni_2(OH)_2$(formazan)$_2$ complexes **5** and **6**. $R_1 = R_3 =$ mesityl, $R_2 = NO_2$ and $R_1 = R_3 =$ 2,6-dimethylphenyl, $R_2 = CN$, respectively.

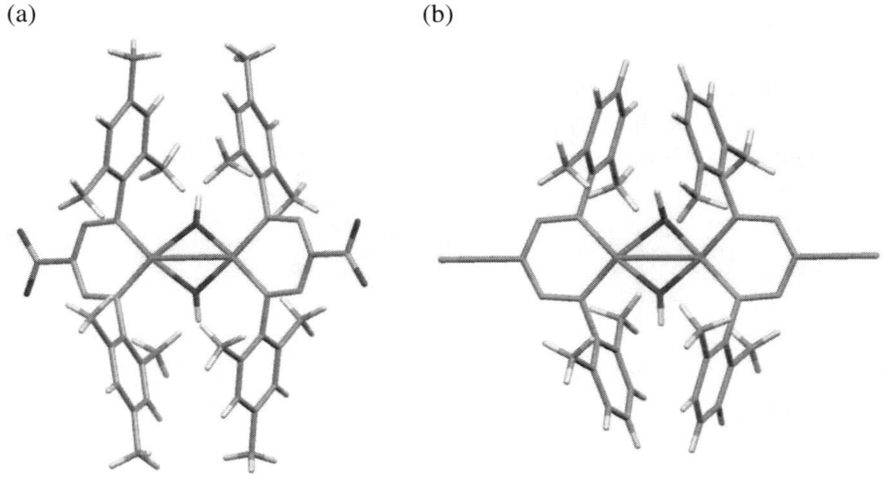

Figure 3. Crystal structures of **5** (a) and **6** (b).

6.2. Complexes of Tridentate Formazans

An additional coordination possibility can be adjoined to formazan ligands by the introduction of, e.g., benzathiazole. This and other similar substituents possess an extra nitrogen donor atom, and are able to coordinate the metal center in tridentate fashion. Thus, a series of nickel(II), iron(II), cobalt(II) and palladium(II) benzathiazolyl formazanates were prepared by relatevly simple straightforward procedures (Scheme 19) [54-57].

In the case of nickel(II), iron(II) and cobalt(II), formazans act as tridentate ligands supporting the formation of ML_2 type complexes with the MN_6 coordination site (**7-14**). They chelate the metal ion in a similar fashion forming two five-membered metallacycles [MN-C=N-N]. The metal ions exhibit a distorted octahedral coordination sphere that is filled by six nitrogen atoms of two formazan ligands, i.e. by N1 and N4 nitrogen atoms of formazan backbone and by the benzathiazolyl nitrogen atom (Scheme 19A). The crystal structures of some of the above-mentioned metal formazanates are represented in Figure 4.

A two step synthetic procedure towards dinuclear palladium(II) 1,3-diphenyl-5-(benzthiazol-2-yl)formazanate was reported recently (Scheme 19BC) [57]. It was assumed that dinuclear palladium(II) formazanate is formed in the first stage (Scheme 19B, **15a**), while in the second stage it undergos transformation towards the other dimer (Scheme 19C, **15b**) [57]. In **15b** the ligand chelates the paladium ion via backbone nitrogen atoms in bidentate fashion, thus forming a six-membered metallocycle [MN=N-C=N-N], while the benzthiazolyl nitrogen atom is coordinated to the collateral palladium ion supporting the formation of the dimer (Figure 5).

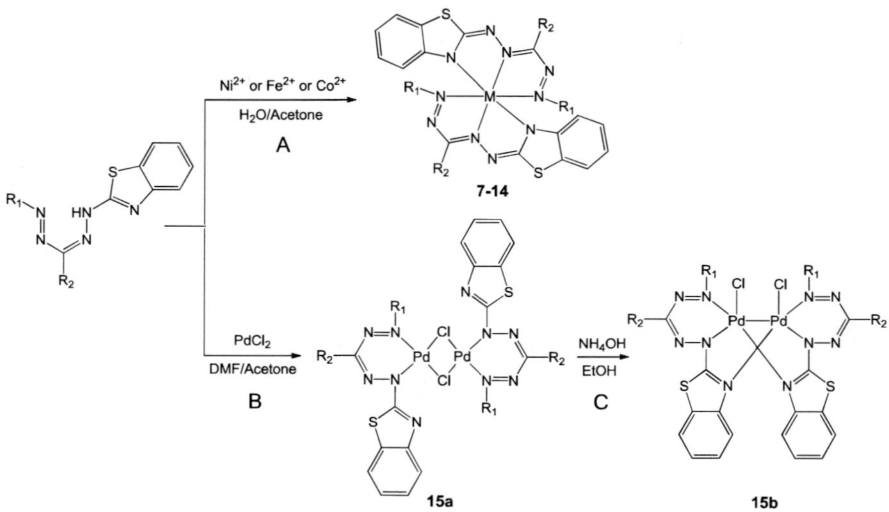

Scheme 19. Synthesis of nickel(II)/iron(II)/cobalt(II) and Pd(II) benzathiazolyl formazanates: for Ni(II) R_1 – 2-chlorophenyl / 4-butylphenyl and R_2 – *n*-butyl (**7** and **8**); for Fe(II) R_1 – 2-chlorphenyl / 4-chlorophenyl and R_2 – *n*-butyl (**9** and **10**), or R_1 – phenyl / 4-methoxyphenyl and R_2 – *i*-propyl (**11-12**), or R_1 = R_2 – phenyl (**13**); for Co(II) and Pd(II) R_1 = R_2 – phenyl (**14** and **15**).

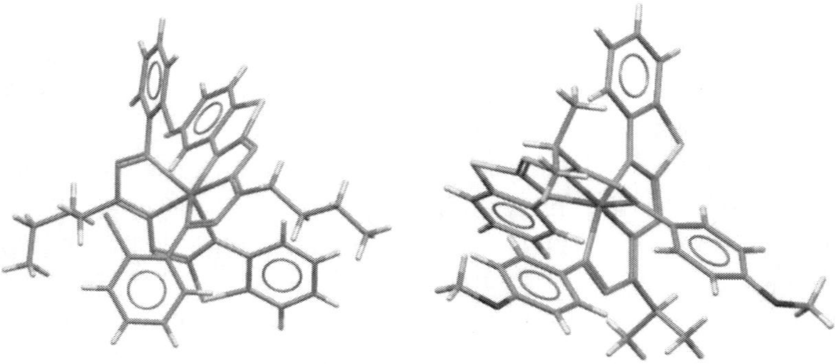

7 **12**

Figure 4. (Continued).

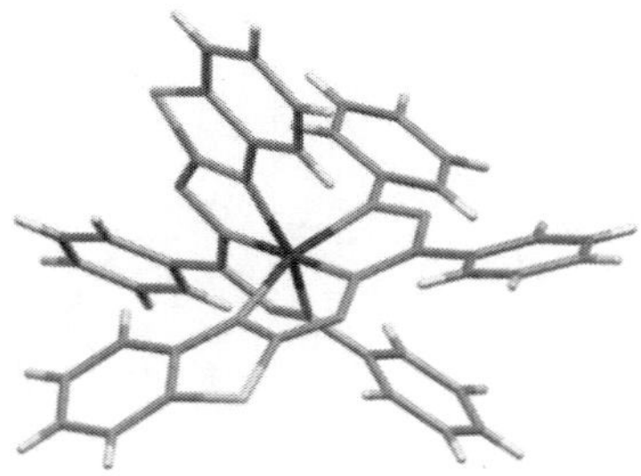

14

Figure 4. Crystal structures of nickel(II) 1-(2-chlorophenyl)-3-(*n*-butyl)-5-(benzthiazol-2-yl)formazanate (**7**), iron(II) 1-(*p*-methoxyphenyl)-3-(*i*-propyl)-5-(benzthiazol-2-yl)formazanate (**12**), cobalt(III) 1,3-diphenyl-5-(benzthiazol-2-yl)formazanate (**14**).

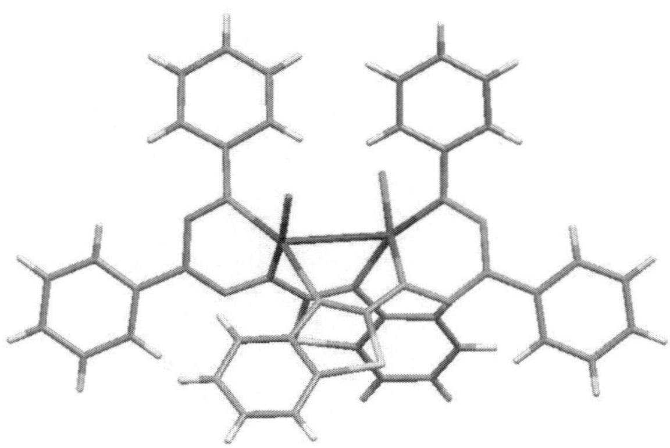

Figure 5. Crystal structure of **15b**.

The introduction of one more functional group into the benzathiazolyl formazan extends its chelating ability. Thus, the interaction of 1-(2-hydroxy-4-nitrophenyl)-3-methyl-5-(benzothiazol-2-yl)formazan with cobalt chloride in acetone yields another type of the metal benzothiazolyl formazanate **16** (Scheme 20, Figure 6) [56]. The formazan acts as a tridentate chelating ligand. However, in contrast to the previously mentioned tridentate benzathiazolyl formazan, it provides the formation of CoL_2 complex with an MN_4O_2 coordination core. The cobalt atom exhibits a distorted octahedral coordination sphere and is coordinated by two formazan backbone nitrogen atoms and one alkoxy oxygen atom of each ligand, thus forming five- [MN-N=C-N] and six-membered [MN=N-C=C-O] metallocycles (Scheme 20), while the benzathiazolyl moiety remains uncoordinated.

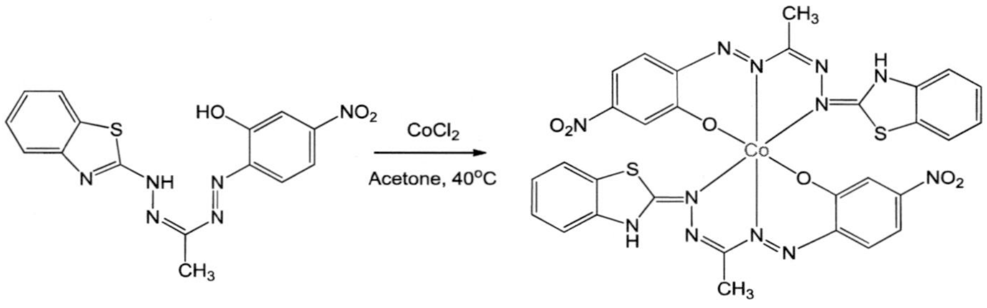

Scheme 20. Synthesis of cobalt 1-(2-hydroxy-4-nitrophenyl)-3-methyl-5-(benzothiazol-2-yl)formazanate **16**.

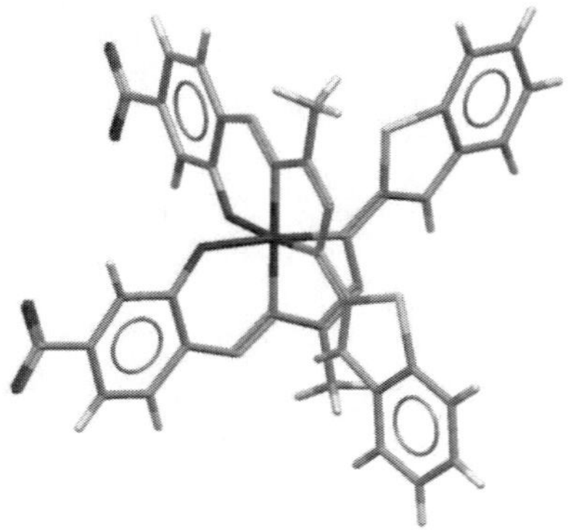

Figure 6. Crystal structure of **16**.

6.3. Complexes of Tetradentate Formazans

Formazans with two additional donor groups can act as tetradentate ligands. Thus, the reaction of Cu^{2+}, Fe^{3+} or Co^{2+} salts with bis(*o*-methoxyphenyl)- or bis(*o*-hydroxyphenyl) cyanoformazans leads to the mononuclear ML type complexes Cu^{II}(1,5-bis(*o*-methoxyphenyl)-3-cyanoformazan)Cl **17**, Fe^{III}(1,5-bis(*o*-hydroxyphenyl)-3-cyanoformazan)$(py)_2$ **18** and Co^{III}(1,5-bis(*o*-hydroxyphenyl)-3-cyanoformazan)$(py)_2$ **19**, respectively (Scheme 21) [53]. The reactions were performed in two steps: in the first step the corresponding cyanoformazan was deprotonated with sodium hydride and then reacted with the metal salt. Reactions of cobalt(II) and iron(III) salts with the formazan ligand were performed in the presence of pyridine. Moreover, the oxidation of cobalt(II) to cobalt(III) was observed.

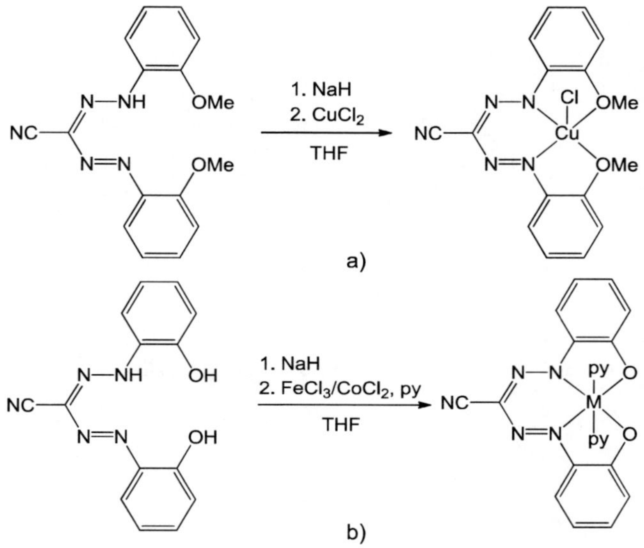

Scheme 21. Synthesis of metal complexes of 1,5-bis(*o*-methoxyphenyl)- (a) and 1,5-bis(*o*-hydroxyphenyl)-3-cyanoformazans (b).

The planar positions of the distorted square pyramidal coordination sphere of the copper ion in **17** (Figure 7) are filled by two nitrogen atoms, one methoxy oxygen atom of the ligand and one chloride atom, while at the apical position is the second methoxy oxygen atom of the ligand; thus, one six- and two five-metallocycles are formed. In both the structures of **18** and **19** (Figure 7) the apical positions of the octahedral coordination sphere are filled with pyridine nitrogen atoms, while two nitrogen and two alkoxy oxygen atoms of the formazan ligand occupy equatorial positions (Figure 7b,c) [53].

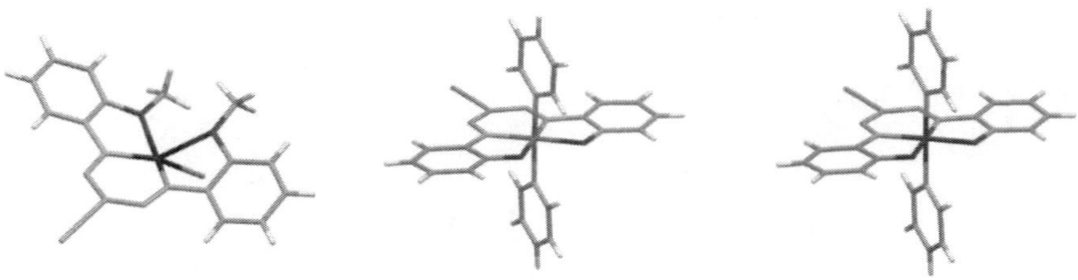

17 **18** **19**

Figure 7. Crystal structures of **17**, **18** and **19**.

Tetradentate 1-(2-carboxyphenyl)-3-isopropyl-5-(benzimidazole-2-yl)formazan reacts with nickel acetate, providing the formation of ML type complex **20** (Scheme 22) [58]. The ligand in **20** exhibits another coordination fashion, chelating the nickel(II) center through two backbone nitrogen atoms and thus forming a five-membered metallocycle. The structure is essentially planar with a low deviation of atoms from the plane (Figure 8).

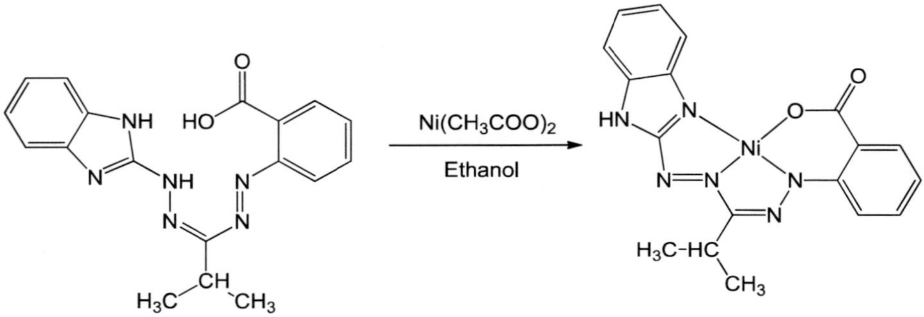

Scheme 22. Synthesis of Ni(II) 1-(2-carboxyphenyl)-3-isopropyl-5-(benzimidazole-2-yl)formazanate (20).

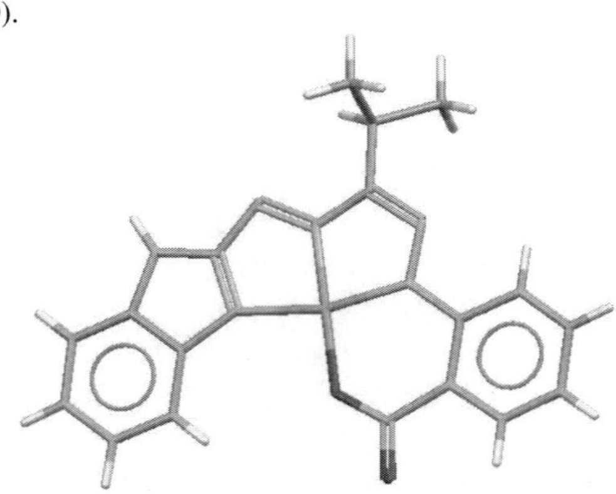

Figure 8. Crystal structure of **20**.

CONCLUSION

Effective methods of the synthesis of formazans were developed a long time ago, while the specific physico-chemical properties ensured a wide range of applications of these interesting compounds. The introduction of additional functional groups and structural elements into the molecules of formazans allowed the expansion of the spectrum of their properties and applications. Thus, antibacterial, antifungial and other useful pharmaceutical activities are described in the literature. Moreover, formazan-based ion-selective electrodes and other analytical materials for express monitoring were introduced. Metal formazanates are used as dyes, data recording media, as thermosensitive elements and active catalysts or catalyst precursors. The molecular magnetic properties of the complexes of formazans are also described. Formazans and their derivatives are useful as tools for the detection and study of Ox-Red, photo- and thermochromic reactions in biological and organic systems.

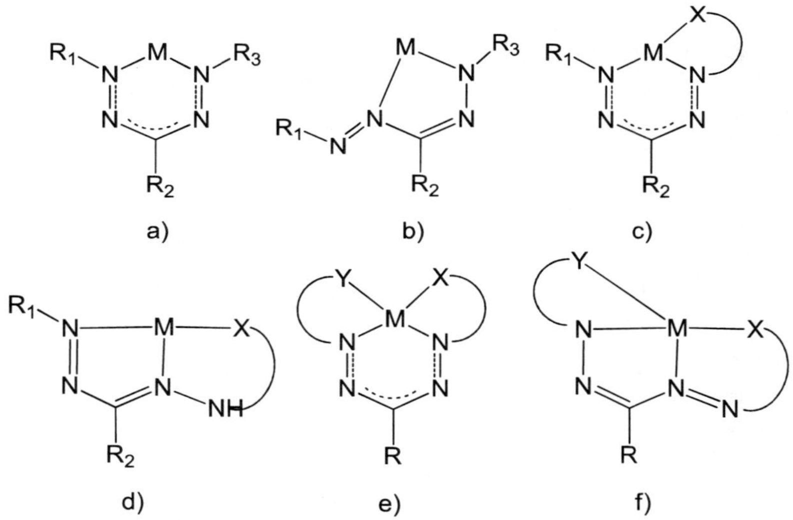

Scheme 23. Some typical formazan coordination modes (X= -OH, -OMe, -COOH, etc).

Concerning coordination chemistry, the simple bidentate formazans usually exhibit the *N,N*-coordination modes, thus providing the formation of six-membered or five-membered metallacycles (Scheme 23ab). However, the addition of functional groups (such as -OH, -COOH, etc.) into the adjacent moiety increases the coordination ability of formazans and allows them to act as tri- and tetradentate ligands (Scheme 23cdef). On the other hand, bulky substituents, due to their steric hindrance, can provide the formation of "half sandwich" ML or polynuclear M_nL_n complexes.

REFERENCES

[1] Nineham, A.W., *Chem. Rev.* 55 (1955) 355.

[2] Buzykin, B.I., Lipunova, G.N., Sysoeva, L.P., Rusinova, L.I., *Khimiya Formazanov (The Chemistry of Formazans)* Moscow: Nauka (1992).

[3] Sigeikin, G.I., Lipunova, G.N., Pervova, I.G., *Uspekhi Khimii* 75 (2006) 980.

[4] Twentyman, P.R., Luscombe, M., *Br. J. Cancer.* 56 (1987) 279.

[5] Gilroy, J.B., Otieno, P.O., Ferguson, M.J., McDonald, R., Hicks, R.G., *Inorg. Chem.* 47 (2008) 1279.

[6] Bednyagina, N.P., Serebryakova, N.V., Lipunova, G.N., *Chem. Heterocyclic Comp.* 3 (1969) 268.

[7] Frolova, N., Vatsadze, S., Stash, A., Rakhimov, R., Zyk, N., *Chem. Heterocyclic Comp.* 42 (2006) 1444.

[8] Ibrahim, Y.A., Abbas, A.A., Elwahy, A.H.M., *J. Heterocyclic Chem.* 41 (2004) 135.

[9] Dziomko, V.M., Ostrovskaya, V.M., Zhukova, T.E., *Chem. Heterocyclic Comp.* 15 (1979) 848.

[10] Abbas, A.A., *Tetrahedron* 54 (1998) 12421.

[11] Rybakova, Y.A., Bednyagina, N.P., *Chem. Heterocyclic Comp.* 7 (1971) 906.

[12] Lipunova, G.N., Gulemina, N.N., Kourova, L.V., Bednyagina, N.P., *Chem. Heterocyclic Comp.* 13 (1977) 367.

[13] Sutherland, M.W., Learmonth, B.A., *Free Radical Research* 27 (1997) 283.

[14] Xu, C., Liu, S., Liu, Z., Song, F., Liu, S., *Analyt. Chim. Acta* 793 (2013) 53.

[15] Mao, Z., Liu, Z., Chen, L., Yang, J., Zhao, B., Jung, Y.M., Wang, X., Zhao, C., *Analyt. Chem.* 85 (2013) 7361.

[16] Hussain, R.F., Nouri, A.M.E., Oliver, R.T.D., *J. Immunol. Met.* 160 (1993) 89.

[17] Verma, A., Prasad, K.N., Singh, A.K., Nyati, K.K., Gupta, R.K., Paliwal, V.K., *J. Microbiol. Met.* 81 (2010) 175.

[18] Williams, C., Espinosa, O.A., Montenegro, H., Cubilla, L., Capson, T.L., Ortega, B.E., Romero, L.I., *J. Microbiol. Met.* 55 (2003) 813.

[19] Stevens, M.G., Olsen, S.C. *J. Immunol. Met.* 157 (1993) 225.

[20] Berridge, M.V., Herst, P.M., Tan, A.S. In *Biotechnology Annual Review*; El-Gewely, M.R., Ed., Elsevier 11 (2005) 127.

[21] Plumb, J.A., Milroy, R., Kaye, S.B., *Cancer Res.* 49 (1989) 4435.

[22] Green, L.M., Reade, J.L., Ware, C.F., *J. Immunol. Met.* 70 (1984) 257.
[23] Barltrop, J.A., Owen, T.C., Cory, A.H., Cory, J.G., *Bioorg. Med. Chem. Lett.* 1 (1991) 611.
[24] Ngamwongsatit, P., Banada, P.P., Panbangred, W., Bhunia, A.K. *J. Microbiol. Met.* 73 (2008) 211.
[25] Berridge, M.V., Tan, A.S., *Arch. Biochem. Biophys.* 303 (1993) 474.
[26] Sims, J., Plattner, R., *Cancer Chemother. Pharmacol.* 64 (2009) 629.
[27] Lü, L., Zhang, L., Wai, M.S.M., Yew, D.T.W., Xu, J., *Toxicol. in Vitro* 26 (2012) 636.
[28] Haselsberger, K., Peterson, D.C., Thomas, D., Darling, J.L., *Anti-Cancer Drugs* 7 (1996) 331.
[29] Vistica, D.T., Skehan, P., Scudiero, D., Monks, A., Pittman, A., Boyd, M.R., *Cancer Res.* 51 (1991) 2515.
[30] Malich, G., Markovic, B., Winder, C., *Toxicology* 124 (1997) 179.
[31] Buttke, T.M., McCubrey, J.A., Owen, T.C., *J. Immunol. Met.* 157 (1993) 233.
[32] Goodwin, C.J., Holt, S.J., Downes, S., Marshall, N.J., *J. Immunol. Met.* 179 (1995) 95.
[33] Ukeda, H., Kawana, D., Maeda, S., Sawamura, M., *Bioscience Biotech. Biochem.* 63 (1999) 485.
[34] Tan, A.S., Berridge, M.V., *J. Immunol. Met.* 238 (2000) 59.
[35] Henriques, C., Moreira, T.L.B., Maia-Brigagao, C., Henriques-Pons, A., Carvalho, T.M.U., de Souza, W., *Analyt. Methods* 3 (2011) 2148.
[36] Wang, P., Lei, J., Su, M., Liu, Y., Hao, Q., Ju, H., *Analyt. Chem.* 85 (2013) 8735.
[37] Gill, R., Zayats, M., Willner, I., *Angew. Chem. Int. Ed.* 47 (2008) 7602.
[38] Desai, K.G., Desai, K.R., *J. Heterocyclic Chem.* 43 (2006) 1083.
[39] Zhang, Y., Liu, D., *Dyes and Pigments* 29 (1995) 57.
[40] Sokolowska-Gajda, J., Freeman, H. S., Reife, A., *Dyes and Pigments* 30 (1996) 1.
[41] Sokołowska-Gajda, J., *Dyes and Pigments* 36 (1998) 149.
[42] Edwards, L.C., Freeman, H.S., *Color. Tech.* 121 (2005) 271.
[43] Szymczyk, M., El-Shafei, A., Freeman, H.S., *Dyes and Pigments* 72 (2007) 8.
[44] Sokolowska-Gajda, J., Freeman, H.S., Reife, A., *Textile Res. J.* 64 (1994) 388.
[45] Uchiumi, A., Tanaka, H., *Analyt. Sciences* 5 (1989) 425.
[46] Grote, M., Schwalk, A., Kettrup, A.Z., *Analyt. Chem.* 313 (1982) 297.

[47] Grote, M., Schwalk, A., Hüppe, U., Kettrup, A.Z., *Analyt. Chem.* 316 (1983) 247.
[48] Stozhko, N.Y., Lipunova, G.N., Maslakova, T.I., Aleshina, L.V., Brainina, K.Z., *J. Analyt. Chem.* 59 (2004) 179.
[49] Brainina, K.Z., Stozhko, N.Y., Aleshina, L.V., Lipunova, G.N., *J. Analyt. Chem.* 58 (2003) 965.
[50] Stozhko, N.Y., Inzhevatova, O.V., Kolyadina, L.I., Lipunova, G.N., *J. Analyt. Chem.* 60 (2005) 163.
[51] Hoshino, H., Nakano, K., Yotsuyanagi, T., *J. Chromatogr. A* 515 (1990) 603.
[52] Gilroy, J.B., Ferguson, M.J., McDonald, R., Hicks, R.G., *Inorg. Chim. Acta* 361 (2008) 3388.
[53] Gilroy, J.B., Patrick, B.O., McDonald, R., Hicks, R.G., *Inorg. Chem.* 47 (2008) 1287.
[54] Pervova, I., Slepukhin, P., Zaidman, A., Lipunova, G., Lipunov, I., *Rus. J. Coord. Chem.* 36 (2010) 213.
[55] Gorbatenko, Y., Rezinskikh, Z., Lipunova, G., Pervova, I., Maslakova, T., Slepukhin, P., Lipunov, I., *Rus. J. Appl. Chem.* 81 (2008) 2127.
[56] Rezinskikh, Z., Slepukhin, P., Pervova, I., Lipunova, G., Maslakova, T., Lipunov, I., Sigeikin, G., *Rus. J. Inorg. Chem.* 54 (2009) 396.
[57] Lipunova, G., Rezinskikh, Z., Maslakova, T., Slepukhin, P., Pervova, I., Lipunov, I., Sigeikin, G., *Rus. J. Coord. Chem.* 35 (2009) 215.
[58] Rezinskikh, Z., Slepukhin, P., Lipunova, G., Pervova, I., Lipunov, I., Sigeikin, G., *Rus. J. Coord. Chem.* 34 (2008) 653.

In: Ligands
Editors: Paula Gawryszewska et al.
ISBN: 978-1-63117-143-7

Chapter 9

N-ACETYLSALICYLHYDRAZIDATE: A VERSATILE LIGAND FOR THE SYNTHESIS OF HIGHER NUCLEARITY COMPLEXES

Manas Sutradhar **and** ***Armando J. L. Pombeiro***
Centro de Química Estrutural, Complexo I, Instituto Superior Técnico,
Universidade de Lisboa, Lisboa, Portugal

ABSTRACT

N-acetylsalicylhydrazide (H_3ashz) is a triacidic species which undergoes deprotonation and enolization in the presence of a metal ion in solution, where the basic form N-acetylsalicylhydrazidate ($ashz^{3-}$) can act as a pentadentate trinegative ligand. This exhibits a versatile coordination behaviour, depending on the nature and charge of the metal ions. The coordination flexibility and interesting donor properties of the $ashz^{3-}$ ligand lead to the formation of various types of metal complexes, e.g., metallacrowns, 1D chain and mixed-valence complexes, which are illustrated herein along with their applications.

Keywords: N–acetylsalicylhydrazidate, metallacrowns, mixed-valence oxovanadium(IV,V), 1D Ni(II) chain, diorganotin(IV)

* Corresponding author: Centro de Química Estrutural, Complexo I, Instituto Superior Técnico, Universidade de Lisboa, Av. Rovisco Pais, 1049-001 Lisboa, Portugal. Email: manaschem@yahoo.co.in, manaschem@gmail.com (MS).

1. INTRODUCTION

Ligands play a fundamental and diverse role to build high nuclearity complexes. The coordination behaviour of a ligand towards a metal ion depends on various features of its own, such as, electron donor properties, denticity, chelation effect, acidic/basic character, steric requirements, etc., as well as on the reaction conditions, solvent properties, pH of the reaction medium, etc., apart from the metal itself. This work illustrates the coordination ability of N-acetylsalicylhydrazidate ($ashz^{3-}$), a potentially pentadentate trinegative chelator (Scheme 1).

Acetylation of salicylaldehydrazide produces N-acetylsalicylhydrazide (H_3ashz), an important pro-ligand, which undergoes enolization and deprotonation in solution in the presence of metal ions, yielding $ashz^{3-}$. Due to its versatile coordination behaviour, $ashz^{3-}$ is a useful ligand for the syntheses of high nuclearity complexes. It has been vastly used in coordination chemistry since its pentadentate ligating nature is suitable in the construction of metallacrowns, particularly with three-valent metal ions, such as Fe^{3+}, Mn^{3+}, Co^{3+}, Ga^{3+}, etc. [1-8]. The study of metallacrowns has been developed as "host–guest" or supramolecular chemistry [4,9]. Some Fe- and Mn-containing metallacrowns have interesting magnetic and antimicrobial properties, and catecholase activity [5,6]. An hexanuclear manganese metallamacrocycle of $ashz^{3-}$ served as a secondary building unit for the construction of a porous host network [4]. The $ashz^{3-}$ ligand has also been utilized to the synthesis of various structural types of complexes, namely trinuclear [10,11], hexanuclear [12] and 1D-polymeric [13] ones. The presence of two potential chelating moieties with different negative charges of $ashz^{3-}$ shows distinct affinities for different oxidation states and can stabilize a mixed-valence V(IV,V) compound [12], which acts as highly efficient catalyst towards alkane oxidations by aqueous H_2O_2.

2. COORDINATION BEHAVIOUR OF N-ACETYLSALICYLHYDRAZIDATE ($ASHZ^{3-}$)

The potentially pentadentate trinegative chelator $ashz^{3-}$ generally forms two coordination pockets, one bidentate and the other tridentate, and in most of the cases it simultaneously encapsulates two metal ions (Scheme 1b). In one case it was found that it can simultaneously bind to three metal ions due to the

μ_2-oxo bridging mode of the phenolate oxygen (Scheme 1c) [10]. These two pockets play a diverse role in coordination to metal ions, e.g., stabilize different coordination geometries of the metal centre, generate various structural motifs, stabilize different oxidation states of the metal ion, etc. Examples are illustrated below.

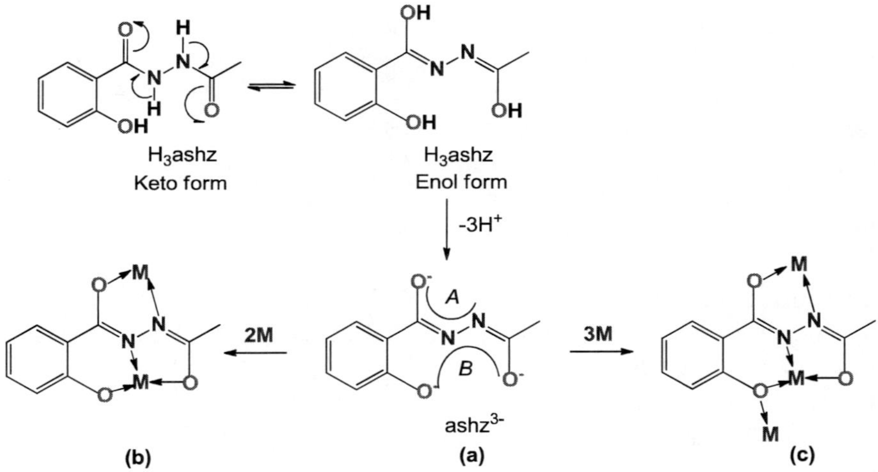

Scheme 1. Coordination modes of $ashz^{3-}$.

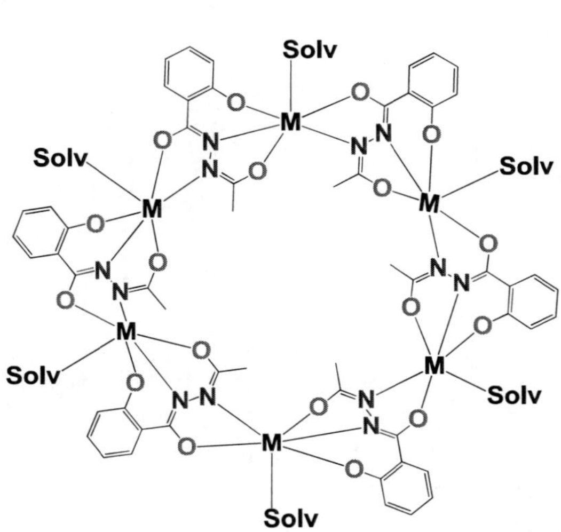

Scheme 2. Hexanuclear metallamacrocycle $[M–N–N]_n$ azametallacrowns.

3. METALLACROWNS

Metallacrown were first reported in 1989 [14], being structurally analogous to crown ethers where methylene carbons [15] are replaced by transition metal ions and nitrogen atoms, sometimes the oxygen atoms being replaced by nitrogen atoms [16]. To obtain the macrocyclic metal cluster, it is important to bridge two metal ions simultaneously in a cyclic way by the ligands. Generally two types of metallacrowns are found, one having repeated [M–N–O] units interlinked to a cyclic structure and the other one with repeated azametallacrown [M–N–N] units. A rich variety of $[M–N–O]_n$ metallacrowns, e.g., 9-metallacrowns-3 [17], 12-metallacrowns-4 [18] and 15-metallacrowns-5 [18d,19], are formed by hydroxamate and/or ketonoximate ligands, whereas higher analogues are found in $[M–N–N]_n$ azametallacrowns, such as, Mn(III) 18-azametallacrown-6 [1-8,16a], Mn(III)/Fe(III) 30-azametallacrowns-10 [16b] and Mn(III)/Ni(II) 36-azametallacrowns-12 [20], by N-acylsalicylhydrazide type ligands. The anionic ligand $ashz^{3-}$ obtained from N-acetylsalicylhydrazide (H_3ashz) has been used to synthesize several [M–N–N]n 18-azametallacrown-6 complexes discussed below [8].

All azametallacrowns built with the $ashz^{3-}$ ligand have the common hexanuclear metallamacrocycle core (Scheme 2) and show two different chiralities (Δ and Λ) around the metal centres (Scheme 3).

3.1. Mn(III)–Azametallacrowns

When N-acetylsalicylhydrazide (H_3ashz) is treated with manganese(II) acetate tetrahydrate in DMF the hexanuclear metallamacrocycle $[Mn(ashz)(DMF)]_6$ (**1**) (Scheme 2, M = Mn, Solv = DMF) results [7]. This complex represents the first structurally characterized azametallacrown with the $ashz^{3-}$ ligand. A similar Mn(III)–azametallacrown, $[Mn_6(ashz)_6(CH_3OH)_3(H_2O)_3]\cdot 3H_2O\cdot 3DMF$ (**2**), is obtained when a methanolic solution of $Mn(ClO_4)_2\cdot 6H_2O$ reacts with H_3ashz in DMF [6]. One more structure with coordinated methanol, $[Mn_6(ashz)_6(CH_3OH)_6]\cdot 6\ CH_3OH$ (**3**), is also known [21]. Both **1** and **2** have the same hexanuclear core and differ only with the coordinated solvent molecules.

The hexanuclear metallamacrocycle has been used as a secondary building block to synthesize new open framework materials. Exchange of the coordinated solvent molecules of the metallamacrocycle by an exobidentate ligand constructs a three-dimensional framework. Reaction of a mixture of the pro-ligand H_3ashz and the exobidentate ligand 1,2-bis(4-pyridyl)ethane (bpea) in DMF-hexane solution with manganese(II) acetate tetrahydrate yields dark brown rectangular crystals of the self-assembled three-dimensional molecular open framework (host network) $[Mn_6(ashz)_6(DMF)_2(bpea)_2]_n$ (**4**) [Scheme (4a)][2].

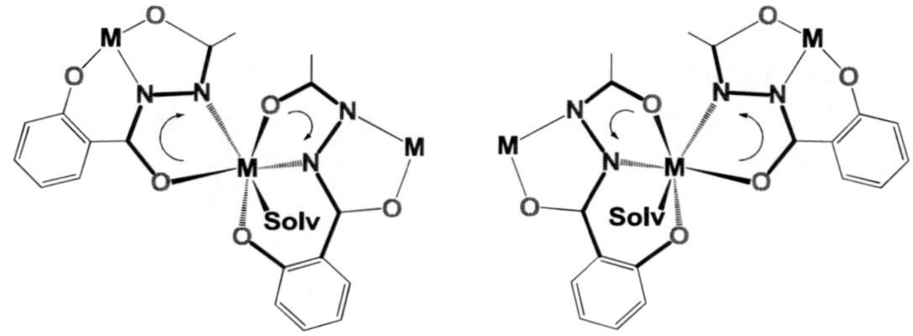

Δ chirality Λ chirality

Scheme 3. Two different chiralities around the metal ion in azametallacrowns.

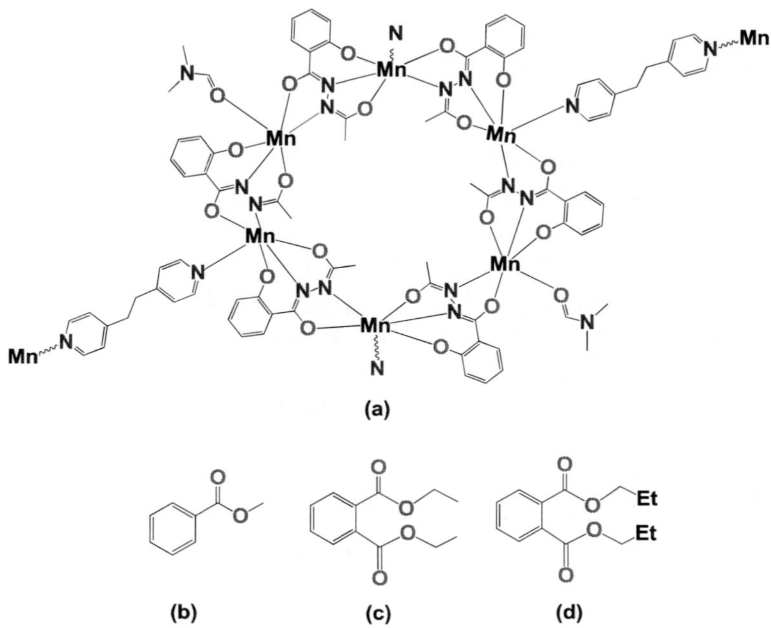

Scheme 4. Three-dimensional molecular open framework of **4** (a), mono-podal guest molecule (b), di-podal guest molecules (c, d).

By utilizing this host network, several other host-guest networks were prepared. An overall methodology similar to that indicated above was also applied to synthesize such networks. Hence, when the above reaction was carried out in the presence of guest molecules a host-guest network is obtained with the general formula $[Mn_6(ashz)_6(bpea)_2(DMF)_2]\cdot x(guest)\cdot y(DMF)$ [4]. Two types of guest molecules were used for the host-guest interactions (Scheme 4), viz., ethylbenzoate as a monopodal guest [Scheme 4(b)] and *ortho* disubstituted alkyl ester benzenes as dipodal guests [Scheme 4(c) and (d)].The monopodal guest molecule (b) and short or medium length dipodal molecules (c) are well fitted at the screw channels of the three-dimensional cavities, but the long length dipodal guest molecule (d) does not fit therein.

Compound **2** shows an effective catechol oxidase activity towards the oxidation of 3,5-di-*tert*-butylcatechol to 3,5-di-*tert*-butylquinone. Variable temperature magnetic susceptibility measurement of **2** shows a typical antiferromagnetic behaviour. In addition, **2** possesses an intense fluorescence at room temperature [6].

3.2. Fe(III)-Azametallacrowns

Reaction of iron(II) chloride tetrahydrate with H_3ashz in DMF yields dark brown crystals of $[Fe(ashz)(DMF)]_6$ (**5**) [1]. Similarly, ferric nitrate decahydrate in methanol yields dark black-violet crystals of the hexanuclear metallamacrocycle $[Fe_6(ashz)_6(MeOH)_6]\cdot 8MeOH\cdot 2H_2O$ (**6**) [5]. The ligand $ashz^{3-}$ effectively stabilizes the hexanuclear Fe(III)–azametallacrown. Changing the precursor iron salts or the reaction conditions does not alter the hexanuclear cyclic structure. It is also obtained when the the oxo-centred trinuclear Fe(III) cluster $[Fe_3(\mu_3\text{-}O)(Piv)_6(H_2O)_3(Piv)]$ (Piv = Bu^tCOO^-) is treated with H_3ashz in ethanol at room temperature, the core of this precursor being destroyed to yield the cyclic Fe(III)-azametallacrown $[Fe(ashz)(EtOH)]_6$ (**7**) with coordinated ethanol (Scheme 5) [8]. The reactivity of the cyclic Fe(III)–azametallacrown **7** concerns only the replacement of the loosely bound coordinated solvent molecule by a N-donor Lewis base to produce a similar cyclic core structure (Scheme 5) [8].

The structures of Fe(III)–azametallacrowns are very similar to that of Mn(III)–azametallacrowns, possess a 18-membered hexanuclear core ring system with an $[Fe–N–N]_n$ repeating unit having alternating Λ /Δ chiralities. **6** shows an antiferromagnetic coupling between the Fe(III) spin 5/2 centres via the –N–N– bridging group [5].

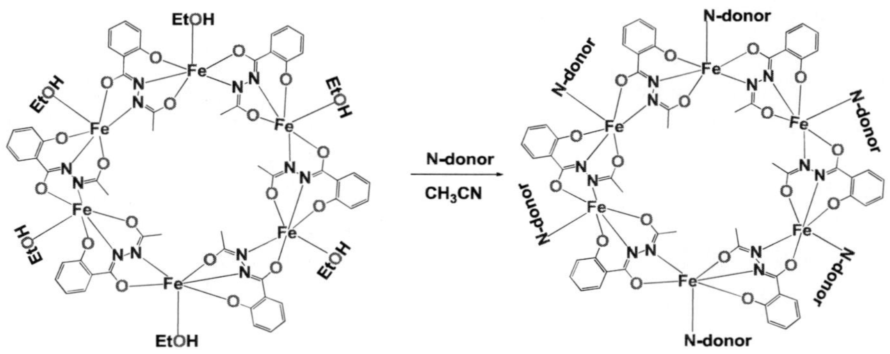

Scheme 5. Reactions of **7** with N-donor Lewis bases (pyridine, pyridine derivatives, imidazole, etc.).

3.3. Co(III)-Azametallacrowns

As Fe(III) and Mn(III) metal ions, Co(III) also forms azametallacrowns with H_3ashz. Reaction of triacetylacetonato cobalt(III) with H_3ashz in DMF-MeOH produces a dark green solution from which quite air-sensitive black–brown rhombohedral crystals of $[Co_6(ashz)_6(MeOH)_6]\cdot 6MeOH$ (**8**) [3] were isolated. A similar microcrystalline azametallacrown, $[Co(ashz)(MeOH)]_6$ (**9**), is also obtained by the reaction of cobalt(II) acetate tetrahydrate with H_3ashz in MeOH [1]. The molecular structure of **8** possesses a non-crystallographic pseudo-C_{3i} symmetry and crystallographic inversion centres where the cobalt atoms exhibit alternate Λ and Δ chiralities [3].

3.4. Cr(III)-Azametallacrowns

The first metallacrown complex of Cr(III) was synthesized using H_3ashz by hydrothermal synthesis. Red prismatic crystals of $[Cr_6(ashz)_6(H_2O)_6]\cdot 11H_2O$ (**10**), a hexanuclear azametallacrown, is obtained from the reaction of a water suspension of $Cr(ClO_4)_3\cdot 6H_2O$ and H_3ashz in an autoclave at 150 °C [22]. The structure of **6** is similar to those of other 18-azametallacrown-6 complexes and exhibits alternative Λ and Δ chiralities.

Significant antiferromagnetic coupling through the Cr–N–N–Cr unit is observed by magnetic susceptibility measurements of **10** [22].

3.5. Ga(III)-Azametallacrowns

The basic structural features of the Ga(III)-azametallacrown are similar to the other corresponding hexanuclear metallamacrocycles illustrated above. Transparent crystals of the metallacrown $[Ga(ashz)(MeOH)]_6$ (**11**) [1] are obtained by the reaction of H_3ashz and gallium (III) nitrate hydrate in methanol. X-ray crystal structure analysis shows that the size of the hexanuclear gallium metallamacrocycles lies between those of the cobalt and manganese analogues.

4. Other Types of Complexes

4.1. Ni(II) Complexes

Bivalent metal ions may not produce azametallacrowns upon reaction with H_3ashz. When $Ni(OAc)_2{\cdot}4H_2O$ reacts with H_3ashz in DMF, in the presence of pyridine, red block shaped crystals of the trinuclear Ni(II) complex $[Ni_3(ashz)_2(py)_2(DMF)_2]{\cdot}2DMF$ (**12**) are obtained (Scheme 6) [10].

Its structure exhibits a trinuclear arrangement, where three nickel ions are linked by two $ashz^{3-}$ ligands. The two terminal nickel ions exhibit the same coordination environment with a square planar geometry and they occupy the tridentate pocket of the $ashz^{3-}$ ligand. The fourth coordination is satisfied by the ligated pyridine. The central nickel ion adopts a six-coordinate octahedral-type geometry, being coordinated by two $ashz^{3-}$ ligands in their bidentate pockets. The axial positions are occupied by coordinated DMF molecules.

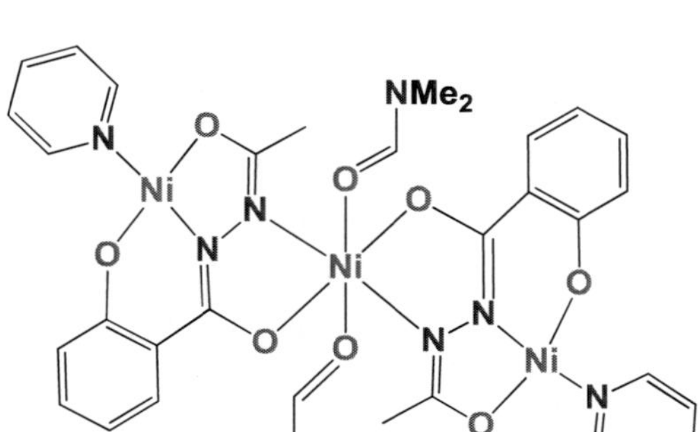

Scheme 6. Trinuclear Ni(II) complex **12**.

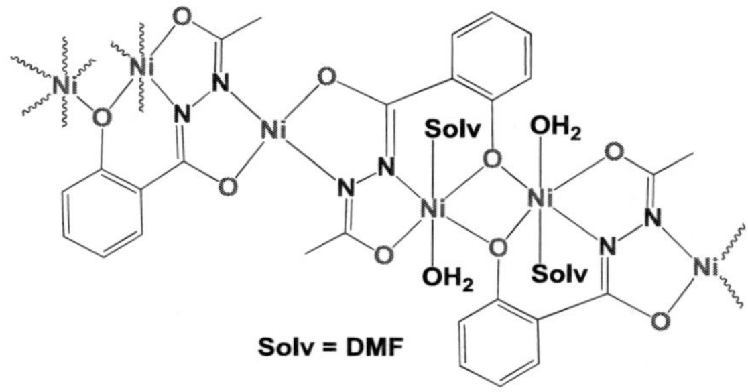

Scheme 7. Polynuclear 1D Ni(II) chain of **13**.

Interestingly, the reaction of H_3ashz with $Ni(OAc)_2.4H_2O$ (in 1:2 mole ratio), in the absence of pyridine, in DMF produces light green single crystals of a polynuclear 1D Ni(II) chain (Scheme 7) having $[Ni^{II}_3(ashz)_2(DMF)_2(H_2O)_2]_n$ (**13**) as the basic unit [13], with an unusual binding mode of the $ashz^{3-}$.

Generally the trianionic ligand $ashz^{3-}$ simultaneously binds to two metal ions in its two coordination pockets (Scheme 1a,b), but in the case of **13** it simultaneously binds three metal ions due to the μ_2-oxo bridging mode of the phenolate oxygen (Scheme 1c). The coordination environments around the Ni(II) centres in **13** are the reverse than those of **12**, the octahedral and square-planar arrangements being stabilized by the tridentate and the bidentate pockets, respectively. The presence of the μ_2-oxo bridging mode of the phenolate moiety gives rise to the formation of a 1D chain without any additional spacer ligand. This structure contains a crystallographically imposed inversion centre passing through the Ni_2O_2 core centre, where the $O_{phenolate}$ atom acts as a bridge between the Ni(II) centres and leads to an edge-sharing bioctahedron. The axial positions of the octahedral Ni(II) centres are satisfied by water and DMF molecules, which are placed oppositely of the Ni_2O_2 core. Magnetic studies of the coordination polymer **13** having octahedral and square-planar environments shows that the Ni(II) centres are antiferromagnetically coupled [13].

4.2. Mixed-Valence Oxovanadium(IV,V) Complex

As mentioned above, N–acetylsalicylhydrazidate ($ashz^{3-}$) bears two potential chelating moieties [coordination pocket A and B, Scheme 1(a)] with different negative charges, which exhibit distinct affinities towards different vanadium oxidation states (IV and V). The mono negative NO coordination pocket A stabilizes the +4 oxidation state, whereas the dianionic NO_2 coordination pocket B stabilizes more the +5 oxidation state. Reaction of H_3ashz with $VOSO_4{\cdot}3H_2O$ in ethanol under reflux yields the hexanuclear mixed-valence $V^{V/IV/V/V/IV/V}$ complex $[V_3O_3(OEt)(ashz)_2(\mu\text{–}OEt)]_2$ (**14**) (Scheme 8)[12].

X–ray analysis reveals that the molecular structure of this complex contains two trinuclear vanadium assemblies bound by two exogenous ethoxide bridging groups forming a central V_2O_2 core, and possesses a crystallographically imposed inversion centre at this core centre, ultimately leading to the hexanuclear arrangement. **14** has three different sets of VO centres with distinct donor groups around the metal cations and its structure can be represented as an edge-sharing $V_2O_{10}N_2$ bioctahedron, with two terminal oxide ligands placed on opposite sides (anti) of the V_2O_2 core. The existence of $V^{IV}O$ centres is supported by EPR study and the presence of mixed-valence state is confirmed by electrochemical study [12].

It is found that **14** acts as an efficient catalyst towards the oxidation of alkanes in the presence of 2–pyrazinecarboxylic acid (PCA) by aqueous H_2O_2 in CH_3CN at 50°C to the corresponding alcohols and ketones, via formation of alkyl hydroperoxides as primary oxidation products. Remarkably high TON (turnover number, moles of products/mole of **14**) ca. 25,000 were achieved with a product yield of 27%. The mechanism of the catalytic process can be described by a radical pathway (with involvement of $HO^{\bullet}$ radicals) [12].

4.3. Diorganotin (IV) Complex

The ligand N–acetylsalicylhydrazidate ($ashz^{3-}$) is also found in an organometallic tin(IV) complex. Reaction of H_3ashz and dimethyltin(IV) dichloride $[Sn(CH_3)_2Cl_2]$ in the presence of EtONa in dry ethanol produces a white solid of the trinuclear diorganotin(IV) complex $[(CH_3)_2Sn]_3(ashz)_2$ (**15**), which upon recrystallization from dichloromethane/ethanol gives colourless crystals [11].

The structure of **15** (Scheme 9) contains two types of tin atoms. The terminal ones are five-coordinate with a distorted trigonal bipyramid geometry, whereas the central Sn six-coordinate has a distorted octahedral geometry. The axial positions of the central Sn are occupied by two *trans* methyl groups and lies on a crystallographic mirror plane of the molecule.

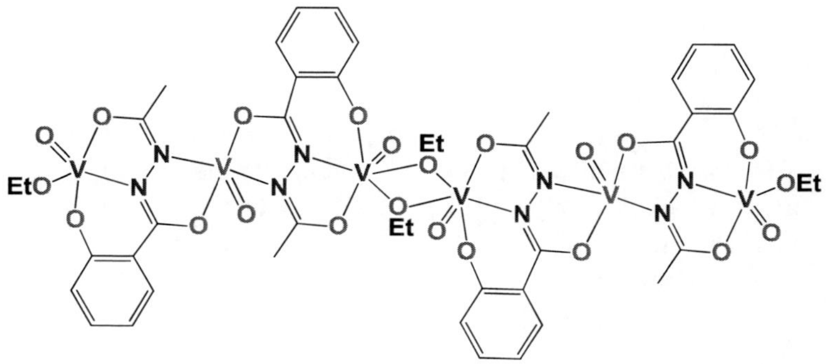

Scheme 8. Mixed-valence oxovanadium(IV,V) complex **14**.

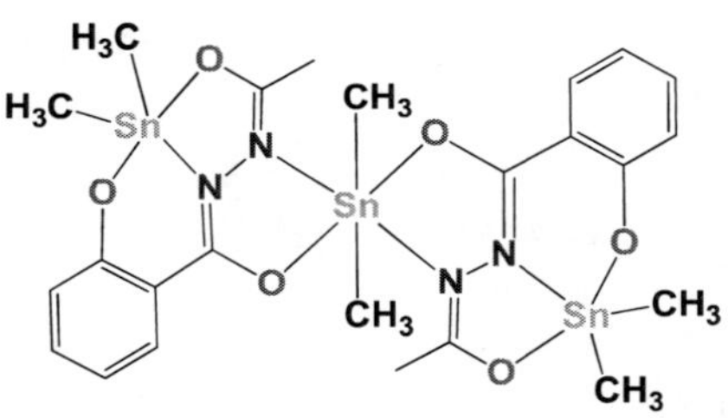

Scheme 9. Trinuclear diorganotin(IV) complex **15**.

CONCLUSION

The coordination versatility of N-acetylsalicylhydrazidate ($ashz^{3-}$) is accounted for by its potential pentadentate nature, with two coordination pockets with different charges and denticities (one being bidentate and the other one tridentate). As a result, it can simultaneously encapsulate two or more metal ions. The coordination flexibility and interesting donor properties of the $ashz^{3-}$ ligand allow to construct a variety of structures. The leading structural pattern is the hexanuclear metallamacrocycle 18-azametallacrown-6 with trivalent metal ions.

However, different types of molecular structures are obtained with divalent metal ions. Hence, in Ni(II)-complexes, two different coordination geometries viz., square-planar and octahedral, are observed around the Ni(II) centres. The ability of $ashz^{3-}$ to form a 1D polymeric chain has also been documented, which deserves further exploration for the synthesis of coordination polymers.

Interestingly, on account of the different charges of the two coordinating pockets of $ashz^{3-}$, capable of stabilizing distinct metal oxidation states, a hexanuclear mixed-valence vanadium(IV,V) complex is known. Furthermore, it acts as an efficient catalyst towards the oxidation of alkanes under mild conditions. Therefore, $ashz^{3-}$ is a suitable ligand in catalysts for oxidation reactions, e.g., of alkanes.

A trinuclear diorganotin(IV) complex is also found with the $ashz^{3-}$ ligand, in which Sn(IV) exhibits two different coordination geometries, a distorted trigonal bipyramid and a distorted octahedron.

Though a versatile coordination chemistry is already well established with $ashz^{3-}$, there are still many areas (such as, concerning non-transition metal complexes and heterometallic complexes) that need to be explored.

ACKNOWLEDGMENT

Authors are grateful to the Foundation for Science and Technology (FCT; projects PTDC/EQU-EQU/122025/2010 and PEst-OE/QUI/UI0100/2013, and postdoc fellowship SFRH/BPD/86067/2012), Portugal, for financial support.

REFERENCES

[1] Kim, I., Kwak, B. & Lah, M. S. (2001). *Inorg. Chim. Acta*, *317*, 12–20.
[2] Moon, M., Kim, I., Lah, M. S. (2000). *Inorg. Chem.*, *39*, 2710–2711.
[3] Lin, S., Liu, S-X., Huang, J-Q. & Lin, C-C. (2002). *J. Chem. Soc., Dalton Trans.* 1595–1601
[4] Moon, D. & Lah, M. S. (2005). *Inorg. Chem.*, *44*, 1934–1940.
[5] Lin, S., Liu, S-X. & Lin, B-Z. (2002). *Inorg. Chim. Acta*, 328, 69–73.
[6] Majumder, A., Goswami, S., Batten, S. R., El Fallah, M. S., Ribas, J. & Mitra, S. (2006). *Inorg. Chim. Acta*, *359*, 2375–2382.
[7] Kwak, B., Rhee, H. & Lah, M. S. (2000). *Polyhedron*, *19*, 1985–1994
[8] Sutradhar, M., Guedes da Silva, M. F. C., Pombeiro, A. J. L. (2013). *Inorg. Chem. Commun.*, *30*, 42–45.

[9] Saalfrank, R. W. & Demleitner, B. (1999). in *Perspectives in Supramolecular Chemistry*, ed. Sauvage, J. P., Wiley-VCH, Weinheim, *5*, 1–51.

[10] Xiao, F-P. & Jin, L-F. (2008). *Z. Anorg. Allg. Chem.*, *634*, 397-400.

[11] Cui, J-C., Qiao, Y-L., Ding, L-H., Yin, H-D., *Chinese J. Struct. Chem.*, 29 (2010) 1784–1787.

[12] Sutradhar, M., Kirillova, M. V., Guedes da Silva, M. F. C., Martins, L. M. D. R. S. & Pombeiro, A. J. L. (2012). *Inorg. Chem.*, *51*, 11229−11231.

[13] Sutradhar, M., Guedes da Silva, M. F. C., Nesterov, D. S., Jezierska, J. & Pombeiro, A. J. L. (2013). *Inorg. Chem. Commun.*, *29*, 82–84.

[14] Lah, M. S., Kirk, M. L., Hatfield, W. & Pecoraro, V. L. (1989). *J. Chem. Soc. Chem. Commun.*, 1606–1608.

[15] (a) Lah, M. S. & Pecoraro, V. L. (1991). *Inorg. Chem.*, *30*, 878–880; (b) Lah, M. S., Gibney, B. R., Tierney, D. L., Penner-Hahn, J. E. & Pecoraro, V. L. (1993). *J. Am. Chem. Soc.*, *115*, 5857–5858; (c) Dendrinou-Samara, C., Psomas, G., Iordanidis, L., Tangoulis, V. & Kessissoglou, D. P. (2001). *Chem. Eur. J.*, *7*, 5041–5051; (d) Noord, A. D. C. -V., Kampf, J. W. & Pecoraro, V. L. (2002). *Angew. Chem. Int. Ed.*, *41*, 4667–4670.

[16] (a) Kwak, B., Rhee, H., Park, S. & Lah, M. S. (1998). *Inorg. Chem.*, *37*, 3599–3602; (b) Liu, S. X., Lin, S., Lin, B. Z., Lin, C. C. & Huang, J. Q. (2001). *Angew. Chem. Int. Ed.*, *40*, 1084–1087.

[17] (a) Pecoraro, V. L. (1989). *Inorg. Chim. Acta*, *155*, 171–173; (b) Gibney, B. R., Stemmler, A. J., Pilotek, S., Kampf, J. W. & Pecoraro, V. L. (1993). *Inorg. Chem.*, *32*, 6008–6015.

[18] (a) Gibney, B. R., Wang, H., Kampf, J. W. & Pecoraro, V. L. (1996). *Inorg. Chem.*, *35*, 6184–6193; (b) Gibney, B. R., Kessissoglou, D. P., Kampf, J. W. & Pecoraro, V. L. (1994). *Inorg. Chem.*, *33*, 4840–4849; (c) Halfen, J. A., Bodwin, J. J. & Pecoraro, V. L., *Inorg. Chem.*, *37* (1998) 5416–5417; (d) Cutland, A. D., Malkani, R. G., Kampf, J. W. & Pecoraro, V. L. (2000). *Angew. Chem. Int. Ed. Engl.*, *39*, 2689–2692; (e) Stemmler, A. J., Kampf, J. W. & Pecoraro, V. L. (1995). *Inorg. Chem.*, *34*, 2271–2272.

[19] Kessissoglou, D. P., Kampf, J. & Pecoraro, V. L. (1994). *Polyhedron*, *13*, 1379-1391.

[20] (a) Rohith, P. J., Lee, K. & Lah, M. S. (2004). *Chem. Commun.*, 2660–2661; (b) Bai, Y., Dang, D. D., Duan, C. Y., Song, Y. & Meng, Q. J. (2005). *Inorg. Chem.*, *44*, 5972–5974.

[21] Lin, S., Liu, S-X. & Lin, B-Z. (2002). *Chin. J. Inorg. Chem*, *18*, 1205–1210.
[22] Chen, G-J., Gao, C-Y., Gu, W., Liu, X., Yan, S-P., Tian, J-L., Liao, D-Z. (2011). *Z. Anorg. Allg. Chem.*, *637*, 374–376.

EDITORS' CONTACT INFORMATION

Dr. Paula Gawryszewska, D.Sc. Eng.
Faculty of Chemistry
University of Wroclaw
14 F. Joliot-Curie Str., 50-383
Wroclaw, Poland
paula.gawryszewska@chem.uni.wroc.pl

Dr. Piotr Smoleński, D.Sc. Eng.
Faculty of Chemistry
University of Wroclaw
14 F. Joliot-Curie Str., 50-383
Wroclaw, Poland
piotr.smolenski@ chem.uni.wroc.pl

INDEX

D

E

F

G

H

I

K

L

M

N

O

P

Q

R

S

T

V

X

Y

Z

β